INFANT CARE AND FEEDING
IN THE SOUTH PACIFIC

Food and Nutrition in History and Anthropology
A series edited by John R.K. Robson

Additional volumes in preparation

ISSN 0275-5769

This book is part of a series. The publisher will accept continuation
orders which may be cancelled at any time and which provide for
automatic billing and shipping of each title in the series upon
publication. Please write for details.

INFANT CARE AND FEEDING
IN THE SOUTH PACIFIC

Edited by
LESLIE B. MARSHALL
University of Iowa

GORDON AND BREACH SCIENCE PUBLISHERS
New York   London   Paris   Montreux   Tokyo

Gordon and Breach Science Publishers

P.O. Box 786
Cooper Station
New York, NY 10276
United States of America

P.O. Box 197
London WC2E 9PX
England

58, rue Lhomond
75005 Paris
France

P.O. Box 161
1820 Montreux 2
Switzerland

14-9 Okubo 3-chome,
Shinjuku-ku,
Tokyo 160
Japan

Chapters 2–9, 12, and 14–16 were originally published in Volumes 14–16 of the journal *Ecology of Food and Nutrition.*

Library of Congress Cataloging in Publication Data
Main entry under title:

Infant care and feeding in the South Pacific

(Food and nutrition in history and anthropology,
ISSN 0275-5769; v. 3)
    Based on papers presented at a Symposium on
Infant Care and Feeding in Oceania, 1983, in New Harmony,
Indiana and on papers previously published in
Ecology of food and nutrition.
    Includes bibliographies and index.
    1. Infants — Oceania — Nutrition — Congresses.
2. Infants — Care and hygiene — Oceania — Congresses.
I. Marshall, Leslie B.   II. Symposium on Infant Care
and Feeding in Oceania (1983: New Harmony, Ind.)
III. Ecology of food and nutrition.   V. Series.
[DNLM: 1. Ethnic Groups — Pacific Islands — congresses.
2. Infant Care — congresses. 3. Infant Nutrition —
congresses. WS 113 143 1983]
RJ216.1496   1985   306   85-12648
ISBN 2-88124-037-2 (Gordon and Breach Science)

CONTENTS

PREFACE TO THE SERIES

Two decades ago, interchanges between nutritionists and anthropologists were negligible, and mutual disregard and mistrust were rife. A few of us working (usually alone) on nutrition problems in remote areas of the world inevitably encountered anthropologists and we learned the value of their specialized knowledge of human behavior. As a result of this interaction, the vagaries of human populations became more understandable to us, and the need for nutritionists to achieve a better comprehension of human behavior became more apparent. The ensuing years, which provided a phylogeny of nutritional anthropology, have been extremely fruitful.

In an attempt to bring about improved working relationships between nutritionists and anthropologists, the international journal *Ecology of Food and Nutrition* published its first issue in 1971. Its objective was to provide scientists with a forum for the discussion of the remote, as opposed to the immediate, or etiological, causes of malnutrition. The journal published studies conducted throughout the world showing how culture could affect nutritional status. At the same time, scientists in the two disciplines started to recognize that the quality of nutritional assessment could be improved if the two disciplines worked together. One area much in need of improvement was nutrition survey techniques. Too often these had ignored cultural practices. For example, nutrition contributions in the form of gleanings of indigenous foods were overlooked. Nutritionists had also failed to recognize that food was frequently eaten outside of meals and that even those animal and plant materials not acceptable as food to the Western world could make valuable, sometimes life-saving, contributions to the diet.

A step was made toward rectifying this situation in 1975 when the American Society of Physical Anthropology decided to hold a "Symposium on Nutrition in Physical Anthropology" that included a presentation on how anthropologists could contribute to nutrition surveys. Three years later, the Federation of American Societies for Experimental Biology opened their doors to nutritional anthropology by devoting a section at their annual meeting to a review of studies relating to nutrition and human behaviour. This appeared to signal an acceptance of nutritional anthropology as a scientific speciality. In the next few years, the scientific literature was further enhanced by numerous articles that showed the global interest in, and tremendous breadth of, the science. As a result of these studies, numerous foods previously unrecognized as contributors to nutrient intake by the nutrition establishment were shown to have significant nutritional value as well as cultural importance. A better understanding was also achieved of nutrition and childbirth, food taboos in pregnancy and lactation, and other cultural phenomena of nutritional importance in many parts of the world, and there was a gradual recognition of the need for more in-depth studies. In response to this, one of the Gordon Conferences in 1984 focused on diet in human evolution. The physical anthropologists' interest in body size and eating capacity and the archeologists' interest in early humans, other primates, and diet, served to bring together a large number of scientists to discuss a large number of topics with the central objective of learning more about the human past, including food-related behavior and, by implication, the effects of diet on health.

Ecology of Food and Nutrition has continued to encourage the publication of papers of anthropological and nutritional interest and, as the breadth of the science of nutritional anthropology became increasingly recognized, the publishers decided to republish certain selected papers in monograph form. The first monograph in the series Food and Nutrition in History and Anthropology was *Food, Ecology and Culture.* It was followed by *Famine — Its Causes, Effects and Management.* In this second volume, the published papers were supplemented by invited papers in order to provide a more comprehensive in-depth treatise. This second work was awarded the World Hunger Award as the best scientific book on famine in 1982.

The present book, *Infant Care and Feeding in the South Pacific,* is a further step in the development of our nutritional anthropology series. It provides information on a topic that was previously considered the domain of the pediatrician or the nutritionist. Inevitably, the topic was then presented through the eyes of an ethnocentric Western world. This is an attempt to broaden the perspective. This book, which includes invited papers as well as papers published in *Ecology of Food and Nutrition,* is a specialized publication that we are confident will have a beneficial influence on infant feeding practices in both the Western world and less-developed countries.

John R.K. Robson
Charleston, South Carolina

PREFACE

During the past decade, a great deal of international attention has been focused on methods of infant feeding, especially breastfeeding. Since War on Want published *The Baby Killer* in 1974, most public concern has been directed toward the increased use of the feeding bottle to supplement or replace breastfeeding in the Third World. During this time, the World Health Organization and UNICEF have become more visibly involved in the promotion of breastfeeding and the regulation of the marketing of infant formula. A comprehensive campaign to promote breastfeeding is part of a recent UNICEF six-point approach to improving child health. Pediatric and public health associations have adopted policy statements that strongly endorse breastfeeding. Some national governments, like that of Papua New Guinea, have acted to make bottlefeeding more difficult. Although the topic of weaning foods has not captured the public interest to the same extent as the breastfeeding issue has, it has been of deep concern to nutritionists and health care practitioners. As field workers have repeatedly discovered, programs for promoting breastfeeding or improving the weaning diet in any community must be based on a detailed knowledge of extant infant feeding practices in their social and cultural context.

A large and growing literature exists on the feeding of infants. Biomedical research has centered around the physiology of lactation, the biochemical composition of human milk and its substitutes, and the effect of various nutrients on human growth and development. Others have been particularly interested in the relationships between breastfeeding and fecundity or fertility. Economists have analyzed the costs in money or time associated with breast- and bottlefeeding at the levels of both individuals and nations. Historians have documented changes in feeding apparatus and medical pronouncements and have shown the recurrent nature of debates over how best to feed babies. Innumerable surveys — among small groups, encompassing a nation, or multinational in scope — have been conducted by epidemiologists, nutritionists, psychologists, sociologists, and anthropologists to examine the prevalence of different types of milk feeding and their association with demographic characteristics or morbidity and mortality rates. Many general ethnographies contain at least brief descriptions of infant care and feeding practices. Detailed ethnographic accounts of these practices in social, cultural, and physical contexts have now begun to appear more frequently.

Infant care and feeding practices form an interrelated set of specific behaviors that can be observed and discussed. They do not, however, occur in a vacuum. They are influenced by culturally shaped beliefs and attitudes, the social and physical environment, and the past experiences of the individual. Thus, a study of infant feeding may provide new insights into certain aspects of social organization or may clarify particular cultural concepts. Likewise, familiarity with the broader context makes the individual practices more understandable.

This monograph is intended to present detailed descriptions of infant care and feeding practices, and associated attitudes and beliefs, within their social, cultural, and physical contexts for a number of social groups in the South Pacific region; this includes the tropical islands of Melanesia and Polynesia lying south of the Equator (see maps for specific sites). By focusing on a relatively small geographical region, variations in physical environment, population size, and social organization have been limited. Despite the geographical limitation, however, a great deal of diversity in practice and belief is evident — even in those groups occupying adjacent territories (see Chapters 9 and 10, 12 and 13). A brief summary that would fairly characterize all of the groups is not possible. The study of infant care and feeding is particularly relevant in this region, given the great importance of food and of infants in Pacific Island societies.

In addition to providing an ethnographic account, each contributor uses that information to address particular topics in the literature on infant care and feeding. The four commentators then summarize ethnographic details from all of the contributors to address several general issues.

This monograph had its origin in the slight unrest felt in doing a study of self-reported infant care and feeding practices among a heterogeneous group of urban women (see Chapter 2). Because they came from such diverse cultural backgrounds, the only context in which they could be placed was their work situation. The generally strong probreastfeeding attitudes expressed by them and so many others in Papua New Guinea was intriguing and led to speculation on how widespread this might be in the nation and the region. Over 20 researchers responded to requests for information with abstracts of papers for a proposed symposium. Many others sent detailed answers to specific questions asked. A consideration of these questions formed the backbone of all the papers included in this monograph: What were the methods of feeding milk, other fluids, and solids? What special problems were encountered with any of these methods? What were the causes and timing of the separation of mothers from their infants? How was the infant cared for and fed in her absence? What were the general attitudes toward the separation? Were there any changes in these practices in the recent past? Were there restrictions on activities or diet for mother or infant? What was the physical, social, and cultural environment of the lactating woman?

The group of researchers — representing cultural and physical anthropology, nursing, nutrition, psychology, and education — convened twice during the annual meetings of the Association for Social Anthropology in Oceania. We met first as a working session at Hilton Head, South Carolina, in March of 1982 and then as a Symposium on Infant Care and Feeding in Oceania a year later in New Harmony, Indiana. The topics addressed by the group were diverse: women's conflicting roles and their social support systems, the symbolism and meaning of foods, infant feeding as early socialization, fostering of social and physical independence, breastfeeding and sexual activity, social change (planned and unplanned), and health and nutritional status.

At the working session, ethnographic particulars were presented and basic issues were identified. During the symposium, however, discussion focused on general themes and underlying problems; reflections on these discussions have been incorporated into many of the contributions to this monograph. The variability in practice between and within societies made it difficult to draw any generalizations for this geographical region other than that breastfeeding is, indeed, very common and of long duration by current industrialized world standards.

Differences were also noted between the normative or ideal pattern constructed (by informant or ethnographer) for the group as a whole and the real behavior of individuals. Some of the intragroup variability could be attributed to change over time. The baseline often used to determine change has been "traditional practice," which has often simply represented an ideal or general pattern rather than historical reality. Some of the contributors had done longitudinal studies of sufficient duration to actually document changes; others made extensive use of historical documents. Assuming that a change in practices could be demonstrated, some of the questions felt to be important included: What has caused the change or influenced its course? What have been the consequences — social and biological — for the people involved? These questions have important policy implications that were addressed by several contributors.

xiii

Some of the variability in practices within and between groups was attributed to differences in women's roles, activities, and support systems in the first year after birth. The individual strategies and social mechanisms for resolving the conflict between reproductive and non-reproductive demands on a new mother's time were carefully delineated by many of the contributors.

Because the importance of making this information available to those in disciplines other than our own was felt strongly, appropriate research methodology and clear communication were major topics of discussion at the symposium. The benefits and drawbacks of observation, informal questioning, standard surveys, pure description, and quantification were discussed in the context of how one can best use limited fieldwork time. The use of emic versus etic categories for measuring age (relevant to growth and development measures and to feeding practices) was also considered as we grappled with the problem of defining and measuring malnutrition.

It became apparent that, despite all the differences noted, food is a core concept in all the groups described. The taboos on food for parents (prospective or real) and infants indicated that food is believed to be something important to or for the individual. The act of feeding in these societies also establishes and defines critical social relationships (usually parenthood) between adults and an infant and demonstrates caring.

The composition of our group changed over the three years of exchange preceding the completion of this monograph. Some of the working session or symposium participants did not continue and others joined in after one or the other of the meetings. Eventually, fifteen contributed chapters to the monograph. Of these, twelve also appeared in Volumes 14, 15, and 16 of *Ecology of Food and Nutrition* by special agreement with the editor.

The commentators for the monograph were chosen because of their research interest in infant feeding or early development. All have worked in the tropical Third World — two specifically in Papua New Guinea. Two of the commentators — Bambi Schieffelin and Judith Gussler — were symposium discussants. Schieffelin is an Anthropologist and assistant professor of education at the University of Pennsylvania, has carried out fieldwork in rural Papua New Guinea for over three years, and has published extensively on early socialization and language learning. Gussler is the former manager of medical anthropology research at Ross Laboratories, has done fieldwork with women and infants primarily in the Caribbean, and has written and lectured extensively on infant feeding issues. The third commentator, John Biddulph, is professor of child health at the University of Papua New Guinea and senior pediatrician for that nation's Department of Public Health. He has worked there for twenty-four years, twenty-two of them teaching at the country's medical school, and has played a major role in the strong breastfeeding promotion programs in that country. The final commentator, Penny Van Esterik, is now assistant professor of anthropology at York University (Ontario) after several years as part of the Cornell University research group conducting a four-country study of infant feeding. She has done extensive fieldwork in rural and urban Thailand, including research on infant feeding and women's activities. All of these commentators have discussed the ethnographic reports from their own particular perspectives and have presented their own opinions.

The intended audience for this monograph is even more diverse than its contributors. The material should prove valuable to nurses, nutritionists, doctors, other health care practitioners, and health planners. The general message for all health care workers, repeated frequently throughout the monograph, is that behaviors cannot be taken out of their sociocultural context and that understanding the context is essential for planning any intervention. In addition, for those working in the South Pacific region, there is a great deal of information relevant to specific areas.

The material should also be of interest to those anthropologists and other social scientists involved in research — basic or applied — on women's roles, early childhood, or health-related issues. Because the chapters contain data and arguments on general issues as well as good regional ethnography, they should appeal to more than just the community of Pacific scholars.

It is hoped that this book will also find its way into the hands of the lay public — in the Pacific islands and elsewhere — who care about the place of women, infants, or food in societies other than their own. Different solutions to very common problems may be found here.

The basic layout of the book is as follows: An introductory commentary summarizes the major themes in the ethnographic reports that follow it. Because many authors discuss more than one of the themes, the descriptive chapters have not been grouped by topic. The first one concerns urban women, the others have rural settings. The chapters on Papua New Guinea are followed by those on the neighboring Solomon Islands, then Fiji, and finally Western Samoa — an easterly progression through the islands. The first of the three concluding commentaries emphasizes infant health and nutrition, the second stresses the relationship between infant feeding and women's other activities, and the final one directs the reader's attention to some general concepts in anthropology to which these papers relate.

Leslie B. Marshall
University of Iowa

ACKNOWLEDGMENTS

Many people and institutions contributed to this project in special ways during its three-year gestation. It began while I was a research associate funded by the World Health Organization at the Papua New Guinea Institute for Applied Social and Economic Research (IASER). I am particularly grateful to the administrators and staff of that institution for their kindness and generous support. After I returned to the University of Iowa College of Nursing, support for this project continued unabated. I am greatly indebted to the excellent secretarial staff and to the administration for all their help. Mary Jane Benson and Julie Duncan, my graduate research assistants at the College, did a painstakingly thorough job of preparing the index. The University of Iowa Graduate College funded the preparation of the regional maps by the Graphics Unit. For lively discussion and encouragement in continuing the project, I thank all the participants, including those whose work does not appear in this monograph — the late Edwin Cook, Helen Doan, Christine Duval, Donna Foster, Jonathan Friedlaender, Eulalia Harui-Walsh, Brigit Obrist, Susan Pflanz-Cook, Nancy Pollock, Rebecca Stephenson, and Stanley Ulijaszek. John R.K. Robson and Ann Tamariz of *Ecology of Food and Nutrition* provided invaluable assistance with the editorial tasks. Finally, I owe a very special thank you to Mac Marshall for convincing me to convene an ASAO Symposium and to try my hand at editing; to Terry Hairoi, R.N., and Reia Taufa, M.D., for early and continued encouragement in studying women and infants in Port Moresby; and to numerous other Pacific Island women community leaders, professionals, and students who, by emphasizing the need to get research findings into a form useful for their communities and nations, prompted me to assemble this collection.

MAPS OF THE RESEARCH SITES

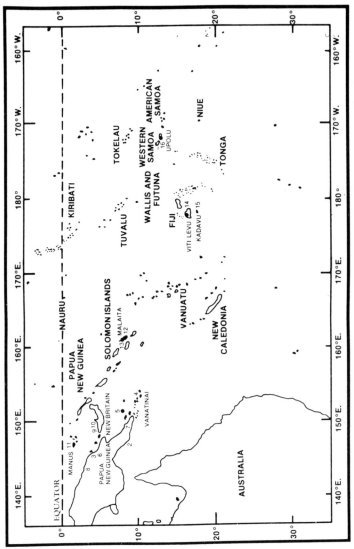

MAP 1   Partial map of the islands of the South Pacific region plus Australia. The numbers indicate approximate locations for the research reported in the chapters so numbered. Adapted from The New Pacific map, Hawaii Geographic Society.

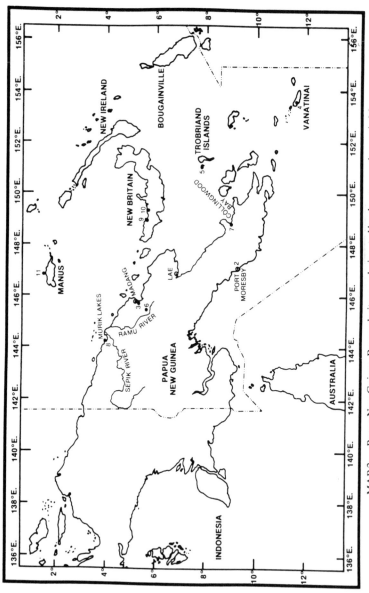

MAP 2   Papua New Guinea. Research sites are designated by chapter numbers as in Map 1. Adapted from Crown Series map no. 520, Robert Brown and Associates Pty Ltd.

XX

CONTRIBUTORS

K. Gillogly Akin, B.A.,
Department of Anthropology,
University of Hawaii, Manoa,
Honolulu, Hawaii 96822, U.S.A.

Kathleen Barlow, Ph.D.,
Department of Anthropology,
University of California,
San Diego,
La Jolla, California 92093,
U.S.A.

John Biddulph, M.D., F.R.C.P.,
F.R.A.C.P.,
Faculty of Medicine,
University of Papua New
Guinea,
Boroko, Papua New Guinea

Achsah H. Carrier, B.A., Dip.
Soc. Anth.,
Department of Anthropology,
University of Papua New
Guinea,
University, Papua New Guinea

Ann Chowning, Ph.D.,
Department of Anthropology,
Victoria University of
Wellington,
Wellington, New Zealand

Leslie Conton, Ph.D.,
Department of Anthropology
and Fairhaven College,
Western Washington University,

Bellingham, Washington 98225,
U.S.A.

Dorothy A. Counts, Ph.D.,
Department of Anthropology,
University of Waterloo,
Waterloo, Ontario, N2L 3G1,
Canada

David Welchman Gegeo, M.S.,
Graduate School of Education,
Harvard University,
Cambridge, Massachusetts
02138, U.S.A.

Judith D. Gussler, Ph.D.,
Ross Laboratories,
Columbus, Ohio 43216, U.S.A.

Peter F. Heywood, Ph.D.,
Institute of Medical Research,
Madang, Papua New Guinea

Carol L. Jenkins, Ph.D.,
Institute of Medical Research,
Goroka, Papua New Guinea

Mary Maxwell Katz, Ed.D.,
Graduate School of Education,
Harvard University,
Cambridge, Massachusetts
02138, U.S.A.

Maria A. Lepowsky, Ph.D.,
M.P.H.,
Public Health Nutrition

xxi

Program,
School of Public Health,
University of California,
Berkeley, California 94720,
U.S.A.

Leslie B. Marshall, Ph.D.,
College of Nursing,
The University of Iowa,
Iowa City, Iowa 52242, U.S.A.

Susan P. Montague, Ph.D.,
Department of Anthropology,
Northern Illinois University,
De Kalb, Illinois 60115, U.S.A.

Janice M. Morse, R.N., Ph.D.,
Faculty of Nursing,
University of Alberta,
Edmonton, Alberta T6G 2G3,
Canada

Bonnie A. Nardi, Ph.D.,
IntelliCorp,
Menlo Park, California 94025,
U.S.A.

Alison K. Orr-Ewing, M.S.,
Institute of Medical Research,
Madang, Papua New Guinea

Bambi B. Schieffelin, Ph.D.,
Graduate School of Education,
University of Pennsylvania,
Philadelphia, Pennsylvania
19104, U.S.A.

Anne Marie Tietjen, Ph.D.,
Department of Psychology,
University of Washington,
Seattle, Washington 98195,
U.S.A.

Penny Van Esterik, Ph.D.,
Department of Anthropology,
York University,
Downsview, Ontario M3J 1P3,
Canada

Karen Ann Watson-Gegeo,
Ph.D.,
Graduate School of Education,
Harvard University,
Cambridge, Massachusetts
02138, U.S.A.

CHAPTER 1

COMMENTARY:
THE IMPORTANCE OF CULTURAL
PERSPECTIVES ON INFANT CARE AND
FEEDING[†]

BAMBI B. SCHIEFFELIN

INTRODUCTION

These ethnographic studies represent a unique collection documenting one of the most critical concerns for all social groups, the care and feeding of infants and children. Of particular importance is their emphasis on cultural factors in this process. This includes not only the beliefs that shape people's understandings of infant care, but also the contingencies generated by particular social and cultural contexts in which infants and children are cared for. The chapters focus on a number of different societies in the Pacific. Through the examination of a wide range of topics, the authors raise methodological and theoretical issues around which future studies of infant care and feeding might be organized. Additionally, these studies offer information that will be of great relevance to policy makers by demonstrating the importance of social and cultural variables for developing and recommending programs concerned with improving maternal and child health.

METHODOLOGICAL ISSUES

These studies raise important methodological issues. The relevance of collecting in-depth case studies that include ethnographic detail is obvious. It is only when we have a sufficiently fine-grained

[†]My sincere appreciation goes to Kathleen Barlow and Buck Schieffelin for their insightful comments which have been incorporated into this chapter.

1

description and analysis of feeding and caregiving practices as they are integrated within their cultural and historical contexts that we can begin to compare these data with regard to such topics as health, nutrition, and work, which are themselves culturally organized systems. The ethnographic perspective sees culture not as a patchwork of different ideas and practices, but as the system that integrates them all. Thus, one of the more important methodological and theoretical issues is the *integration of different types of data* to get at the underlying system of organization. For example, Montague (ch. 5) points out the value of this when she notes that Trobriand Islanders do not accurately report infants' and childrens' actual feeding patterns in brief survey interviews. Instead they provide data which describe locally ideal adult patterns. Thus, she suggests that interview data alone are not enough to understand what Trobrianders are doing. Interviews must be supplemented by detailed systematic observations of actual practice. This is necessary in many Pacific societies where ideals about what people eat and what they actually eat vary considerably. This difference, together with intra-group variation in behavior, potentially makes it difficult to characterize the dietary practices of any particular society (let alone comparing practices between societies) on the basis of interview data alone.

The major questions here are not only *what* is the relationship between what people say they do and what they actually do, but also *why* there is such a difference and *what it means* for the way people feed and care for their children. Inquiry should proceed on the assumption that there is a relationship between different cultural statements and practices even if people say there is not. Lepowsky (ch. 4), Montague (ch. 5), Conton (ch. 6), and others point out the underlying interaction of different systems with each other and their functional interrelations even when these are not articulated by the people themselves.

It is also important to understand the ways in which the people of a given society act according to explicit cultural norms and knowledge. For example, it is the case that most Pacific peoples have their own ways of perceiving and evaluating developmental progress. Standard clinical weights and bodily measurements do not capture their relevant categories. Jenkins, Orr-Ewing and Heywood (ch. 3) argue that it is important to understand how the stages of children's growth are conceptualized emically as well as etically (according to Western ideas) in order to grasp the relationship of children's growth to feeding practices in any society.

Several sets of original findings in this book demonstrate the importance of integrating different research perspectives. Lepowsky (ch. 4) provides an excellent example of the integration of cultural and biological approaches by combining medical and comparative ethnographic literature in a particular case study. Jenkins, Orr-Ewing and Heywood (ch. 3) show the relevance of undertaking nutritional studies in the context of ethnographic observation. In all of these chapters, it is clear that researchers must continue to draw on a variety of disciplinary perspectives on the study of infant care and feeding.

THE ROLE OF IDEOLOGY

A major theme discussed in many of the chapters is the role of ideology in shaping infant feeding practices. Only by understanding the cultural ideology that underlies a set of practices can we make sense of the meaning that actions and events have for participants. These studies include data focusing on people's cultural ideas about what constitutes a healthy child and proper feeding practices. Montague's research on ideology and nutrition in the Trobriand Islands (ch. 5) details the local cultural expectations about proper body configuration. This type of information, in turn, helps make sense of what otherwise might seem to be maladaptive feeding practices. Since parents positively value light body weight for their young children, they make sure that children eat enough to be muscled, but visible fat deposits are undesirable. Furthermore, in the Trobriands, the emphasis is on the musculature of the legs of children, not the arms.

Jenkins, Orr-Ewing and Heywood (ch. 3) found that the Amele beliefs that breastmilk increases in value as children age and that liquid foods are most suitable for small children, result in little supplementary food being offered to infants in the first year of life.

According to the Kaliai (ch. 9), breastmilk is the only female effluvium that is not polluted. It may be contaminated, however, especially by male substance (semen). Thus, the Kaliai must make sure a wetnurse is not sexually active. The neighboring Kove (ch. 10) share this belief and concern. The Kwaio (ch. 12) believe that the quality of breastmilk changes over time. This, too, is an important consideration in matching the needs of the nursling to a wetnurse.

Morse (ch. 14) compares infant care and feeding practices in Fijian and Fijian-Indian groups, pointing out that withholding the breast for three to four days, a custom practiced by both groups, is done for very different cultural reasons. This underlines the importance of understanding the beliefs behind apparently identical types of behavior, rather than just describing the acts themselves. Policy makers and health care workers in particular must be alert to these differences in meaning because they indicate that apparently similar health care interventions may have quite dissimilar implications in different societies and require alternative approaches for implementation.

Situations of social and historical change present complex problems for assessing the relations of ideological patterns to cultural practices. Counts (ch. 9) shows that the Kaliai have undergone considerable social change, including giving up their traditional procreation theory and changing their ideology regarding sexuality. In addition, they have introduced new foods into the diets of older children and adults. These changes do not, however, include new social or economic activities which compete with women's traditional domestic responsibilities, and thus there are no new incentives of this sort for changing the age of weaning or introducing food supplements. Ideological changes in this case were not reflected in changes in infant feeding practices.

Among the Kwara'ae (ch. 13), where traditional taboo systems have been abolished and Christianity is being accepted, men and women no longer practice ritual separation and now live together in the same house. As a result of this and of the abandonment of infanticide, the fertility rate and family size are increasing. At the same time there are critical changes in available subsistence resources. This has led to changes in Kwara'ae family structure, work organization and infant care. However, as Gegeo and Watson-Gegeo point out, these changes are not merely a result of changing population characteristics and concomitant social structural accommodations. Many of the changes result from conscious decision-making by the Kwara'ae and a determination to maintain their traditional central values while participating in a modernizing society.

VARIATION

Another important theme to emerge from these chapters concerns variation in infant care and feeding practices. We expect a great deal of variation among Melanesian societies; what is quite impressive is the amount of variation within each social group.

Some sources of variation are imposed or enabled by external ecological conditions or by pressing social factors that parents do not themselves control. For example, in many places there is a seasonal variability in the food supply. As Barlow (ch. 8) points out for the Murik, there is variability in the foods and other resources to which different families have access. Katz (ch. 15) points out that in the outer islands of Fiji there are several additional things that contribute to variations in infant care practices: parental age, and social and economic factors. She found that feeding practices were related to whether or not the husband or male kin grew green vegetables in addition to essential root crops, and how much cash was available to the family to buy food. Family composition also had an effect on the individual child's diet, because animal foods received high value and were given first to the adult men in a family. Katz found factors associated with early infant weaning in a traditional context which permitted a mother to return to work. She found that mothers whose consanguineal kin resided in the same village ended breastfeeding earlier, suggesting that a plentiful food supply and/or greater availability of infant caregivers enabled this kind of flexibility.

Conton (ch. 6), Counts (ch. 9), Chowning (ch. 10), Akin (ch. 12), Gegeo and Watson-Gegeo (ch. 13), and Nardi (ch. 16) are all concerned with variation in traditional patterns of infant care and feeding due to socio-historical and ideological changes brought about by contact with outside influences, especially Western religious dogma, trade practices, cash, and business.

Individual decision making is an important basis for variation in some societies. This point is made by Conton (ch. 6) for the Usino people who make their choices individually as to when to introduce semisolid and solid foods to infants. Conton argues that there is a lack of consensus in both theory and practice among the Usino people regarding food restrictions for nursing mothers and their children and for practices regarding supplementary feeding. There are no universally agreed upon restrictions, nor is there a common body of knowledge shared by all mothers about them. However, for the Usino people, almost all lactating mothers follow some food taboos to protect the health of the nursing offspring and all must avoid the ingestion of animal protein.

Gegeo and Watson-Gegeo (ch. 13) present the case for there being no clear agreement — even traditionally — within the Kwara'ae community about what women should eat during pregnancy except for one category of foods. In this society, individuals have a great deal of choice, not only in deciding dietary practices they themselves will follow, but what their young children may or may not eat. Variation extends to the choice of a child's first foods, the times solid foods are introduced, and the age at which children are weaned. For the Kwara'ae, the variation is due to a complex interaction of individual choice and the birth order of the child. The role of birth order in determining the food habits of both mother and child is also discussed by Lepowsky (ch. 4) for Vanatinai society, where there are variations in mothers' food taboos according to stages in their life cycle and the numbers of children that they have already borne.

In contrast to some peoples who have well-formulated explicit ideologies about infant care and feeding (as do the Murik, ch. 8), Usino people (ch. 6) do not talk much about these things. This apparent freedom of choice, however, must not be taken to mean that in some societies feeding practices and their underlying conceptual bases are applied strictly idiosyncratically rather than in a culturally normative way. Even Usino mothers' range of personal choice must fall within the range of cultural definitions of food versus non-food and the meaning of feeding. There is always the possibility that a practice is culturally shaped even when the people do not (or cannot) articulate reasons for it. While it is important to document variation among and within cultures, it is even more important to look further into the cultural sources for the variability. We must not lose sight of the fact that variation derives from many different sources. For example, seasonal variation in Murik food supply (ch. 8) is related to and integrated with the ritual cycle of trade. This is a different kind of variability than Conton (ch. 6) or Carrier (ch. 11) describe: variation due to general acceptance of the idea that one eats what is available or variation due to social class difference or cultural change. To continue with the Usino case, eating whatever is available is a very different cultural attitude toward food than a situation in which people are suspicious of exotic foods (ch. 8) or animal substances (ch. 5), or in which certain foods have prestige value or are taboo. Thus, what seems to be a lack of Usino people's ideology about food is itself a cultural phenomenon. Given the importance of feeding and the complexes of meaning attached to it in other Pacific societies, the Usino people's laissez-faire attitude is quite striking, culturally distinctive, and probably related to other aspects of the culture and current situation. A great deal of free variation for multiple reasons is something that needs further study. This investigation of variation should uncover other kinds of information about feeding practices and their relation to other aspects of the culture. This is not to say that everything will turn out to be meaningful or well-integrated, but a focus on the diversity of surface variation must not be allowed to undermine the systematic underlying cultural sources for it across Pacific societies if the real implications of such variations are to become clear. In this effort, how explicitly a given cultural group articulates their norms and expectations regarding feeding practices is only one consideration among many, even though explicit norms frequently shape actual practice.

SOCIAL CONTEXT AND MEANING

Another theme developed in these ethnographic studies is the relation between the social context and the meaning of giving and receiving food. In many Pacific societies, including these, the meaning of giving food has less to do with nutritional value than with affective associations. Breastmilk is especially important. As Gegeo and Watson-Gegeo (ch. 13) point out, "Even when an infant seems completely satisfied with cow's milk, caregivers say that it is dissatisfied; any signs of discomfort are attributed to the absence of maternal love associated with breastfeeding".

For most of the peoples described in this volume, dietary prohibitions as well as the prescription of special foods in pregnancy and child bearing years can be seen as part of a larger pattern of dietary changes and social meanings which mark the critical points in the life cycle.

Among the Murik (ch. 8), feeding is a way of negotiating membership in social groups for whom the recruitment of children is essential. For these people, a major means of claiming a child for adoption is through feeding both child and mother. A similar set of beliefs and practices is documented by Counts (ch. 9) for the Kaliai, who also maintain that sharing food is an important aspect of adoption, as food establishes kinship relationships before and after birth. See Carroll (1970) and Brady (1976) for a more detailed consideration of adoption in the Pacific region.

Counts suggests that for the Kaliai, early feeding practices set up enduring personal assumptions and expectations. The breast is offered to young children at the first sign of discomfort, and the

8 B.B. SCHIEFFELIN

practice of using food to provide comfort continues throughout life. Furthermore, Kaliai emphasize the importance of breastmilk as the substance of maternal kinship. They see the sharing of breastmilk as critical to nurturance.

Tietjen (ch. 7) also emphasizes the social nature of food giving among the Maisin, pointing out that infant feeding is an important vehicle for introducing very young members of a society to the ideology and appropriate behaviors of that social group. Beliefs and practices important to Maisin adults are reflected in the care and feeding of their infants. Further investigations of what can or cannot occur when one is eating or feeding someone, and what is the general mood or the range of moods of the feeding situation in any society would also be valuable for our understanding of the socialization which occurs with infant feeding.

CULTURAL ADAPTATION AND NUTRITION

In her discussion of cultural adaptation and infant feeding in Vanatinai, Lepowsky (ch. 4) poses the following questions: Why do areas with apparently abundant food resources (such as Vanatinai) have large numbers of children between the ages of one and three years measured as malnourished? Why are there large differences in the reported rates of child malnutrition for different parts of the Louisiade Archipelago? In Vanatinai, children below weaning age eat no animal protein foods, fruits, greens or storebought food. Reviewing the medical literature, Lepowsky suggests that an antagonistic relationship exists between nutritional deficiencies and malaria. She concludes that these customary restrictions on animal protein in the diets of infants and lactating mothers may be a "long-standing cultural adaptation to an environment of endemic malaria". Lepowsky cites evidence from ethnographic literature elsewhere suggesting that similar taboos against consumption of animal protein are said to have similar consequences. The strict reliance on breastmilk means that the child is dependent on mother's milk for virtually all of its protein intake. This provides the necessary antibodies to a variety of diseases and assists in the avoidance of possibly contaminated water and of fatty foods, both of which might bring on diarrhea. The suggestion that the more poorly nourished individuals seem to be more resistant to malaria is provocative, and opens up an important line of research.

Lepowsky also argues that some cultural practices or courses of action, such as limiting protein intake, may appear arbitrary or maladaptive because their connection to survival and reproduction has not been carefully studied. As she makes clear, it is not enough to study the indigenous dietary systems for their symbolic value. One must examine the interactions of different systems for their effects on one another.

CHANGING PATTERNS OF BREASTFEEDING

For groups such as the Usino people (ch. 6), there is a compatibility of women working while continuing traditional patterns of prolonged breastfeeding because of extended kinship support systems. Among the Kaliai also (ch. 9), the cultural ideology supports traditional breastfeeding patterns. In most Pacific societies discussed here, however, patterns of breastfeeding are changing and several authors examine the reasons for and consequences of these changes.

One reason for changing the age of weaning relates to changes in the domestic relations between husband and wife. For example, on Ponam Island, Carrier (ch. 11) describes the general patterns of infant care and feeding within the context of husband-wife and adult-child relations. She suggests that less segregation between the sexes has resulted in a decrease in birth interval, which in turn has had a significant effect on the age of weaning. Counts (ch. 9), Chowning (ch. 10), and Akin (ch. 12) also consider the relationship between weaning, postpartum sex taboo, and next conception. In many societies, weaning is supposed to occur before the resumption of sexual relations, as semen or pregnancy is thought to spoil breastmilk. However, weaning often takes place once the mother knows she is pregnant. Therefore, some women resume sexual relations while still nursing and this raises the important question of the extent to which nursing limits conception.

Another reason for early termination of breastfeeding is women's greater workload associated with their increased involvement in the cash economy. Nardi (ch. 16) suggests that in Western Samoa this reason, rather than other factors (such as the availability of infant formula for bottle feeding, women's participation in the wage labor force, insufficient lactation, medical intervention or breast eroticism) has resulted in a decline in the age of weaning. Samoan women initiate earlier weaning so as to attend to other responsibilities.

Chowning (ch. 10) documents the difficulties of breastfeeding for the Kove, who seek to emulate what they have thought to be European practices. She points out the burdens of long breastfeeding for nursing mothers experiencing ill-health, work demands, and restrictions on sexual intercourse, which contribute to a less than idyllic situation for these women. (See also Akin, ch. 12). The Kove also feel that arrangements made for feeding babies whose mothers have to leave the village to work are unsatisfactory, placing excessive demands for milk on those women who stay at home and wetnurse. In addition, there is the danger to the baby of pollution of breastmilk by pregnancy, which is a more serious source of pollution for the Kove than semen and sexual intercourse. Chowning sees the major problem with changing traditional feeding patterns to be the weaning of infants onto a limited range of storebought foods of dubious value.

Another factor that can change traditional breastfeeding patterns is wage employment of the mother outside of the home. Marshall (ch. 2) examined the effect of this among public health nurses in Port Moresby, finding that personal commitment, knowledge of infant feeding options, and social support from employers and co-workers contributed to the high prevalence and long duration of breastfeeding among this urban group. In addition, policy decisions supported more traditional feeding patterns through the assignment of nurses to posts close to home.

Marshall's study shows the complexity of factors affecting the choice to breastfeed, and the importance of direct investigations of these variables. Her research has application for changing work situations and for increasing policy makers' awareness of the consequences of their decisions.

IMPLICATIONS FOR POLICY

A final theme addressed in these chapters concerns the policy implications of these studies on changing patterns of infant care and feeding. Both Lepowsky (ch. 4) and Conton (ch. 6) argue for caution on the part of government and health personnel when considering intervention and change in terms of traditional adaptations and food taboos. If such traditional food taboos do protect young children against malarial disease, any change in feeding patterns without accompanying malaria control programs could be devastating to the infant population. Conton urges the monitoring of the impact of economic development programs on the health of people after establishing a baseline of observations made before the programs are instituted. Self-sufficient economic systems and traditional value systems frequently give way to systems that emphasize the acquisition of cash. Such changes in economic organization, traditional marriage and residence can lead to radical changes in infant feeding strategies. The potential consequences of departure from more traditional practices which Conton, Chowning (ch. 10), and Akin (ch. 12) outline are certainly negative, and well worth the attention of health care individuals. However, as Marshall (ch. 2) points out, change does not always lead to negative outcomes if proper planning takes place beforehand.

CONCLUSION

The chapters in this volume present ethnographic studies that raise key issues concerning infant care and feeding. In particular, the importance of contextualizing mother-infant interaction within the broader family and community environment that support them is emphasized. Demands on the mother that are directly related to the baby, as well as demands that less directly affect the infant-caregiver relationship, are examined in detail. The cultural context, and not just biological necessity, makes comprehensible the expectations that guide the roles of mother, father and baby. The ideological system gives meaning to the symbols and value of food and mother's milk, including the quantity and quality of the milk. The effects of outside agents that influence local customs, such as government, business, and church are analyzed in their full complexity.

Throughout these studies, variation among cultural groups is documented. We see variation in terms of who cares for the child in the early years; in when the breast is offered, who can offer it, and when it is taken away; in what supplementary foods may be used and when they are given. There is variation in contraception, abortion, and infanticide practices across the cultures described. The sources of this variation are diverse. In some societies it is due to social change from outside influences (Western and/or from other indigenous groups). Changes may affect a society rather homogeneously. However, there is a great deal of intra-group variation, especially in societies with a system of individual options on how to care for children. In addition, families and individuals have differential access to resources such as

B

land, food, and labor assistance. This degree of variation due to diverse internal as well as external sources suggests a dynamic interaction in which many factors contribute to what individuals eat. The combination of within-group variation in the context of rapid social change from outside forces creates particular problems for research methodology and analysis.

From a cultural perspective, how do we characterize what constitutes the infant care and feeding situation in a given society? These chapters suggest many good ways: through an ethnographic approach, viewing individuals through the life cycle, documenting the relevant details not only of what people say they should do and what they say they do, but also of observations on what they actually do every day. Using a variety of research methods in an integrated fashion is imperative to making sense of these complex situations. Not only must we take account of variation, but we must also determine the cultural, biological, and ecological patterning of this variation. Finally, the interactions between the cultural-symbolic and biological-ecological realms remain extremely interesting and, from a practical health policy point of view, potentially the most rewarding to pursue.

REFERENCES

Brady, I. (Ed.) (1976). *Transactions in Kinship: Adoption and Fosterage in Oceania.* ASAO Monograph No. 4, University Press of Hawaii, Honolulu.
Carroll, V. (Ed.) (1970). *Adoption in Eastern Oceania.* ASAO Monograph No. 1, University Press of Hawaii, Honolulu.

CHAPTER 2

WAGE EMPLOYMENT AND INFANT FEEDING: A PAPUA NEW GUINEA CASE[†]

LESLIE B. MARSHALL

INTRODUCTION

For a mother, the situation in which the conflict between the demands of one's baby and the demands of one's other responsibilities is most clearly defined is that of wage employment outside the home. Working hours usually are rigidly set and must be strictly observed. The workplace is often far from where the baby is kept. The tasks to be completed and the general work environment are such that a baby's presence would be disruptive if not dangerous. Any regular, frequent aspect of childcare — such as breastfeeding — is simply not possible during working hours when mother and infant lack easy access to one another. Even the passage of legislation based on the International Labor Organization Maternal Protection Convention Number 3, which prescribes breastfeeding breaks in the working day (Richardson, 1975), does not necessarily guarantee employers' compliance or alert women to their rights. Furthermore, the problem of providing a desirable environment for the baby in a convenient location for the rest of the working hours is rarely addressed.

Many recent studies in tropical Asia and Africa have shown a decreased prevalence or duration of breastfeeding among women who work for wages outside their homes, especially in urban areas (Knodel and Debavalya, 1980; Pongthai, Sakornrattanakul, and Chaturachinda, 1981; Pathmanathan, 1978; Koh, 1980; Akin *et al.*,

[†]The study was conducted while the author was a Research Associate at the PNG Institute for Applied Social and Economic Research in Port Moresby. This investigation received financial support from the World Health Organization.

13

1981; El-Mougi *et al.*, 1981; Harfouche, 1981; Mudambi, 1981;
Igun, 1982; Uyanga, 1980; Power, Willoughby and de Waal, 1979;
WHO, 1981). This demonstrated relationship between wage work
and decreased breastfeeding is not a simple one and may be affected
by the mother's ethnic background, educational level, income and
parity, as well as by the nature of her employment. For example, farm
laborers and market traders actually have a higher prevalence of
breastfeeding than do housewives. Increased distance between home
and workplace has also been associated with a decline in breastfeeding (Popkin and Solon, 1976).

It is debatable whether or not return to wage employment postpartum causes termination of breastfeeding or introduction of a
feeding bottle. In some studies, employment was stated by a substantial portion of the interviewees to be their reason for decreased
breastfeeding (Biddulph, 1975; Overseas Education Fund, 1979;
Mohrer, 1979); in other studies, it was mentioned infrequently
(Zurayk and Shedid, 1981; West, 1980; Van Esterik and Greiner,
1981; Overseas Education Fund, 1979). In a prospective study of
breastfeeding practices, Sjolin, Hofvander and Hillervik (1979) found
that 11 out of 46 women curtailed breastfeeding because of such
conflicting demands on their time as employment, studies or the care
of other children. Unfortunately the proportion of women employed
and their particular occupations were not specified in these studies,
even though these factors would have an important influence on the
results. Obviously, return to employment would not be given by a
housewife as a reason for using a feeding bottle or weaning from the
breast.

A few recent studies of breastfeeding practices have focused on the
wage employed mother. In a survey of Yoruba women in Ibadan, Di
Domenico and Asuni (1979) found that 30% of those in a modern
working environment, but only 15% of those in a traditional working
environment such as the market, gave "work" as a reason for
terminating breastfeeding. Of the Hyderabad working mothers interviewed by Thimmayamma, Vidyavati and Belavady (1980), 23%
cited "work" as their reason for weaning and 81% felt a conflict
between the time demands of their job and the time required for
proper childcare. Eighty-six percent of Malaysian nursing personnel
gave "work" as their reason for discontinuing breastfeeding and most
of these women stopped at six weeks when their maternity leave
ended (Sinniah, Chon and Arokiasamy, 1980). Only 14% of the
Israeli nurses surveyed by Bergman and Feinberg (1981) returned to

work within six months postpartum while still breastfeeding. Inconvenience due to working hours was cited as the reason for sevrage by 56% of the nurses and 29% of the teachers in urban Nigeria surveyed by Ojofeitimi, Elegbe, and Etuknwa (1982), and all of these had introduced a supplementary bottle in the first month postpartum. The major reason for bottlefeeding given by Cameroonian nurses (62%) was their academic or employment activities (Garrett and Ada, 1982). Although supplementary bottles usually were introduced soon after birth by the Thai professional women described by Van Esterik (1982), most of these women did not terminate breastfeeding until their 45-day maternity leave was over and they returned to work.

Worldwide, the infant feeding bottle has become a widely used and convenient substitute for mother's breast. Its exclusive use, however, has been linked to increased prevalence of infant morbidity and mortality, especially in the developing world (Jelliffe, 1976; Marshall and Marshall, 1980). As a result of concern for the health of young children, many countries recently have attempted to promote breastfeeding (WHO, 1981, p. 133) or to restrict the availability of feeding bottles (Biddulph, 1981). Papua New Guinea (PNG) has attempted to do both.

BACKGROUND

In 1977, the PNG government enacted legislation to make feeding bottles available only by prescription from authorized health personnel (medical officers and nursing sisters). For a number of years previous to that, various educational programs promoted breastfeeding. Training programs for health personnel emphasized breastmilk as proper infant food and treated feeding bottles as pieces of hospital equipment. The Native Employment Ordinances of the colonial period established the rights of employed breastfeeding mothers to two half-hour nursing breaks daily during working hours (Richardson, 1975). Thus, in PNG, access to a convenient substitute for breastfeeding has been limited and breastfeeding — even among the wage-employed — has been strongly encouraged. Yet many of the difficulties of combining all the other aspects of childcare with wage employment remain.

The present study was designed to investigate the attitudes of one group of working women in PNG — the urban public health nursing staff — toward certain infant feeding practices and the personal experiences they had in reconciling the demands of infant care and feeding and wage employment. As working mothers, nursing staff in the urban clinics and hospitals in PNG were in a particularly awkward situation. They were well-educated town-dwellers, facing all the demands of wage-employment on their time and energy, and thus were among the group of mothers in a developing country who should be most likely to bottlefeed. In addition, they were well-qualified and well-situated to obtain prescriptions for bottles once those became legally necessary. At the same time, they had been trained to encourage breastfeeding and were expected to provide such encouragement as part of their job. They were in a position to influence other mothers by giving advice, by discussing how they personally were coping with the demands of motherhood and employment and by controlling access to feeding bottles through prescriptions.

The daily tasks of most nurses or nurse aides in PNG were incompatible with continuous or even regular mother-infant contact. The urban health clinics usually were open from 8 a.m. to noon and from 1 p.m. to 4 p.m. The staff arrived approximately a half hour before opening to prepare for the morning's activities and often continued work part way through the lunch hour. Upon occasion, in the busier clinics, some of the staff would continue working until 5 or 6 p.m. Busy times in each clinic varied with the time of day, the day of the week and the season. Generally, staff were involved in treating trauma or infections, drawing blood, analyzing urine, interviewing or giving medicines or injections. All of these tasks required continued concentration and could not be interrupted easily. Other employment situations often were more demanding of nursing staff than were the urban clinics. Hospital nursing staff on the wards were responsible for monitoring patients and carrying out routine and emergency treatments. Days off occurred sporadically and shifts often switched. Staff members assigned to urban domiciliary care or mobile clinic units made house calls or travelled to neighboring villages all day long. Rural clinics had fewer patients, but also fewer staff members and thus less chance of finding anyone to take over extra tasks while a nurse slipped out to breastfeed her baby. Mobile clinic units in rural areas sometimes were gone for days or even weeks at a time.

METHODS

In June and July, 1981, those nurses and aides who were mothers and also fulltime staff at public health care facilities in Port Moresby, the major urban area and capital of PNG, were interviewed in English by the author or the project assistant regarding their infant care and feeding practices. The project assistant was a well-educated local woman who was active in community affairs and who had five children of her own. The interviews took less than 30 minutes and were conducted in a private, quiet area of each woman's place of work. The sample group consisted of 88 of the 96 nurses and aides who worked fulltime at the 11 public clinics in Port Moresby plus nine of the 59 staff who worked fulltime in the Port Moresby General Hospital wards designated for women or children. Of the eight clinic staff not interviewed, seven were on-leave and one declined to participate. Only nine ward staff were interviewed because of great scheduling difficulties. The responses from ward and clinic staff were first tabulated separately and then combined when no obvious differences were found in the patterns of responses of each group. Informal observations of clinic and ward activities were carried out at intervals from September, 1980, through July, 1981.

The women in the sample ranged in age from 21 to around 50 years old, and were from all over PNG (49 from the Papuan coast, 20 from the New Guinea coast, 19 from the New Guinea islands and 9 from the Highlands), primarily from rural areas or small towns. All had received the standard nurse or nurse aide training and most had experience in a variety of health care settings throughout the country. While the number of their liveborn children varied from one to seven, the mean was just under three and few women had more than four.

Data on mode of milk feeding, birthdate and postpartum employment were obtained for all 260 children of all 97 sample group women. In addition, more detailed information was elicited from each of the sample mothers on feeding practices and specific problems they encountered with their youngest child only. This group of children included 94 singlets and 3 pairs of twins.

In the course of the interview, all 97 sample women were asked about the timing of sevrage (termination of breastfeeding) and introduction of bottles, cups, spoons and solid food for their youngest child. The accuracy of these memories obviously cannot be determined. However, because the interview questions were designed to check for internal consistency in each respondent's replies, because nursing staff have been trained to attend to the timing of events and because some nurses reported that they relied on memories of their own experiences in giving advice to the clinic visitors, these data are considered reliable and relevant and thus are reported here. In order to examine only the most recent experiences and yet to include all those who were still breastfeeding, information on the timing of events obtained from the 77 mothers of only those last-born children who were under $5\frac{1}{2}$ years old at the time of the study (thus, were born after January, 1976) was tabulated. The cut-off point for this subgroup was selected because all children older than this age already had been taken off the breast (at ages ranging from six months to five years) and one five-year-old child was still nursing. This age was also a point at which inconsistencies began to show up in answers.

For purposes of analysis, each child was classified as breastfed (breast with or without supplementary use of a cup and a spoon to give fluids), mixed fed (use of a supplementary bottle to give fluids for one week or more) or bottlefed (no breastmilk offered, use of a bottle with or without the cup and spoon to give fluids).

RESULTS AND DISCUSSION

Feeding Routines

As shown in Table I, the predominant mode of milk feeding reported by the sample of nursing staff was breastfeeding. Over 50% of the mothers breastfed *all* of their children, 20% of the mothers used mixed feeding for *all* of their children and one mother bottlefed her only baby. The rest did not feed all their children in one particular fashion — usually breast alone for some of their children and mixed feeding for the others, in patterns unrelated to parity. No statistically significant change in milk feeding practices occurred among the interviewees after mid-1977, when prescriptions became required for purchase of infant feeding bottles.

When only the youngest children of these sample women are

TABLE I

Mode of milk feeding by birthdate for all children of sample of public health nurses

| Birthdate | Number of Children Mode of Feeding | | | |
	Breast only	Mixed	Bottle only	Total
Before July 1977	132 (74%)	44 (24%)	3 (2%)	179 (100%)
After July 1977	53 (65%)	28 (35%)	0	81 (100%)
Total	185 (71%)	72 (28%)	3 (1%)	260 (100%)

considered, a similar pattern is seen. Of these 100 children, 69 were breastfed, 30 were mixed fed and 1 was bottlefed. Over 80 % of this group were put to the breast within 24 hours postpartum. In a majority of cases in which there was a delay, either the mother or the baby had experienced particular difficulties associated with the birth. Breastfeeding of this group continued from several months to five years. The foods most commonly reported as first solids were traditional ones: mashed sweet potato, pumpkin, banana and papaya. Egg, green vegetables, commercial baby food, rice, white potato and tinned fish also were mentioned occasionally. This pattern of long breastfeeding and starchy *beikost* (weaning food) is found in many areas of PNG (Oomen, 1961; Shaw, 1979; Obrist, unpublished observations, 1981;[†] Pflanz-Cook, Foster and Cook, unpublished observations, 1982;[‡] Barlow, ch. 8; Carrier, ch. 11; Conton, ch. 6; Counts, ch. 9; Jenkins, Orr-Ewing and Heywood, ch. 3; Lepowsky, ch. 4; Tietjen, ch. 7).

The timing of changes in feeding routines is shown in Table II. The age ranges and standard deviations for each event were wide, indicating a diversity in individual practice. Breastfeeding was of long duration, as only five children were weaned from the breast before the end of the first year.

Two patterns emerged in the use of the feeding bottle and the cup. Some women in the sample followed PNG Health Department recommendations and supplemented their breastmilk with bottle-feeding for only a few months, from the time they returned to work until the time their baby could first drink from a cup. The majority of mixed feeders, however, offered the bottle until the baby was on an adult diet and could handle a cup completely on his own. Rarely were the bottles used to give anything other than milk. The cup (and sometimes the spoon) were used for all fluids.

Combining Motherhood and Employment

Whether or not a mother returned to a staff position or to a nurse training course within the first year postpartum was significantly

---

[†]Obrist, B. (1981). *Shadow Pigs and Sweet Potato Vines: A Study of Food in its Cultural Context among the Chimbu.* Master's thesis. University of Basel. pp. 26-28.

[‡]Pflanz-Cook, S.M., D.L. Foster and E.A. Cook (1982). Lactation and the reproductive cycle among the Manga, Jimi District, Western Highlands Province, Papua New Guinea. Paper presented at the eleventh annual meeting of the Association for Social Anthropology in Oceania. Hilton Head, SC.

TABLE II

Termination of breast milk and introduction of other foods by age for all youngest children born after January, 1976, to sample of public health nurses

Event	No. of Children	Age (Months) Mean ± 1 S.D.	Range
Termination of breast	44[a]	21 ± 11	2–56
Introduction of bottle	28	2 ± 1	0–6
Termination of bottle	24[b]	12 ± ·7	1–36
Introduction of cup or spoon	77	3 ± 2	0–9
Introduction of mashed solids	74[c]	4 ± 2	½–12
Introduction of juices	75[d]	3 ± 2	½–12
Introduction of adult diet	67[e]	9 ± 4	3–24

[a]32 children were not yet weaned from breast at the time of the interview; 1 child was not breastfed.

[b]4 children were not yet weaned from the bottle at the time of the interview.

[c]1 child was not yet introduced to solids; 2 mothers could not recall this information.

[d]2 mothers could not recall this information.

[e]6 children were not yet on an adult diet; 4 mothers could not recall this information.

associated with her decision to use a feeding bottle, as is indicated in Table III for all children of the sample women and in Table IV for their last-born children only. Very few mothers used bottles at all for milk feeding if they had not returned to gainful employment or to training within six months after birth. Of the 25 mothers who had not used the same modes of milk feeding for every one of their children, 17 had changed feeding modes when there was a change in their employment situation: An earlier return to work after the birth of a particular baby or an increased distance from home to work was associated with at least partial bottlefeeding for that child. It should be noted at this point that not all of the children, or even all of the youngest children, were born while their mothers were working in Port Moresby. The mothers returned to work postpartum in the full gamut of public health facilities described briefly in the background section.

Over 80% of the entire group of mothers in the study said that they began to leave their youngest babies regularly when they returned to work or to nurse training, and 80% of these returned within the first three months for a variety of reasons. Many of these mothers did turn to bottles for supplementary feeding at that time; however, a surprising number (41 out of 67) did not use bottles at all. These latter ones arranged for breastfeeding sessions frequently

TABLE III
Mode of milk feeding by postpartum employment for all children of sample of public health nurses[a]

| Length of postpartum absence from work | Number of Children Mode of Feeding | | | |
	Breast only	Mixed	Bottle only	Total
≤6 months	90 (58%)	63 (40%)	3 (2%)	156 (100%)
7–12 months	13 (87%)	2 (13%)	0	15
>12 months	82 (92%)	7 (8%)	0	89
Total	185 (71%)	72 (28%)	3 (1%)	260 (100%)

[a]For return to work within 6 months postpartum versus use of bottle, $\chi^2 = 34.43$, df. $= 1$, and p $<.001$. For return to work within 12 months postpartum by use of bottle, $\chi^2 = 29.02$, df. $= 1$, and p $<.001$.

TABLE IV
Mode of milk feeding by postpartum employment for all youngest children of sample of public health nurses[a]

| Length of postpartum absence from work | Number of Children Mode of Feeding | | | |
	Breast only	Mixed	Bottle only	Total
≤6 months	47 (62%)	28 (37%)	1 (1%)	76 (100%)
7–12 months	3 (100%)	0	0	3 (100%)
>12 months	19 (90%)	2 (10%)	0	21 (100%)
Total	69	30	1	100

[a]For return to work within 6 months postpartum versus use of bottle, $\chi^2 = 6.25$, df. $= 1$, and p $<.02$. For return to work within 12 months postpartum by use of bottle, $\chi^2 = 4.53$, df. $= 1$, and p $<.05$.

through the day or had the childminder use a cup and spoon in their absence.

One alternative to the use of the bottle or the cup (and spoon) for providing milk to an infant during the mother's absence is the wetnurse. The custom of wetnursing has been reported to be widespread, though infrequently practiced, in PNG (Pflanz-Cook, Foster and Cook, unpublished observations, 1982; Carrier, ch. 11; Counts, ch. 9; Jenkins, Orr-Ewing, and Heywood, ch. 3; Montague, ch. 5; Tietjen, ch. 7). Most interviewees said it was customary in their villages; however, only 12% of them actually had used a wetnurse for their youngest baby. Three others had breastfed the babies of other

22   L.B. MARSHALL

women. The most common rationale offered by the mothers in the study for not exchanging breastfeeding was that there was no appropriate person, generally a close kinswoman, with whom to share this. In many PNG societies, there is concern that breastmilk becomes polluted by intercourse (Carrier, ch. 11; Counts, ch. 9; Montague, ch. 5). In addition, some women expressed concern about the health status of a potential wetnurse. Thus, selection of a wetnurse was a serious matter. Very few expressed great enthusiasm for it.

It was Health Department policy to allow staff to transfer to clinics as near to their homes as possible. Many sample women said this facilitated continued breastfeeding. Either the childminder brought the child to the mother, or the mother had enough time to go home to breastfeed during her mid-morning and afternoon nursing breaks, as well as at lunch.

Over 20% of the group described situations in which they were able to take the baby along to work with them, especially in the first three to six months. For example, the government hospital in Lae, the second largest city in PNG, has had a childminding center in which breastfeeding by staff is strongly encouraged. In contrast, the child-minding center at Port Moresby General Hospital had been defunct for around two years at the time of the study. Certain clinics and health centers did have a "back room" which was not used for patient care and thus could serve as a safe place to keep a very young baby sleeping in a makeshift cradle or a traditional net bag. Allowing mothers to keep their young babies at work made it possible for many new mothers to return to work when alternative childminding was not yet available.

Childminding for their youngest child was an important problem for these women. If they could not find appropriate help they said they did not return to work. Most of the childminders were relatives of the woman or her husband who lived with them in their home. Very few women took their babies to childminders outside their home. In four of the nine situations in which the babies stayed at another house, it was the home of a close relative. This preference for relatives as child caretakers has been reported for numerous societies within PNG (Barlow, ch. 8; Carrier, ch. 11; Conton, ch. 6; Counts, ch. 9; Lepowsky, ch. 4; Pflanz-Cook, Foster and Cook, unpublished observations, 1982; Tietjen, ch. 7) and for other groups of educated PNG women as well (Harper, 1974). More than 55% of the women reported preparing food ahead for the childminder to give their children. Three women reported leaving expressed breastmilk for their mother or cousin to give the baby. In most cases where the mother did not fix food in advance for the child, the childminder was a close relative to whom the mother had taught food preparation.

Since nearly all the mothers interviewed continued to breastfeed their babies at least in the evenings and on weekends, they had the problem of keeping their milk supply adjusted to their work schedule as well as to the changing demands of their growing infants. Forty percent of them reported not having enough milk to satisfy their young child at all times. Primary symptoms of this — for the mother — were that she felt empty and/or the baby sucked too much or cried after its feed. Most ascribed this to their not taking good enough care of their own bodies (for example, too little food and rest, too much work). Some noted that the problem arose only on weekends when they had to feed their babies much more frequently than during the work week. A few felt that the baby's appetite had simply grown too fast, or they blamed the problem on contraceptive drugs they were taking. The standard remedy for this was to continue breastfeeding and to increase non-milk fluids and solid foods for the baby. This may have reduced the suckling stimulus, but was probably quite important in maintaining the health of the baby. All but two of the cases of insufficiency occurred after the baby was six months old and already being supplemented with solid food. The symptoms used by the mothers to indicate insufficiency were similar to those reported by Gussler and Briesemeister (1980), although the prevalence was lower in this study than in the cases they cited.

Advocacy of Breastfeeding

The opinions on proper milk feeding for infants which were reported by this group were strongly in favor of breastfeeding, with bottles viewed as appropriate only under highly specific conditions. When asked under what circumstances it was appropriate for a mother to use a bottle to feed her baby, 16% of the sample group replied, "Never." There was some debate about whether working for wages was a proper reason, but the majority felt it was. Other reasons offered included the inability of the real mother to feed the baby (due to illness, insufficiency, death or multiple birth), adoption by another woman or another pregnancy. In a great many cases, the answers were qualified by stipulating that the woman who uses a bottle must be able to do so properly and should do so only if the baby cannot use a cup and spoon. Many elaborated with personal accounts of sick babies brought in to the clinic by bottlefeeding mothers who often were described as educated working women. Several of the nurses mentioned the techniques they had employed to dissuade mothers from bottlefeeding, usually confiscating the bottle and scolding its owner. Only 4 of the 97 health staff said that a woman needed no special excuse for bottlefeeding, if she could do so properly.

SUMMARY AND CONCLUSIONS

Because of the differences in sample selection and definitions of what constitutes a given mode of milk feeding, a comparison of studies of prevalence rates for breast or bottlefeeding among particular groups of people in different societies is often difficult and should only be taken as suggestive. Compared to other wage-employed urban women surveyed in Sarawak (Koh, 1980), Malaysia (Sinniah, Chon and Arokiasamy, 1980), Thailand (Van Esterik, 1982), Philippines (Burgess, 1980), India (Thimmayamma, Vidyavati and Belavady, 1980), Israel (Bergman and Feinberg, 1981), Yemen (Harfouche, 1981), Cameroons (Garrett and Ada, 1982) and Nigeria (Di Domenico and Asuni, 1979; Ojofeitimi, Elegbe, and Etuknwa, 1982), higher proportions of the PNG nursing staff breastfed their babies and continued to do so for much longer periods of time. The rate of initiation of breastfeeding by the nursing staff in this study (98%) was higher than that reported by Biddulph (1975) for women delivering at Port Moresby General Hospital (45% in the semiprivate ward, 91% in the public ward). However, the prevalence of at least partial bottlefeeding among the working mothers in the present study was approximately the same as that (30%) for a group of Port Moresby women of varied background in 1976 (Lambert and Basford, 1977) and much higher than that (8%) for a second group of Port Moresby women surveyed in 1979 (Benjamin and Biddulph, 1980). Only 5-15% of the mothers in the latter surveys were employed. The cup and/or spoon were used by all the nursing staff, but by only 4-5% of the women in the Port Moresby surveys. Compared to a cross-section of other mothers in the same urban area, then, the women in this study relied somewhat more heavily on feeding bottles (and much more on the cup and spoon) for giving milk to their babies. This relationship between the prevalence of bottlefeeding and wage employment is consistent with that found in all the studies cited above.

Several factors appeared to influence the infant feeding practices of these nurses. The traditional practices of breastfeeding for several years and using the major local starch staple as a weaning food were being continued. These were encouraged by information gleaned from mothers and other relatives in the village setting and by experience with the health services, which usually emphasized the use of locally available foods. Breastfeeding, the early introduction of solid foods and the use of cup rather than bottle to give fluids received strong support from both mission and government health care providers all over the country. The recent government nutrition education programs and legislation made all health workers responsible for discouraging bottlefeeding, so they were forced to become aware of the possible health hazards of bottle use. On the other hand, these women were living in an urban environment with easy access to commercially prepared foods and often without the advice and encouragement of the traditional village-based female support network. The availability of traditionally appropriate wetnurses — an alternative to the use of cup or bottle — also was limited by living away from the majority of female kin. As working women, they had rigidly scheduled demands on their time which conflicted with those of traditional breastfeeding practices.

In order to return to work or to training in the first 6 to 12 months postpartum, these women employed a variety of strategies. All were dependent upon having a satisfactory childminding situation, either in the home or at their place of work. If distance, transport and the disposition of the childminder allowed, the mother could continue to breastfeed her baby even during working hours so long as she and the baby could get together easily at fairly regular times. There was a great deal of individual variability in the obstacles that had to be overcome in order to continue breastfeeding. When mother's breast was not available, a number of substitutes were given: milk or other fluids in a bottle or cup, gruels or solid foods or occasionally breastmilk from a closely-related wetnurse. In commenting upon the appropriateness of bottlefeeding, the majority — whether bottle users or not — indicated that the wage employed woman should be able to bottlefeed. However, a significant proportion strongly opposed bottle use by working women and could offer advice from personal experiences on feasible alternatives to the feeding bottle.

As is obvious from even a cursory perusal of the literature, there are many reasons for supplementing or replacing breastmilk in an infant's diet. Wage employment of the mother is only one of these. It is probably not the most important factor in the gradual decline in prevalence of breastfeeding that has occurred worldwide (Van Esterik and Greiner, 1981). However, for the individual mother who might wish to continue breastfeeding, returning to work — especially to urban wage employment — *does* present a number of difficulties. As is well demonstrated in the case of the public health nurses in Port Moresby, these difficulties are *not* insurmountable. Their return to work while breastfeeding was encouraged by certain government policy features: Increasing the proximity between mother and baby during working hours by providing creches or reassigning employees to workplaces close to their homes and increasing the opportunities for mothers to suckle their babies during working hours by providing regular breastfeeding breaks. Furthermore, the personal commitment, knowledge about infant feeding options and social support for breastfeeding from co-workers and employers so obvious among this group of working women undoubtedly contributed to the high prevalence and long duration of breastfeeding.

ACKNOWLEDGEMENTS

The author gratefully acknowledges the cooperation of the PNG Department of Health and the Matrons for the Port Moresby urban clinics and general hospital in permitting the study; the assistance of Madia Geita in data collection; and the kindness of the clinic and hospital nursing staff in consenting to be interviewed.

REFERENCES

Akin, J., R. Bilsborrow, D. Guilkey, B.M. Popkin, D. Benoit, P. Cantrelle, M. Garenne, and P. Levi (1981). The determinants of breast-feeding in Sri Lanka. *Demography* **18**(3), 287-307.

Benjamin, A., and J. Biddulph (1980). Port Moresby infant feeding survey, 1979. *Papua New Guinea Med. J.* **23**(2), 92-96.

Bergman, R., and D. Feinberg (1981). Working women and breastfeeding in Israel. *J. Adv. Nurs.* **6**, 305-309.

Biddulph, J. (1975). Every baby deserves the breast. *Nutr. Devel.* **1**(2), 29-34.

Biddulph, J. (1981). Promotion of breast feeding: Experience in Papua New Guinea. In D.B. Jelliffe and E.F.P. Jelliffe (Eds.), *Advances in International Maternal and Child Health.* Oxford University Press, New York. Vol. 1, pp. 169-174.

Burgess, A.P. (1980). Breastfeeding: The knowledge and attitudes of some health personnel in metropolitan Manila. *J. Trop. Pediatr.* **26**(5), 168-171.

Di Domenico, C.M., and J.B. Asuni, (1979). Breastfeeding practices among urban women in Ibadan, Nigeria. In D. Raphael (Ed.), *Breastfeeding and Food Policy in a Hungry World.* Academic Press, New York. pp. 51-57.

El-Mougi, M., S. Mostafa, N.H. Osman and K.A. Ahmed (1981). Social and medical factors affecting the duration of breast feeding in Egypt. *J. Trop. Pediatr.* **27**(1), 5-11.

Garrett, N.R. and V. Ada (1982). Infant feeding beliefs and practices: A study of Cameroonian health care personnel. *J. Trop. Pediatr.* **28**(4), 209-215.

Gussler, J.D., and L.H. Briesemeister (1980). The insufficient milk syndrome: A biocultural explanation. *Med. Anthropol.* **4**(2), 145-174.

Harfouche, J.F. (1981). The present state of infant and child feeding in the Eastern Mediterranean Region, *J. Trop. Pediatr.* **27**(6), 299-303.

Harper, J. (1974). Educated women in Niugini. *Aust. and NZ J. Sociol.* **10**(2), 90-95.

Igun, U.A. (1982). Childfeeding habits in a situation of social change: The case of Maiduguri, Nigeria. *Soc. Sci. Med.* **16**, 769-781.

Jelliffe, D.B. (1976). World trends in infant feeding. *Amer. J. Clin. Nutr.* **29**, 1227-1237.

Knodel, J., and N. Debavalya (1980). Breastfeeding in Thailand: Trends and differentials, 1969-79. *Stud. Fam. Plann.* **11**(12), 355-377.

Koh, T.H.H.G. (1980). Breastfeeding in Sarawak. *Brit. Med. J.* **280**(6207), 95-96.

Lambert, J., and J. Basford (1977). Port Moresby infant feeding survey. *Papua New Guinea Med. J.* **20**(4), 175-179.

Marshall, L.B., and M. Marshall (1980). Infant feeding and infant illness in a Micronesian village. *Soc. Sci. Med.* **14B**(1), 33-38.

Mohrer, J. (1979). Breast and bottle feeding in an inner city community — an assessment of perceptions and practices. *Med. Anthropol.* **3**(1), 125-145.

Mudambi, S.R. (1981). Breast-feeding practices of mothers from Mid-Western Nigeria. *J. Trop. Pediatr.* **27**(2), 96-100.

Ojofeitimi, E.O., I. Elegbe and U.T. Etuknwa (1982). Knowledge and breastfeeding practices among nurses and teachers in Ile-Ife, Nigeria. *Pediatr. Nurs.* **8**(6), 400-402.

Oomen, H.A.P.C. (1961). The Papuan child as a survivor. *J. Trop. Pediatr.* **6**(4), 103-121.

Overseas Education Fund (1979). *Child Care Needs of Low Income Mothers in Less Developed Countries.* Overseas Education Fund of the League of Women Voters, Washington, D.C., pp. 36-42.

Pathmanathan, I. (1978). Breast feeding — a study of 8750 Malaysian infants. *Med. J. Malaysia* **33**(2), 113-119.

Pongthai, S., P. Sakornrattanakul and K. Chaturachinda (1981). Breast feeding: Observation at Ramathibodi Hospital. *J. Med. Assoc. Thai.* **64**(7), 324-327.

Popkin, B., and F.S. Solon (1976). Income, time, the working mother and child nutrition. *J. Trop. Pediatr.* **22**, 156-166.

Power, D.J., W. Willoughby and R.H. de Waal (1979). Breast feeding in Cape Town. *South African Med. J.* **56**(18), 718-721.

Richardson, J.L. (1975). Review of international legislation establishing nursing breaks. *J. Trop. Pediatr.* **21**(5), 249-258.

Shaw, B. (1979). *Human Milk as a National Resource: Economic and Demographic Aspects of Infant Feeding in Papua New Guinea.* Development Issues Seminar Series, Research School of Pacific Studies, Australian National University, Canberra.

Sinniah, D., F.M. Chon and J. Arokiasamy (1980). Infant feeding practices among nursing personnel in Malaysia. *Acta Paediatr. Scand.* **69**(4), 525-529.

Sjolin, S., Y. Hofvander and C. Hillervik (1979). A prospective study of individual courses of breast feeding. *Acta Paediatr. Scand.* **68**(4), 521-529.

Thimmayamma, B.V., M. Vidyavati and B. Belavady (1980). Infant feeding practices of working mothers in an urban area. *Indian J. Med. Res.* **72**, 834-839.

Uyanga, J. (1980). Rural-urban differences in child care and breastfeeding behaviour in Southeastern Nigeria. *Soc. Sci. Med.* **14D**(1), 23-29.

Van Esterik, P. (1982). Infant feeding options for Bangkok professional women. In M.C. Latham (Ed.), *The Decline of the Breast: An Examination of its Impact on Fertility and Health, and Its Relation to Socioeconomic Status.* Cornell International Nutrition Monograph Series No. 10, Cornell University, Ithaca, NY., pp. 56-79.

Van Esterik, P. and T. Greiner (1981). Breastfeeding and women's work: Constraints and opportunities. *Stud. Fam. Plann.* **12**(4), 184-197.

West, C.P. (1980). Factors influencing the duration of breast-feeding. *J. Biosoc. Sci.* **12**(3), 325-331.

World Health Organization (1981). *Contemporary Patterns of Breast-feeding. Report on the WHO Collaborative Study on Breast-feeding.* WHO, Geneva.

Zurayk, H.C. and H.E. Shedid (1981). The trend away from breast feeding in a developing country: A women's perspective. *J. Trop. Pediatr.* **27**(5), 237-244.

CHAPTER 3

CULTURAL ASPECTS OF EARLY CHILDHOOD GROWTH AND NUTRITION AMONG THE AMELE OF LOWLAND PAPUA NEW GUINEA[†]

CAROL L. JENKINS, ALISON K. ORR-EWING and
PETER F. HEYWOOD

INTRODUCTION

The first attempt to take a broad look at nutrition in Papua New Guinea was made in 1947 when the Australian Department of Health carried out a general survey in five different areas of the country (Hipsley and Clements, 1950). The general growth pattern revealed by the cross-sectional anthropometric data showed a marked departure from the Harvard growth curves for both weight and height. The 1950's and 1960's saw further extensive documentation of growth retardation in a number of areas of the country by Scragg (1955), McKay (1960), Oomen and Malcolm (1958), Venkatachalam (1962), Bailey (1964) and Malcolm (1970). These studies showed that breastfeeding was universal and occurred for extended periods of time up to four or five years of age. In those in which food intake was measured, it was frequently reported that protein intake was inadequate and that in some cases energy was, too, leading to the general conclusion that the traditional diet was responsible for the slow growth and small size of the population. However, more recent studies in which both food intake and growth were measured have not been able to explain the observed patterns of growth (Sinnett, 1972; Ferro-Luzzi, Norgan and Durnin, 1975). Several authors have stressed what appears to be a nonchalant attitude toward infant and child feeding on the part of Papua New Guinea mothers. Malcolm

[†]This work was supported in part by World Health Organization Grant No. 1638.

30   C.L. JENKINS, A.K. ORR-EWING AND P.F. HEYWOOD

(1975) refers to this as "demand" or "opportunity feeding", implying that the traditional modes of feeding are basically responsible for inadequate food intake.

Other factors common in many developing countries do not appear to be responsible for poor growth in Papua New Guinea. Breastfeeding up to at least one year of age remains nearly universal. Diarrheal disease does not occur on the same scale as in some other countries as shown by its relatively small contribution to mortality (Stanhope, 1967; Sturt, 1972). Land alienation, which underlies poverty and malnutrition in many countries, is not a major problem.

The diets of the rural population, which comprises approximately 90 % of the nation, are still predominantly traditional. Where dietary shifts have taken place, they do not appear to contribute to poor nutrition. In fact, where, as a consequence of economic development, rural groups now consume significant amounts of imported foods, particularly rice and tinned meat or fish, improved growth is evident (Harvey and Heywood, 1983).

Therefore, partly by exclusion and partly due to the demonstration of improved growth where substantial dietary change has occurred, traditional modes of feeding infants and young children, as suggested by early observers, remains an important possible explanation of the commonly documented childhood growth patterns. Although Papua New Guinean cultures have received wide coverage by anthropologists, information on foodways, especially as they affect infants and young children, is lacking. Although ethnographers and nutritionists have noted food taboos in a number of areas, the information is difficult to dovetail with growth outcome.

The purpose of this study is to begin correcting this deficit with reference to a single, well-documented population using the combined methods of ethnography and nutrition.

METHODS

The selected study population is the Amele, an ethnolinguistic group of about 6000 persons speaking a non-Austronesian language who reside in the sub-coastal region of Madang Province. The Amele are included in on-going demographic, malarial and nutritional surveillance projects conducted by the Madang branch of the Institute of Medical Research (I.M.R.).

Growth

The birthweights of 788 children born between 1974 and 1981 were obtained from the Yagaum Health Center, a Lutheran-supported medical center established in 1950 with an emphasis on obstetrical care (Braun, 1967; MacDonald, 1972). Cross-sectional anthropometric data were obtained by I.M.R. staff, including the authors, for 792 Amele children with recorded birthdates whose ages at the time of measurement ranged from less than one month to three years. Longitudinal growth data at ten age points (± two weeks) between birth and two years of age were obtained on a separate sample of 22 Amele children during 1981-1982 by one of us (A.O-E) as part of a study on lactation and growth. The weight and length of children were measured without diapers in both clinic and non-clinic village settings on calibrated triple beam balances[†] and locally constructed infant measuring boards. Inter- and intra-observer errors were checked and found insignificant.

Food Intake

Qualitative dietary patterns were documented during visits of 8 to 24 hours' duration with 24 lactating mothers and their infants in six Amele villages. Mothers were informed that breastfeeding was under observation during these visits. In addition, the 22 mothers in the longitudinal growth study set aside, under supervision, duplicate portions of food served their children over a three-day period each time growth data were collected. These food portions were placed in plastic bags, subsequently weighed and partially analyzed.

Ethnography

Although the Amele have been in contact with Europeans for nearly a century, they have received no ethnographic coverage. Using current malaria survey census data, the village of Ohuru, having an average population size (275) and a typical location (on a road), was chosen as the primary field site. Housing was arranged for the anthropologist (C.J.) and her family at Yagaum, a 15-minute walk from the main Ohuru hamlet and a central location for the whole Amele area.

[†]CMS Weighing Equipment, London; Model MPS 120.

Initial fieldwork was conducted in Neo-Melanesian Pidgin in which nearly all Amele adults are fluent. Study of the Amele language commenced immediately and eventually the investigator was able to conduct a village-wide survey, including a standard infant feeding interview, in the Amele language. Participation-observation focused on food, its acquisition, preparation, consumption and distribution. The ethnographic information reported here was gathered during 12 months of fieldwork from May, 1982, to May, 1983.

AMELE HABITAT AND HISTORY

The Amele occupy a region of lowland hill forest ranging in altitude from sea level to 200 m. Their territory is bounded by the Gum and Gogol Rivers and the Pacific Ocean, marking an area of approximately 200 sq. km with a mean population density of about 30 persons/sq. km. Due to increasing population and bush-fallow-swidden-farming, the original forest is now almost totally replaced by gardens and secondary growth. Rainfall averages 3533 mm per year with a marked dry season from June to October. The rapid loss of soil nutrients is well-recognized, but many garden areas are replanted following a fallow period no longer than four years. Low fertility areas are planted in coconuts, cocoa and sometimes manioc, a common food for pigs.

Unlike highland New Guinea, the northeast coast has been exposed to Christianity and foreign commercial interests for about a century. The Amele's first encounter with a European took place when Mikloucho-Maclay, a Russian naturalist, walked through the area around 1872 (Mikloucho-Maclay, 1975). His remarkable early scientific experience was followed by the commercial endeavors of the German New Guinea Company. The Amele and their neighbors saw their first police, enforced labor, copra plantations and cargo during the earliest decades of the twentieth century. At the same time Lutheran missions were established at several distant locations and Amele big men[†] sought to secure their own. Land was granted the mission, which established schools and a health center. Customary practices such as infanticide, bride-capture, sorcery and warfare were suppressed. During World War II the Japanese camped throughout

---

[†]Status of respected elder acquired by men who informally compete in rhetoric, generosity and other qualities of leadership.

the area and most people hid in caves to escape bombing. After the war, Yali, a war-time hero from the Rai coast, visited the area bringing messages of Australia's compensation payments and encouragement of traditional initiation rites (Lawrence, 1964). Many Amele joined Yali's cargo cult[†] and began initiating their sons in a newly-adopted rite originating in Siassi called the *mulung garab.* This initiation rite, involving penile incision, a month-long seclusion in the bush, fasting and a complex food taboo system, grew in popularity. Today most lower altitude villages have their own *bal mei yo* or sacred father's houses, and boys from all over the area are sent to be initiated during the Christmas school holiday. Since Independence in 1975, both mission and cargo cult activities have lessened and cash-oriented economic activity has increased.

The Amele have been exposed to Western medicine through Lutheran-supported health centers since 1934. Yagaum Hospital, which was established after the war, was a well-equipped hospital with the capacity to train medical personnel, conduct major surgery and provide long-term in-patient treatment. The Amele's first experience with Western medicine stands in marked contrast to the minimally equipped aid-post and local orderly with which most Papua New Guineans first became acquainted. The Amele health center utilization rate rose through the years and, reportedly, infant and maternal mortality rates have decreased.

HEALTH STATUS

There are no accurate estimates of overall or age-specific mortality rates for the study area. However, Serjeantson (1975) obtained reproductive histories from 286 married women in the Gum language family and the adjacent Astrolabe Bay area. At the time of interview, 22% of all live births had died. Of these, 27% died in childhood. In a less privileged lowland Sepik population of 3500, Sturt (1972) found an infant mortality rate over a ten-year period of 104/1000 live births and a toddler mortality rate of 19.8/1000.

As with mortality rates, there are no reliable data on causes of childhood deaths in the study area. However, the overall pattern is likely to be the same as in other lowland populations where acute

---

[†]Millenarian religious cult, several of which developed in Melanesia during the first century post-contract, offering the promise of plentiful European-style material wealth.

lower respiratory tract infections and malaria are the leading causes of death in early childhood. Again, the data of Sturt (1972) are pertinent. He found that 24% of infant deaths were due to malaria and 12.5% to acute lower respiratory tract infections. As causes of death, these are also the major contributors of morbidity. Malaria is hyperendemic in the study area and preliminary results of work carried out by the Malaria Research Program at I.M.R. indicate that most infants have experienced at least one malarial infection by 12 months of age. Young children also experience repeated upper and lower respiratory tract infections.

As expected in a malarious region, many children are anemic. Hemoglobin estimates made on a random sample of the population in 19 Amele villages indicate that 42% of children under 12 years of age are anemic using the hemoglobin standards of APHA (1973). Skin infections, particularly *tinea imbricata*, scabies and infected sores and cuts are another common cause of morbidity.

There is little detailed information on the health status of women. Immunosupression during pregnancy has been observed elsewhere (Gilles *et al.*, 1969) and is likely to be important, particularly with respect to malaria. Apart from its direct effect on maternal health, malaria during pregnancy has also been shown to decrease birthweight (Jelliffe, 1968) which, in turn, is an important factor influencing postnatal growth.

GROWTH OF AMELE CHILDREN

Mean birthweights of Amele children are, on average, below those in well-nourished populations. In our sample, 20.4% of the children have birthweights at or below standard (2500 gm). Male birthweights ($N = 409$) average 2.91 ± 0.49 kg with a range of 1.30 to 4.64 kg. Female birthweights ($N = 379$) average 2.79 ± 0.51 kg and range between 1.16 and 4.28 kg.

Figure 1 presents the cross-sectional pattern of weight-for-age among male and female Amele children from birth to three years of age, in relation to the 50th and 5th NCHS centiles. It is evident that, despite low average birthweights, Amele infants initially catch up to international standards but subsequently fall below them again. Males begin to fall away from the median somewhat earlier than females, at 1.5 months compared to 4.5 months. Both sexes continue a low weight-for-age trajectory until nearly three years of age, when a slight recovery seems to begin.

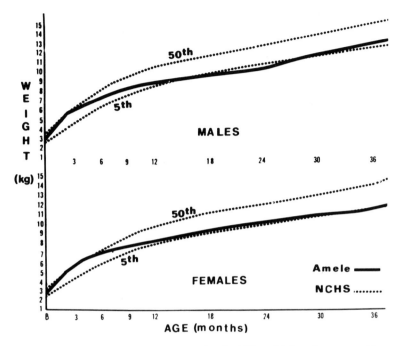

FIGURE 1   Weight-for-age among Amele children, birth to three years old.

Length-for-age patterns are similar, in that males drop away from the median more rapidly, reaching the 5th centile at 12 months, whereas females reach the 5th centile at 18 months. After 24 months, neither sex makes any substantial recovery in length-for-age.

Figure 2 shows Amele weight-for-length curves in relation to standards. Both sexes begin life above the 50th centile in weight-for-length but exhibit steady declines with a notable decrease between 6 and 12 months of age. The male pattern shows a lower dip in the curve but a recovery by 36 months to 94% of standard. Females both gain and lose relative to standards more slowly.

Table I presents the mean percent of NCHS standard values attained by Amele children from birth to three years old for weight-for-age, length-for-age and weight-for-length. The decline in weight-for-length between 6 and 12 months of age is striking and thereafter appears to stabilize until at least three years of age.

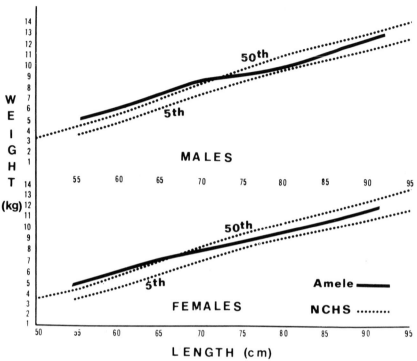

FIGURE 2   Weight-for-length among Amele children, birth to three years old.

AMELE SUBSISTENCE AND ECONOMICS

The principal staples of the Amele are taro (*Colocasia esculenta*), bananas (*Musa spp.*) and yams (*Dioscorea spp.*). Coconuts (*Cocos nucifera*) are also very important and provide the major source of fat. During the rainy season, when taro and yams are unavailable, people depend on bananas, Chinese taro (*Xanthosoma sagittifolium*), pitpit (*Saccharum edule*), winged beans (*Psophocarpus tetragonoglobus*), sago (*Metroxylon sagu*), corn (*Zea mays*), cucumbers and pumpkin (*Cucurbita spp.*), sweet potatoes (*Ipomoea batatas*), manioc (*Manihot esculenta*) and, increasingly, on imported rice. Animal sources of protein are varied but scant and seasonal. A wide variety of green leaves, some of which are high in protein such as *ifer* (*Gnetum gnemon*) and

TABLE I
Mean percent of standards (NCHS) for Amele children, birth to three years.

Age (mos)	N	Sex	Weight/age (Mean %)	Length/age (Mean %)	Weight/length (Mean %)
0- 2.9	18	M	102.4	98.5	106.5
	16	F	102.3	99.2	103.4
3- 5.9	21	M	93.9	96.8	102.2
	15	F	101.6	99.1	102.5
6-11.9	29	M	90.3	93.3	102.7
	31	F	89.5	96.8	96.6
12-17.9	35	M	82.9	94.1	92.0
	24	F	85.3	95.6	91.4
18-23.9	44	M	81.6	93.8	90.5
	41	F	84.6	94.2	92.3
24-35.9	112	M	83.8	93.1	94.1
	79	F	83.6	93.1	92.8

erum (*Abelmoschus manihot*), are regularly enjoyed. Currently, an increased emphasis on cash has brought about greater consumption of tinned meat or fish. Rice, in particular, is becoming the most important store-bought food, followed by sugar, tea and wheat flour products. In many households a substantial portion of available cash is spent on beer, drunk almost entirely by men. Fruits and nuts, including the breadfruit seed (*Artocarpus altilis*) and peanuts, are important snacks.

The daily meal pattern begins with a quick morning meal, usually of bananas or tubers roasted over the fire and a drink, often green coconut water. The mid-day meal depends on the activity and location of the mother. If the children accompany her to the garden, a meal may be cooked there and numerous fresh snacks will be available throughout the workday. Children are sometimes left at home while the mother goes to market or garden and older children or relatives are expected to feed them. Most children can cook something for themselves by ten years of age. In the late afternoon or evening, the main family meal is served. There is little variation in its preparation, although specific ingredients may vary. It is a stew known as *saab ahura ile* or food cooked in coconut milk. The stew always includes different types of tubers and/or bananas, diluted coconut milk and, if possible, tulip leaves, a handful of beans, pieces of pumpkin, pitpit, or whatever else is in season. When fish or meat is available, it is placed on the top of the pot or fried. A proper meal should be freshly cooked because eating stale, cold leftovers is thought to make the body lazy. Each serving should include a substantial amount of liquid as well as solids. The development of this tradition has been dependent on the accessibility of clay cooking vessels obtained from the Bilbil people and elsewhere in trade. Both metal and clay cooking pots are used and, with wooden bowls, form an important part of the brideprice.

About 25% of adult men in most villages work for steady wages or own businesses, such as local trade stores. Many more men work intermittently and the majority of households raise some farm produce for cash. Most earn no more than K100[†] per year from their holdings. Nearly all women sell garden produce at the town market which they attend one to two times weekly earning from K3 to K10[†] each time. Despite the increasing dependence on cash, families continue to take great pride in producing almost all of their own food, probably about 85% by bulk. In lower altitude villages, yams and Chinese taro may be stored for nearly a year, but no other foods are preserved beyond a week. Traditional mechanisms of food exchange, both within and between villages, are vital to all families. The need for cash is less pressing for food than for school fees, transport costs, kerosene, and more importantly, health care, compensation and brideprice payments. Thus, the Amele may be characterized as subsistence farmers with a gradually increasing consumption of selected store-bought foods. This level of consumption may increase as a by-product of the greater effort, time and land invested in cash-earning activities.

FOOD CLASSIFICATIONS AND CHILD DEVELOPMENT

Foods are classified according to their source and several different qualities. *Saab* refers to major staples and anything grown in the garden annually. *Ehiro* includes fruits and nuts growing on trees or vines. *Uhun* (meat) from *dor* (animals) includes mammals, lizards, fish, birds, eggs, crustaceans and insects. Each variety of food may further be classified as *gagadi* (strong) or *bado'e* (soft) and *dain'a* (hot) or *aï be* (cold). The quality of strength is widely recognized and subsumes such characteristics as hardness, density, stringiness and

---

[†]K = Kina. One kina = U.S. $1.14.

dryness. Strong foods are inappropriate for young children without teeth and sick persons. They may cause constipation and swollen bellies, including a syndrome known as *nenege* or swollen spleen. Dryness and maturity are equivalent states (*meg*). Watery foods such as juice (*muhu*) and broths (*we*) represent the extreme of softness and are excellent for infants.

An additional mode of classifying food derives from the philosophy of the *mulung garab*. This hot-cold system is less widely known and is explicitly taught only to initiates, although the accompanying food taboo system is also taught their parents. Medicinal plants, physiological states including stages of development, body fluids and ritual objects are also classified as hot or cold. Heat is associated with life and potency; cold with weakness and death. The logic of this philosophy is used to explain the action of medicinal plant juices, love magic and the causes of sickness. While numerous cold and soft foods are taboo to initiates (who are hot), infants must have soft, watery and preferably warm foods.

Child development proceeds through a series of named stages marked by changes in motor development and tooth eruption. Newborn infants (*momodo*) are cold and soft, like their mothers, and must be strengthened by the application of warm hands heated over a fire. Only liquids, preferably lukewarm, may be given to *momodo*. As strength develops and the infant can hold up its head, it is known as *momo memen*, literally "infant becomes stone." When it is able to sit alone, it is called *biberen* (one who sits). This is a major hallmark of development and signals the possibility of introducing mashed foods. The child is considered still too young to eat strong, dry foods such as those roasted over the fire or very cold foods, such as those which are slimey or bloody. Food cooked in coconut milk becomes somewhat slimey and should not be given children of this age as it can cause a cough. A child who crawls is called *o'o'obona* or sometimes *ho o'o'bon* meaning "he walks like a pig." The first signs of standing are greeted with joy and fathers build cross-bars for the children to use to steady themselves. Eventually the child walks by himself, holding on to the bar or his parent. At this stage he is called *da'a'en*. When able to walk without holding on, the child is referred to as *ji o'obona* meaning "he walks the road." The next stage is called *gudugudu'ena* and is marked by the ability to run about, play with other children and verbally request food and water. All types of food may be given a child by this stage. Dental eruption is also used as a guide to feeding but stages are less well marked. Three types of teeth are recognized, *orugum* (incisors), *sihona* (canines) and *bad gugna* (cheek teeth). There is considerable variability in which foods are given when particular teeth appear, but in general, parents agree it is permissible to give *saab ahura ile* (food cooked in coconut milk) when at least two incisors appear and sago and meat when cheek teeth emerge.

The Amele food classification system dictates behavior strictly only for initiates who adhere to its rules lest they sicken, lose vigor and fail to attract a wife. Married men are obliged to adhere to fewer food taboos and maintain them through fear of sexual impotency. Women and girls have no special food taboos, with the exception of the necessity to avoid hot soups, ginger and capsicum during pregnancy. So-called Mongoloid spots and congenital cataracts are attributed to the burning quality of these foods. The postpartum and lactating mother may eat anything. But idiosyncratic and less systematized food avoidances are common. Many women claim to avoid eating leaves with small hairs on them as they cause pimples. If a child develops skin infections after drinking a soup of a particular green leaf, the leaf will be held responsible. A large belly or herniated umbilicus is attributed to eating strong foods, such as dry, roasted tubers. A commonly followed notion is that sweet liquids such as sweet green coconut water, sugared tea and the juice of sugar cane irritate the throat and cause coughing. If drunk during the day, such liquids may bring on coughing at night. Sugared cough syrups given out by clinic sisters are disliked for this reason. But these beliefs and taboos are more commonly cited as causes of illness after it develops. Should a child greatly desire a food, parents will not deny it and even initiates can have food taboos lifted, or receive a cleansing after breaking a taboo, by ritual experts.

BREASTFEEDING

All women breastfeed their children. Should the milk supply seem insufficient a woman can steam her breasts over the cooking pot. Should this fail or should an elder woman seek to re-lactate (*su gihmudo*) she can obtain a hormone injection from the Yagaum clinic. Rarely, an infant feeding bottle is obtained and used in conjunction with breastfeeding by employed or adoptive mothers. Adoption is very common, but usually the child is breastfed by its biological mother for at least a year before it is transferred to the adoptive parents. If transferred before that time, dry whole milk and green coconut water are fed the child with the aid of the infant feeding cup. Undissolved lumps of dry milk may also cause swollen bellies and should be avoided.

Among the Amele, wetnursing is very rare due to the belief that breastmilk changes from a watery weak fluid to a stronger one as lactation proceeds, in accordance with the development of the child. Thus, breastmilk appropriate for *momodo* is not adequate for the *da' a' en.* If strong breastmilk is given a younger child it will cause constipation and a swollen belly. Since colostrum is more watery than mature milk, elder women consider it appropriate for the newborn, but many younger mothers discard it, thinking its color marks it as a noxious substance. Some mothers believe that only one of their breasts produces good milk, while the other is too cold and causes the child to vomit. Initiation of breastfeeding is rarely delayed beyond one day postpartum.

Shortly after birth, usually within the first week or two, nearly all mothers give their newborns *faila we*, the warmed juice of a ripe papaya. People state that this will keep the child from crying with hunger. The type of papaya most often given is red and known as *maklelika*, named after Mikloucho-Maclay who introduced it to Papua New Guinea. An infant is considered hungry if it cries after being breastfed or chews on its mother's nipples. If this occurs, soups and juices will be introduced. But only in rare cases does this type of feeding continue and lead directly into other supplemental feeding. After about a month, when the mother is convinced that stronger milk has developed and she and the infant have settled into a breastfeeding routine, no other foods will be offered. Occasionally a mother may try giving an egg or other food considered appropriate, but the child is perceived as rejecting it when the food is pushed from its mouth by tongue eversion, which is viewed as equivalent to vomiting. Unless a child cries and points specifically to a food, it will not be given anything but breastmilk, sometimes up to eight or ten months.

A mother rarely refuses her breast to a child, even if it is nearly five years old and she no longer produces milk. Ideal mothering behavior is very indulgent to the demands of the infant, and women rush to feed older children while the infant sleeps or simply claim they cannot cook until the baby allows it. The breast may be bitten, pulled and played with while nursing without negative comment. Attempts at weaning children from the breast are made slowly and half-heartedly. Sometimes a child is sent to a relative for a few days but this is not always effective. Elder siblings may begin teasing a four-year-old for still demanding *su* (breastmilk) and the mother may even passively allow them to place ginger or hot pepper on her breast in order to force the child to give it up. The mother would rarely do this herself, and even while trying to wean the child from the breast, when out of the sight of others, most mothers continue to suckle their children. One factor of importance contributing to lengthy breast-feeding is that husbands are forbidden to sleep with their wives while the child still drinks milk at night. However, this seldom is more than a show for the neighbors, since sexual intercourse may take place elsewhere. Most mothers attempt to hold out against their husbands' demands for sex as often and as long as possible postpartum and can utilize breastfeeding as an excuse. Another important factor is emotional bonding between mother and child. The breastfeeding experience is the first and prototypical food sharing experience in the life of the child. Children frequently suckle one breast and twist the other one up toward the mother's mouth, offering her some of her own milk. To this action mothers always respond positively, smiling and pretending to drink some of the child's gift.

Complete sevrage may not take place until well over four years old if the child is the lastborn or the mother uses contraception. In most cases, however, the mother is forced to speed up the weaning process due to the subsequent pregnancy. Few women know they are pregnant before three months and most continue to breastfeed until their pregnancy is clearly visible to all. It is widely believed that a child who suckles a pregnant mother is drinking the blood of its new sibling (*otig gola je'ena*), which can coagulate to cause a hard swollen belly. The acceptable period of birth spacing is at least two years and, should a child be forced from the breast earlier than this, it is known as *su aga sido* (sent from the breast too soon). This too leads to a swollen belly or *nenege* (literally, spleen) due to the necessity to give the child strong foods prematurely. *Su aga sido* children are subject to some thumb or finger sucking and often play with their mothers' breasts while the new baby suckles. Their distress is obvious and mothers who allow this to come about are severely criticized in village gossip.

FOOD AND GROWTH

The Amele distinguish normal growth (*ben me*, to put big) from retarded growth (*ban me*, to put small) and can discuss the process clearly with respect to trees and plants. Growth in length (*e'ela me*) is also distinguished from weight gain. Children who fail to gain weight are called *aisor na mel*, or children of the tanget (*Cordyline fructicosa*), a narrow leaf commonly used in medicinal preparations. Retardation is seldom directly attributed to inadequate food, although the notion is constantly taught to Amele mothers by clinic sisters. Since growth retardation is nearly always accompanied by repeated bouts of illness, it is the illness which appears paramount to the mothers. The most common explanation for a sickly child is that the mother's brother has cursed it because he has not received the full promised brideprice, particularly the pork component. Fathers frequently rush to secure a pig and to give it to their brothers-in-law when their children's lives appear threatened. Alternatively, chronic and repeated illness is explained by the existence of hidden anger between the parents or the parents and others. This anger frequently concerns food and can be resolved by discussion and the sharing of food.

Despite the continued active maintenance of these beliefs, the Amele simultaneously utilize Western medicine. Taking a child under five years old to the monthly clinic to be weighed and checked is a mother's duty which, in some vague way, guarantees the child's continued growth. At the clinics, sisters perform a short church service followed by a lecture on the causes of malnutrition (*tefur be*, literally, no bones). They show posters of local and imported foods, classifying them in an entirely different way than do Amele mothers. The women are told to introduce food other than breastmilk at four months, but Amele mothers do not consider themselves negligent in this regard since they introduce *we* (juices and soups) very early on. It is perhaps significant that in the Amele language to eat (*yaga*) and to drink (*yaga*) are identical. Betel nut (*Areca catechu*) chewing is also considered to be eating and the nut, especially its soft shell, is given to children less than one year old. By the age of two, and certainly by three, most children frequently chew betel nut with *hur (Piper betel)* and lime derived from roasted and pounded sea shells.

PATTERNS OF DIET IN INFANCY AND EARLY CHILDHOOD

A standard interview concerning the stage of development at which 24 key food items were introduced was administered to 34 Ohuru

C

mothers, referring to 66 children ten years of age and under. The second author (A.O-E) administered the same interview to the 22 mothers in her longitudinal study. Results were substantially the same in both samples, demonstrating no statistically significant differences with regard to either question responses or demographic characteristics of the sample families, with one exception. This exception concerns the frequency at which eggs are said to be introduced during the first two growth stages. This is reported and discussed below. For all food groups the results of the two surveys are combined.

Results reveal a timing sequence which moves from the introduction of total liquids, to mashed solids or semi-solids cooked in water, to solids cooked in water and not mashed, to solids cooked in coconut milk, followed by roasted tubers, nuts, bananas and, finally, to sago and meat. A sub-class of foods appropriate for very young infants, distinguishable by their high moisture content, is delineated by mothers' responses. These include semi-liquid, mashed ripe papaya and bananas, often cooked in water, and the broths of pumpkin, *erum* and sweet potatoes. This sub-set forms a distinct complex of baby foods among the Amele. For ease of analysis, all foods are grouped into four categories: baby foods (as defined above), starchy staples, meat and eggs.

Figure 3 summarizes the growth performance and food introduction schedule of Amele children from birth to two years of age. Children are grouped according to developmental stages as perceived by the Amele. Chronological equivalents are based on known ages of children exemplifying each stage. Since the normal variation in the timing of development milestones may be considerable and sequences among Papua New Guinea children are not yet investigated, the chronological equivalents shown in Figure 3 must be regarded as informed estimates. The bar histograms represent the proportion of children in the longitudinal growth study ($N = 22$) who gained, maintained or lost their position relative to Harvard standards (Jelliffe, 1966) of weight-for-age during one half or more of the time period covered by each developmental stage. Below the histograms are listed the cumulative percentages of children receiving their first tastes of the four food groups by developmental stage.

These data illustrate the relationship between dietary patterns and infant and toddler growth. During the first six months, the *momodo* stage, breastmilk appears adequate in 77% of the children while among the rest, who are losing weight relative to standard, neither

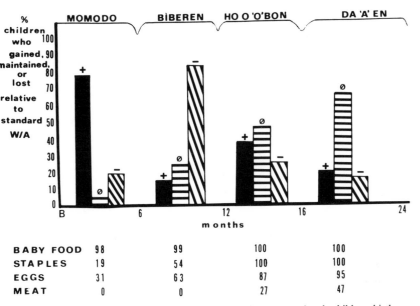

FIGURE 3   Longitudinal growth and food introductions among Amele children, birth
to two years old (+ = gained, ø = maintained, − = lost, weight relative to Harvard
standard).

breastmilk alone or in combination with the type of foods offered is
adequate. Analysis of 16 portions of *we* (pumpkin, sweet potato and
ripe banana broths) demonstrates that the mean portion size is 27.8
gm, 0.37% of which is protein and 91% is water. Although *faila we*
(papaya juice or broth) is the most common supplemental food given
the *momodo*, no sample portions were collected. Nonetheless, it is
clear that the moisture content of this favorite baby food would be
even higher.

Between 6 and 12 months, in the *biberen* stage, the situation
reverses drastically. Relative to standard, 82% of the children lose on
a diet highly dependent on breastmilk. While the average portion size
of *we* increases to 67 gm ($N = 21$), the moisture content remains the
same. Among 54% of the children, boiled mashed banana, yam,
sweet potato, Chinese taro or rice has been introduced. Analysis of
portion size reveals an average of 65 gm ($N = 25$) with an average of
2.5% protein, 75% water and having 137 Cal (572 kJ) per 100 gm.
Therefore, even though mothers report offering the standard staples

(with the exception of taro) at this stage, the prepared foods have a very high water content and can supply a child little energy (less than 200 Cal [836 kJ] daily) or protein of high biological value. While mothers often report giving boiled eggs, observations and further questioning indicate that the frequency of use is low, that the white is usually rejected, and that little of the yolk, being crumbly, is actually swallowed. A few mothers, especially of the more traditional type, boil an egg till firm, then mash it with breastmilk and feed it by spoon to the infant.

The situation improves somewhat in the *ho o' o' bon* stage, when the child begins to crawl and possesses a few teeth. At this stage 27% of the children receive some food of animal origin beside eggs, usually fish. Also at this stage all have been introduced to some staple. Growth has already fallen well below the standard; 45% of the children level off and maintain their reduced trajectory; 36% gain relative to standard, which represents an increase of 22% over the previous stage in that category. Although 23% continue to lose, even this represents an improvement over the previous stage when 82% lost relative to standard.

During the *da' a' en* period, when the child begins to walk, nearly half (47%) of the children are given some meat, nearly all have tasted eggs and all are eating the Amele staples. By this time the child's food may be cooked in coconut milk without danger to his health. Portion size, however, does not appear to keep up with the needs of the growing child. Most children maintain or lose weight relative to standard and few improve their position. The average portion size of starchy staple for the *da' a' en* is 58.1 gm ($N = 29$).

Longitudinal growth data are not yet available for the third year of life, the *gudugudu' ena* stage, but judging from the cross-sectional growth curves in Figures 1 and 2, the third year is a continuation of the last half of the second year when most children maintain a reduced level of growth. During this stage children are far more active than earlier, drink much less breastmilk and actively seek samples of their parents' and siblings' food. During the *gudugudu' ena* stage, 24% of the children receive no new items, although 11% have still not tasted meat and 5% have eaten no eggs.

Adequate frequency, portion size and nutrient value data are not yet available; but longitudinal observations of individual children reveal a common pattern of early introduction to particular foods followed by a cessation of feeding solids of any sort. Mothers unanimously claim that their children refuse anything but breastmilk for some period of time after being introduced to soups or solids. Lack of motivation to persist in offering supplementary foods supported by an indulgent child-rearing ethos may account in part for this feeding pattern. Preliminary analysis of 24 8- to 24-hour observations of Amele children 6 to 18 months old reveals that 64% were offered nothing but breastmilk during the time observed; 21% were fed once and 15% fed twice. None were fed three times daily as clinic sisters encourage and as mothers report when questioned casually. Therefore, since portion sizes and frequency as recorded are inadequate to compensate for the low energy and protein available in Amele foods, the quality and quantity of breastmilk is a vital concern.

It is necessary to remark that the surprisingly high percentage of children introduced to eggs in the *momodo* stage (31%) is highly biased toward children in the longitudinal series. By the next stage (*biberen*) this pattern is reversed and few of these children are newly introduced to eggs. Comparing variables, such as chicken ownership, mother's age, mother's years of schooling and family socio-economic status does not reveal any significant association explaining the difference between samples, although mothers in the longitudinal series, on average, were 4.5 years younger at the time of interview and had 1.9 years more formal schooling than did Ohuru mothers. In addition to this tendency, it is likely that an observer effect operating in the longitudinal growth study was greater than in the ethnographic work. This may bias either the mother's behavior or her reporting. In either case, the nutritional advantage that these figures may imply are amply undermined by numerous other behavioral and food-related factors.

CONCLUSIONS

Although the average birthweight of Amele people is low by international standards, their growth velocity in the first four to six months, when consuming essentially only breastmilk, is high enough for most children to gain in relation to the standards. However, most children do not remain on this favorable growth trajectory. The causes of this growth failure appear to be inadequate nutrient intake from breastmilk and a failure to add significant amounts of supplementary foods of high nutritional value to the diet of the young infant. Because the Amele believe that breastmilk increases in value as the child ages and that liquid foods are most suitable to small children, little supplementary food is offered during the first year of life and that which is given is in such small quantities and/or of such high water content that the contribution to nutrient intake is usually negligible. Little solid food of nutritional value is ever offered the child during its first year of life. Further, there is some evidence from recent studies that soup more effectively suppresses food intake than does equicaloric portions of crackers, cheese or juice indicating a role for temperature in effecting satiety (Kissileff, 1983). Warm soups, the principal food given Amele children up to about one year of age may, in fact, act as Amele mothers claim they do and keep the child from crying with hunger. Nevertheless, this feeding behavior occurs against a background of inadequate lactation beyond four to six months of age, as judged by departure of many children from the growth standards.

The role of beliefs in influencing feeding behavior appears to be very important here. Not only is breastmilk alone believed to be adequate for the newborn but it is also believed to improve as the child ages and thus, implicitly, to continue to meet the child's requirements. Against this background there is little need to introduce supplementary foods. It is possible that those which are introduced, particularly the soups, may lead to decreased breastmilk consumption.

This study shows that conceptualizing growth stages emically, according to the informant, instead of etically, according to the observer, can be a useful method of examining growth in relation to feeding practices. Implicit within the Amele concepts of growth and development are notions about lactation and food qualities which directly influence the timing of food introductions and the continuation of solids in the child's diet. These data also demonstrate the need to observe behavior in addition to asking about it. Food classifications and associated taboos may be more important in the breech as ethnomedical explanations after sickness strikes than as major determinants of feeding behavior.

Although further refinement is necessary, the approach used in this study of combining the methods of nutrition and ethnography, particularly the use of emically derived growth stages, holds promise for addressing the question of the nutritional impact of indigenous food classifications, food taboos, lactation and weaning schedules. This information should then be particularly useful in determining whether it is necessary to change infant feeding behavior and what characteristics of new or old foods would be most likely to ensure their acceptance, and in the design of more appropriate nutrition education programs.

ACKNOWLEDGEMENTS

The authors are grateful to the following members of the I.M.R. staff whose work contributed to this study: Naomi Yupae, Daina Lai, Jacqueline Cattani, Graham Wood and Cecilia Pawe.

REFERENCES

American Public Health Association (1973). Nutritional assessment in health programs. *Amer. J. Pub. Hlth. Suppl.* **63**, 34.
Bailey, K.V. (1964). Growth of Chimbu infants in the New Guinea Highlands. *J. Trop. Med.* **10**, 3-16.
Braun, T.G. (1967). Thirty-seven years of obstetrical and gynaecological experience in New Guinea. *PNG Med. J.* **10**(4), 107-110.
Ferro-Luzzi, A., N.G. Norgan and J.V.G.A. Durnin (1975). Food intake, its relationship to body weight and age, and its apparent nutritional adequacy in New Guinean children. *Amer. J. Clin. Nutr.* **28**, 1443-1453.
Gilles, H.M., J.B. Lawson, M. Sibdas, A. Voller and N. Allan (1969). Malaria, anaemia and pregnancy. *Ann. Trop. Med. Parasitol.* **63**, 245-263.
Harvey, P. and P. Heywood (1983). Twenty-five years of dietary change in Simbu Province, Papua New Guinea. *Ecol. Food Nutr.* **13**, 27-35.
Hipsley, E.H. and F.W. Clements (Eds.) (1950). *New Guinea Nutrition Survey Expedition 1947*. Department of External Territories, Canberra.
Jelliffe, D.B. (1966). *The Assessment of the Nutritional Status of the Community*. World Health Organization, Geneva.
Jelliffe, E.F.P. (1968). Low birth weight and malarial infection of the placenta. *Bull. W.H.O.* **38**, 69-78.
Kissileff, H.R. (1983). Satiety. *Contemp. Nutr.* **8**(1), 1-2.
Lawrence, P. (1964). *Road Belong Cargo*, Melbourne University Press, Melbourne.
McKay, S.R. (1960). Growth and nutrition of infants in the Western Highlands of New Guinea. *Med. J. Aust.* **1**, 452-459.
MacDonald, F. (1972). Good work at Yagaum. *Aust. Nurses' J.* **2**(1), 8-9.
Malcolm, L.A. (1970). Growth and development of the Bundi child of the New Guinea Highlands. *Human Biol.* **42**, 293-328.
Malcolm, L.A. (1975). Some biosocial determinants of the growth, health and nutritional status of Papua New Guinean pre-school children. In E. Watts, F. Johnston and G. Lasker (Eds.), *Biosocial Interrelations in Population Adaptation*. Mouton, The Hague, pp. 367-375.
Mikloucho-Maclay (1975). *New Guinea Diaries*. Translated by C.L. Sentinella. Kristen Press, Madang.
Oomen, H.A.P.C. and S.H. Malcolm (1958). *Nutrition and the Papuan child*. Technical Paper No. 118. South Pacific Commission, Noumea, New Caledonia.
Scragg, R.F.R. (1955). Birthweight, prematurity and growth rate to 30 months of the New Guinea native child. *Med. J. Aust.* **1**, 128-132.

Serjeantson, S. (1975). Marriage patterns and fertility in three Papua New Guinea populations. *Human Biol.* **47**(4), 399-413.

Sinnett, P. (1972). Nutrition in a New Guinea Highland community. *Human Biol. Oceania* **1**(4), 299-305.

Stanhope, J. (1967). Mortality of acute diarrhoea in the lower Ramu Valley 1962-1965. *PNG Med. J.* **10**(1), 15-19.

Sturt, R.J. (1972). Infant and toddler mortality in the Sepik. *PNG Med. J.* **15**, 215-226.

Venkatachalam, P.S. (1962). *A study of the diet, nutrition and health of the people of the Chimbu area.* Monograph No. 4, Department of Public Health, Port Moresby.

CHAPTER 4

FOOD TABOOS, MALARIA AND DIETARY CHANGE: INFANT FEEDING AND CULTURAL ADAPTATION ON A PAPUA NEW GUINEA ISLAND[†]

MARIA A. LEPOWSKY

INTRODUCTION

On Vanatinai (Sudest Island), Papua New Guinea, a remote, culturally conservative island in the Coral Sea, high rates of child malnutrition have been reported in government surveys despite the island's seemingly abundant food resources (National Planning Office, 1978; Lewis and Henton, 1979; Leonard, 1980; see Figure 1).

Nutrition surveys conducted by the national and provincial governments have shown high rates of child malnutrition as measured by weight-for-age and middle upper arm circumference on Vanatinai and nearby islands, particularly in children aged one to three years. The reported patterns of child malnutrition for the Louisiade Archipelago follow cultural rather than ecological boundaries. Census divisions in the Misima language and cultural area in the northwest showed lower percentages of children aged one to five years with middle upper arm circumferences of less than 14 cm. (See Table I.) The Misima-speaking West Calvados Islands had the lowest percentage in Milne Bay Province of children below the cut-off level, 16.39% (Lewis and Henton, 1979). The East Calvados Islands, immediately adjacent to the West Calvados but culturally and linguistically distinct, had the highest rate in Milne Bay Province, 78.31% (Lewis and Henton, 1979), even though the entire Calvados

[†]Research supported by National Science Foundation, University of California, Berkeley Chancellor's Patent Fund and Department of Anthropology, and National Institute of Child Health and Human Development Public Health Service Fellowship.

FIGURE 1.

THE LOUISIADE ARCHIPELAGO

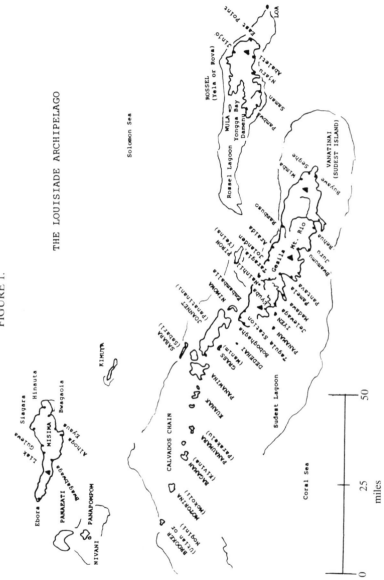

TABLE I

Percentage of children with middle upper arm circumference (MUAC) values less than 14 centimeters in the Louisiade Archipelago, Papua New Guinea.[a]

Census division	Misima	Deboyne-Renard (Panaeati)	West Calvados	East Calvados	Sudest (Vanatinai)	Rossel
12-59 months	29.86	22.86	16.39	78.31	59.80	36.64
12-35 months	34.83	38.60	20.69	90.63	91.94	54.92

[a]after Lewis and Henton (1979) and Leonard (1980)

region provides a similar environment of small, drought-prone islands with poor soil and rich fishing grounds. Vanatinai (Sudest Island), just southeast of the East Calvados, had the third highest rate of children measuring less than 14 cm of the 23 census divisions in Milne Bay Province, 59.80%. Ninety-two percent of Vanatinai children aged one to three were below the 14 cm cut-off level, and mean middle upper arm circumference of one to three year olds was 12.23 cm, the lowest value for this age group in Milne Bay Province (Lewis and Henton, 1979). Virtually all children under five in the province were included in this survey.

A survey of weight charts from Maternal and Child Health clinic records in certain districts conducted by the Milne Bay Provincial Nutritionist showed even higher rates of apparent undernutrition in parts of the Louisiade Archipelago than did the National Nutrition Survey, which also used clinic weight charts. In a sample of 498 children aged one to five from both Vanatinai (Sudest Island) and the neighboring East Calvados Islands, 14% were below 60% of mean weight-for-age of Harvard reference standards. Eighty-two percent were between 60 % and 80 %, 2 % between 80 % and 99 % and 2 % above 100 %. (See Table II.) These were the highest rates of low weight-for-age in the province in this survey with 96% of the children below the 80% cut-off level (Leonard, 1980).

Clearly, further research on the assessment of nutritional status in the Louisiade Archipelago is necessary because of the disparities between the reported rates for each region in the different surveys. The question of the appropriateness of using standards derived from American and European children to measure the nutritional status of Papua New Guinea children is not addressed in this paper because in

TABLE II

Weight-for-age of children aged one to five in the Louisiade Archipelago, Papua New Guinea, as a percentage of mean weight-for-age of Harvard standard.[a]

MCH clinic area	N	Less than 60%	60-79%	80-99%	100+%
Misima	211	2	48	45	5
Nimowa: Sudest Island (Vanatinai) and East Calvados Islands	498	14	82	2	2
Rossel	269	13	64	17	6

[a]After Leonard (1980). Data are derived from Maternal and Child Health clinic records.

all of the surveys the mean values of some districts of the archipelago are substantially lower than the means for Papua New Guinea children aged one to five, which have been measured throughout the country according to the same Western-derived standards.

The main questions arising from the existing survey data are:

1) Why do areas with apparently abundant food resources show large numbers of children measured as malnourished?

2) Why are there such large differences between the reported rates of child malnutrition for different parts of the archipelago?

This paper examines infant and child feeding on Vanatinai in its cultural and environmental context. It discusses present environment and subsistence patterns and describes some of the changes in subsistence strategies which have taken place during the last 100 years as cultural adaptations to new technology, new cultigens and new political systems. The hypothesis is advanced that these changes may have negatively affected the nutritional status of children and adults, since the pre-colonial diet based heavily upon collection of wild foods was more nutritious than the current diet which emphasizes cultivated starchy tubers. After examining the island's infant and child feeding patterns and associated customs, a second hypothesis is advanced: That customary food taboos which prohibit consumption of animal protein foods by children under weaning age contribute to child malnutrition but are a cultural adaptation to the local environment, specifically to endemic malaria. The evidence for a relationship between childhood food taboos and malaria is analyzed, and the implications of this hypothesis are discussed.

METHODOLOGY

The data on which this paper is based were collected during fourteen months in 1978 and 1979 and one month in 1981 on Vanatinai, or Sudest Island. In January, 1978, the author was asked by Dr. Colin Lewis, then the Provincial Health Officer of Milne Bay Province, to write a report (Lepowsky, 1979) on health and nutrition issues as affected by local cultural practices, concentrating particularly upon maternal and child health and nutrition. Government and health officials had virtually no information on conditions on this rarely visited island, which had not been previously studied by an anthropologist. In June, 1978, the National Nutrition Survey of Papua New

Guinea was released (National Planning Office, 1978), indicating widespread child malnutrition in the area and further underlining the urgency of Dr. Lewis' request.

Data were obtained through participant observation and informal interviews.[†] I resided in a household with a local family which included a two year old child, and regularly observed the diet of other young children in Jelewaga Village, my home base, and in other settlements during numerous visits to sites throughout the archipelago. Data include extensive information on subsistence practices, their seasonal variation and their changes over time, food habits, beliefs and proscriptions, and cultural beliefs and practices concerning health, illness, pregnancy, childbirth, and the care of infants and young children.

Quantitative assessments of food intake, food production, and nutritional and health status will be made on a later trip to Vanatinai. These are essential to determine the actual rate of child malnutrition, the prevalence of malaria and its impact upon child mortality, and the degree of disparity between cultural ideals of child and adult diet and actual food intakes.

ENVIRONMENT, SUBSISTENCE AND DIET

Vanatinai, 50 miles long and 10 miles wide, is the largest island in the Louisiade Archipelago, lying 220 miles southeast of mainland Papua New Guinea on the boundary between the Coral and Solomon Seas. (See Figure 1.) It is inhabited by about 2,000 people, giving it a very low population density of about four persons per sq. mile. The island is at the eastern end of one of the world's largest lagoons, with extensive coral reef systems that impede the progress of motor vessels, contributing to the island's lack of integration into the world cash economy. Vanatinai consists primarily of a steep central mountain range trending northwest to southeast. The highest point is Mt. Rio (2,645 feet), believed to be the home of the creator spirit and the spirits of the dead. Most of the coastline is fringed by mangrove swamp, and large tracts of freshwater sago swamps are found in lowland areas. The rest of the island is rainforest, with occasional

---

[†] I attained fluency in Vanatinai Ghalingaji, the unwritten language of Sudest Island, and conversational ability in the unrelated East Calvados (Saisai) and Misima languages. There were no other Caucasians living on Sudest Island at the time of the study.

grassland. Rainfall averages 120 inches per year, normally with no pronounced wet or dry season.

The treacherous reefs and swamps, which for six generations have discouraged European penetration of Vanatinai, are highly valued by the islanders as sources of abundant food. They prepare enough sago starch to trade the surplus to inhabitants of the smaller, drier Calvados Chain Islands, primarily for clay cooking pots from Brooker Island (Utian) and baskets of smoked *Tridacna* clam (Lepowsky, 1983). The mangroves are home to large crabs and many oysters and are nurseries for a wide range of marine species. The fringing reefs are rich in fish, *Tridacna* clam and other shellfish. Turtle (*Chelonia mydas*) and dugong (*Dugong dugon*) are numerous but are rarely harvested. Farther out in the lagoon yellowfin tuna, kingfish, shark, barracuda and other larger fish are found. The island's freshwater streams provide eels (*Anguilla sp.*), crabs, fish and crocodile (*Crocodylus porosus*).

Garden land is plentiful on Vanatinai, and islanders plant yams (*Dioscorea esculenta* and *D. alata*), sweet potato (*Ipomoea batatas*), taro and other aroids (*Colocasia esculenta, Alocasia macrorhiza, Cyrtosperma chamissonis* and *Xanthosoma sp.*), manioc (*Manihot esculenta*), bananas (*Musa spp.*), pineapple (*Ananas comosus*), and pumpkin (*Cucurbita pepo*). Plots are prepared by the swidden, or "slash and burn" technique, with fallow periods up to 40 years.

Four types of yams are the most important cultigens. But yam cultivation is significantly less intensive than in the densely populated islands of the Misima language area to the northwest. The Misimans, famous in the archipelago for their gardening skills, always stake their yam vines. The people of Vanatinai never do, but they rely far more on public yam planting ritual and private yam growing and harvesting magic. Nevertheless, Vanatinai gardens yearly produce a surplus of yams for exchange both to Calvados Chain Islanders and internally during their own elaborate series of mortuary feasts.

Vanatinai hamlets are usually situated near garden areas in groves of coconut (*Cocos nucifera*), breadfruit (*Artocarpus altilis*), mango (*Mangifera indica*) and betelnut (*Areca catechu*) trees. Some settlements are located just inland from the mangrove-fringed shore.

The people of Vanatinai gather a wide range of fruits, greens, tubers, nuts and legumes in the forest. They are prized for their taste. Certain wild foods such as the legume known as *kaikai* (see below), the nut called *mwiga*, and the wild green called *tagarugu* (*Gnetum gnemon*), as well as coconuts and fish, become primary staple foods in times of crop failure or drought. The islanders also hunt wild pig (*Sus scrofa*), monitor lizard (*Varanidus indicus*), possum (*Phalanger sp.*), flying fox (*Dobsonia sp.*) and fruit bat (*Pteropus sp.*). They raise domestic pigs, which are rarely eaten except at mortuary feasts, feeding them mainly on sago pith and allowing them to forage.

There is one small tradestore owned by a local family near the government station at the northwest tip of the island, but its stocks are most often depleted shortly after a supply boat arrives a few times each year. Most islanders eat tradestore foods very infrequently. Rice, canned mackerel, "square meat" (canned corned beef), tea and sugar are highly prized as rare luxuries, although some elders refuse to eat any "European" foods. Per capita income on Vanatinai is less than US$20 per year. Lack of money and the one to three day distance from the tradestore effectively limit consumption of storebought foods.

The word for "food," *ghanika,* also means "yam." Yam and other vegetables including sweet potato, manioc, taro, banana or pumpkin boiled in coconut cream are called *ghanika moli,* which literally means "true food." As their name implies, these vegetables boiled in coconut cream, called *tamja,* or "squeezed food", a reference to the expressed coconut, are a staple food on Vanatinai. Sago (*Metroxylon sagu*), the other staple, is considered to be appealing and desirable food and is often eaten, particularly during the months before the yam harvest. Households vary in the amount of sago they consume, mainly by the amount their members produce. The bulk of the labor of sago-making is done by young and middle-aged men, and older people must depend upon the generosity of kin in giving them sago. A flatbread of sago starch mixed with grated coconut is a favorite breakfast. Sago is also added in chunks or balls mixed with grated coconut to ghanika moli. A meal, as opposed to a snack, always includes ghanika moli or sago.

Garden food is carefully washed before cooking. Yams and manioc are peeled with a piece of polished pearlshell or a knife before cooking. The outer brown peel plus the outermost white layer of the manioc must be removed or people say it will cause sickness. Most vegetable food is boiled with coconut cream as tamja, but tubers are occasionally roasted in the fire as a snack. On special occasions such as clearing a garden, roofing a house or holding a mortuary ritual, when larger groups of people must be fed, a stone oven, or *ghumu,* is made above ground with layers of round heated river stones, soft, broad forest leaves, and tubers, bananas, lumps of sago and grated coconut. Animal foods are rarely added to the ghumu, as in other areas of the Pacific, but occasionally freshwater eel is roasted in a stone oven with chunks of sago as a special treat.

Leftovers (*kurea*) are reheated for breakfast. They must never be offered to even the most casual visitor. Hosts apologize if they have no cooked yams to offer a guest and feel embarrassed to serve boiled manioc, referring to it as "pigs' food" (*bobo ghaji*).

Vanatinai is famous throughout Milne Bay Province as a center of sorcery. Placing a malevolent charm on someone's food is said to be one of the easiest ways to kill a person, and children are warned not to accept food from anyone but close kin. A series of mortuary feasts is held after each death, during which yams, sago starch bundles, pigs and cooked foods are exchanged between the kin and affines of the deceased, who must ritually place cooked sago and yam in one another's mouths to demonstrate their interdependence and lack of fear of each other's sorcery. (See Lepowsky, 1981.)

Fish, shellfish, pork, and other animal protein foods are sometimes referred to collectively as *bwarogi*, the word for fish. Bwarogi and wild vegetable foods such as greens, fruits and nuts constitute a flavorful snack or a sidedish with ghanika moli or sago and are not a meal in themselves. Wild vegetable foods may be eaten as a snack or along with these staples, but the islanders believe that eating pork or fish by itself will lead to sickness, especially diarrhea.

CHANGES IN DIET AS A CULTURAL ADAPTATION

According to the middle-aged and elderly people of Vanatinai, there have been significant changes in subsistence patterns on the island since their youth. As recently as 40 or 50 years ago the islanders spent far less time on the cultivation of crops and more time foraging for a variety of wild foods, making sago, fishing and hunting. Yams were grown during this period, but there were not as many varieties as there are now. A tall aroid called *via* (unidentified) was also culti-vated more than now. People say that only one type of banana was cultivated; its stem rather than its fruit was boiled and eaten. This banana is now rarely grown. A white, firm, watery root called *yoronga* (unidentified), said to be a transplanted wild yam and still occasionally grown in gardens, was also more common.

Still, the earliest known European visitors to Vanatinai recorded evidence of significant cultivation in July, 1849. Thomas Huxley, then

a 23 year old assistant ship's surgeon on the British naval vessel
H.M.S. Rattlesnake, recorded in his diary that the ship's company
succeeded in exchanging "about 360 pounds of yams" for "seventeen
or eighteen hatchets" (Huxley, 1935, p. 203). July is the peak of the
present-day yam harvest. The British were also offered "cocoa-nuts"
(*ibid.*, p. 183), "Indian corn, ginger and sugar-cane" (Macgillivray,
1852, p. 257). Huxley (1935, p. 203) observed "about half an acre
covered with bananas" on Dedehai Island just west of Vanatinai.
Yams, bananas and "two other edible roots" were seen in "large
quantities" in Dedehai houses (Macgillivray, 1852, p. 227). Ginger
grows wild now in Vanatinai forests and is sometimes transplanted to
gardens or hamlets for magical or medicinal use. It is not used as food
and is strongly associated with sorcery. "Indian corn" is presumably a
reference to maize (*Zea mays*). Four inch long ears of maize are
grown on Vanatinai but rarely today. (See Jeffreys, 1971, on the con-
troversy over the Old World introduction of maize.) The British saw
domestic pigs and dogs in the Vanatinai region (Huxley, 1935;
Macgillivray, 1852). Today dogs are kept for hunting wild pig.

The discovery of gold on Vanatinai in 1888 triggered both a brief
goldrush and the annexation of British New Guinea. British
pacification and greater availability of metal tools apparently contributed to more extensive horticulture. David L. White, who lived in
the area from 1887 to 1892, wrote that "before the white men came
here during the 1888 goldrush the other islanders used to make raids
on their gardens, which they destroyed and carried off whatever pigs
or human heads they could get. This had the effect of discouraging
them making large clearances, so that they had hardly enough yams,
&c., to satisfy their wants, and had to live principally on sago, which
grows abundantly in the swamps. They are making larger gardens and
have now no fear of their old enemies, and also have better tools for
clearing scrub, &c., which they purchase from the storekeepers with
gold, bêche-de-mer, or shell" (White, 1893, p. 73).

A massive influx of new cultigens began in the 1880s when several
hundred men who had been tricked into boarding ship and working
on Queensland sugar plantations were returned to Vanatinai and
nearby islands by the Queensland government. The islanders say that
when their grandfathers heard that they were to be taken home, they
began surreptitiously to obtain cuttings and seeds of sweet potato,
manioc, pumpkin, papaya, mango, pineapple, sugarcane, maize,
cucumbers, and other crops. These were smuggled back and planted
in local gardens, where over the course of several generations many

were widely planted and became staples, joining yams as the basic constituents of "true food." Sweet potato and manioc are now particularly valued as drought-resistant and easy to grow. After the goldrush miners left, a few white traders and storekeepers remained for several decades to trade goods to the islanders in exchange for gold dust. They are said to have introduced oranges and different strains of domestic pigs. Tongan and Samoan missionaries who lived intermittently on the island before and after World War II introduced the variety of taro still called "Samoa." One large yellow variety of banana, called "Japan," was introduced to the western tip of Misima Island by Japanese who occupied nearby Panaeati Island during World War II. It is now found in gardens throughout the Louisiade Archipelago. New strains of banana and other cultigens have been brought home by workers returning in recent years from other parts of Papua New Guinea.

Shifts in residence patterns may also have affected Vanatinai subsistence strategies. According to local oral tradition, the earliest settlements were near the shore just inland from the mangrove swamp zone. During the late pre-colonial period, however, the islanders lived in dispersed hamlets on top of steep ridges in the interior for defense against raiders, visiting the shore mainly to fish and gather shellfish. In 1943 the Australia-New Guinea Administrative Unit (ANGAU), which administered the country during World War II, ordered all islanders on Vanatinai and nearby islands to move to larger, nucleated villages in designated sites along the coast so that they could be more easily controlled, an attempt to end a recent outburst of interisland raiding. This order has never been rescinded, and the present-day coastal villages are the result. Some islanders still live for most of the year in small inland hamlets near their gardens, returning to the coastal villages for the yearly census and tax-collecting government patrols. Present-day inland hamlets are close to springs and streams used as water sources.

Those living near the coast complain that the soil is poor, the water supply is unreliable, and the mosquitoes from nearby mangrove and sago swamps are a major nuisance. The proximity of settlements to the swamps and streams and the increasing size of settlements may have contributed to an increased prevalence of malaria. Coastal settlement also makes it more difficult to forage for wild foods in the rainforest, as one must travel longer distances past coastal swamps and grassland, since the limited forested area left near coastal settlements is rapidly stripped of gathered foods.

Although coastal settlement means that the islanders are closer to the shore and reef, the people of Vanatinai have not taken full advantage of their new proximity to marine resources and customarily do not eat seafood more than once or twice per week except during the neap tides of July. By contrast, their nearby neighbors in the small and infertile Calvados Chain Islands immediately to the northwest try to eat fish or shellfish every day and complain if these foods are unavailable. Vanatinai people regard fishing and shellfish gathering as a pleasant respite from their most essential subsistence activities: making sago and cultivating yams. They continue to produce a surplus of sago starch, pigs (fed on sago pith) and yams to trade through traditional exchange links to the Calvados Chain Islanders and, less frequently, the Wari Islanders, 200 miles distant, all of whose islands cannot produce enough garden produce or sago to sustain their populations. The traditional inland orientation of Vana-tinai people, which continues in spite of changes in residence patterns, thus contributes to the sustenance of the smaller island populations, who formerly either raided Vanatinai for food or formed exchange relationships with Vanatinai people to obtain food.[†]

The changes in subsistence patterns on Vanatinai are not due to deliberate intervention or pressure by outsiders from government or mission. They are cultural adaptations in response to the introduction of new technology, new cultigens, and new political systems. The cessation of raiding, government preference for nucleated coastal settlements, the spread of metal tools, and the availability of easy-to-grow starchy root crops have resulted in a tendency to live in larger, lowland settlements, to spend more time on gardening, and to spend less time collecting wild foods, making sago, fishing and hunting.[‡]

[†]During the frequent years of drought or crop failure, these small island peoples are reduced to eating an exclusive diet of coconuts and fish, as observed at Brooker Island (Utian) during the drought period of January, 1978, unless they sail their outrigger canoes to Vanatinai or Misima to trade for yams or sago.

[‡]The foods used ceremonially during mortuary feasts, the island's most important ritual events, are mainly uncultivated and reflect earlier dietary traditions (Foster and Anderson, 1978). These include a sago starch and green coconut pudding called *moni*, roasted sago and coconut loaf, and *kaikai* (an indigenous, not Pidgin, term) a large, wild white legume (unidentified) which used to be a staple food. Kaikai must be leached overnight to be edible, pounded with a stone pestle, and roasted on a potsherd. It is then usually combined in layers with sago starch, wrapped in leaves and roasted in a stone oven, or *ghumu*. Roasting food in hot stones was more common in pre-colonial times, when fewer clay pots were obtained through trade.

I hypothesize that the earlier diet based more heavily on foraging was more varied and more nutritious. These subsistence changes may therefore have had a negative impact upon the nutritional status of both children and adults.

The less varied, starchy present-day diet may have negatively affected the quantity and composition of breastmilk. Although breastmilk volume and composition have often been found to be "surprisingly good" in poorly nourished mothers (Jelliffe and Jelliffe, 1978, p. 80; Boediman *et al.*, 1979), poor maternal nutrition may reduce the quantity of breastmilk and its protein, fat and caloric concentration (Bailey, 1965; Jelliffe and Jelliffe, 1978), and lower values of water-soluble vitamins, Vitamin A and calcium (Jelliffe and Jelliffe, 1978; Osifo and Onifade, 1980). Dietary changes on Vanatinai may have also reduced the variety of weaning foods offered to children between about six months and three years of age and thus the range of nutrients in the child's diet. (See Robson and Wadsworth, 1977.)

INFANT FEEDING

Bottlefeeding is completely unknown on Vanatinai. Children are breastfed until about $2\frac{1}{2}$ to 3 years of age, with supplemental feeding beginning at about 6 months. Almost all women are able to breast-feed successfully, although a few women suffer from abscesses in one breast. Wetnurses should only come from the clan of the mother, father or grandparent. If no wetnurses are available among these groups, the infant is given the water from green coconuts, and its chance of survival is poor.

In earlier times a mother who did not want any more children would abandon her newborn infant, unwashed and with its umbilical cord still attached, by having someone place it in the crotch of a tree outside the hamlet. Others were then free to retrieve the infant, wash it, and raise it as their own. People say that any woman who has already borne and nursed a child could begin to lactate again and nurse the foundling. Three 60 year old men were pointed out who had been adopted in this fashion because their mothers had too many older children to care for. The verb "to adopt", *vaghan*, literally means "to feed". Nowadays, unless a mother has died in childbirth, a child is not usually adopted until it is weaned, although it may be claimed at birth by a kinsperson of the mother or father. Weaning normally takes place at about three years. The mother smears the nipple with ginger and/or leaves the child for a few days with a non-lactating kinsperson, such as its grandmother.

First-born infants and their mothers are secluded in a closed house for two to six months after the birth to protect them from scabies, other illnesses, or death due to sorcery. Seclusion may, in fact, reduce the mother's and infant's exposure to some infectious diseases. After this period, a small feast is held at which ceremonial valuables are given to the father and his kin by the mother's kin in gratitude for his having sired a new member of their matrilineage; these valuables flow in the opposite direction from the earlier bridewealth payments. Later offspring and their mothers need only stay inside the house out of the potentially harmful sight of others for a few weeks. Afterwards a mother may choose to return on some days to work in her garden, bringing her infant with her, normally in the care of an older sibling, who keeps it in the wall-less garden house. The mother nurses the infant periodically during the day. If an infant or a young child cries, it is fed, or its needs are attended to, for it is believed to be dangerous to a child's health to be left to cry. People say that a crying child "may become angry and leave us" (die). Infants and young children are normally left only in the care of siblings or close kin of the mother or father.

The top of an infant's head is blackened with burnt coconut husk as a magical protection against disease until it is four to six months of age. Special "growth magic" is often practiced by a knowledgeable kinsperson of the mother or father upon an infant or toddler to make it grow rapidly. A charm is recited as the practitioner chews a betel nut quid and then sprays reddish saliva upon the top of the child's head. Such magic may have side-effects, making the child cry frequently.

The term *tabwa* is both the adjective "fat" or "plump" and the verb "to grow." A fat baby is considered to be a healthy one. Pre-adolescent children are almost never referred to as fat, as they are typically lean and slender. Children from birth to adolescence are called *gama* (plural: *gamagai*).[†] Obesity is virtually unknown, and only an adolescent girl may be approvingly referred to as *tabwa*.

[†]By contrast, the Amele of northern New Guinea recognize seven named developmental stages of childhood, each linked to different dietary rules (Jenkins, Orr-Ewing and Heywood, ch. 3). There is little cultural emphasis on age-grading on Vanatinai (Lepowsky, 1985).

Pregnant women may eat anything but storebought foods. Eating too much fish is said to cause the woman to have a difficult delivery because the baby will grow too large. Eating fish may lead to conception in the local view; when a woman eats a particular clover-like leaf or a bark (both unidentified) believed to be contraceptive, she must also refrain from eating fish or she will conceive.

The mother of a first-born child may eat no fatty foods and no animal protein foods at all except for small clams (*kiwokiwo*), frequently boiled with wild greens, or freshwater crabs (*vwagu*). According to some people these taboos are in effect until the feast releasing mother and child from seclusion (two to six months), but others say they are observed until the baby is "big enough to sleep on its stomach with its arms in front of it" or until it is walking. The new father must also observe these food taboos. He and his wife should eat vegetable foods and drink herbal teas (*mwaoli* and *wadala*, both unidentified leaves) in order to "make themselves strong."

For a subsequent child, most parental food taboos are lifted after three to four weeks, but no mother may eat European (storebought) food, or greasy foods such as crab, eel, possum (*Phalanger sp.*) or pork fat until the baby is walking around habitually. Similarly, she may not eat pumpkin, for it would make the baby's stomach swell. Violation of any of these taboos is said to cause illness in the child, as will violation of the postpartum sex taboo, which lasts until the child is walking. Several times a day nursing mothers may drink mwaoli and wadala to increase their milk supply. These teas are also drunk by men and women as a general tonic, to reduce hunger, and to make one more able to work.

Leonard (1980) has reported that in many parts of Milne Bay Province colostrum is believed to be harmful to the infant and is expressed. On Vanatinai a mother first feeds her infant about one hour after birth, thereby conferring upon it the immunological benefits of colostrum, but she must first drink mwaoli tea and then express the contents of her breasts in order to rid them of the residue of the fats, coconut, fish, wild game or pig which she ate during her pregnancy.

Breastfeeding is on demand, and most women say their infants nurse about four or five times per 24 hours. Both breasts are emptied at each feeding. By one year of age a child nurses three times per 24 hours, with one feeding during the night, and by two to three years it nurses once during the day and once at night unless it demands the breast more frequently, which some children do. If an infant cries and refuses the breast, its mother may eat fish and then nurse to satisfy it once she has been released from the taboo on eating most animal protein foods.

Most parents say that supplemental foods are not introduced until the infant is about six months of age, but a minority opinion holds that semi-solids and liquids may be given from about three months. The traditional first supplemental food was taro which had been premasticated by the mother. Nowadays the infant is given the liquid from boiled root vegetables and shortly afterwards boiled, mashed or premasticated vegetables from the regular adult meal such as yam, sweet potato, manioc, taro, banana or pumpkin. Occasionally papaya juice or mashed papaya is given. Mashed or semi-solid foods are given by the mother or other caretaker with a spoon or with the traditional eating utensil of polished pearlshell. Semi-solids are fed to the infant about three times per day but not at night. The proportion of the diet consisting of mashed or semi-solid food increases with age. Parents and grandparents also feed the infant a variety of fluids starting at about six months of age. These include the water from green coconuts, the juice from sugarcane, hot mwaoli tea, and cold infusions made from the bark of a young areca palm or the root of the wild pandanus tree (*goa*) which bears a red fruit. A cold infusion made from water in which a reddish leaf called *ye* (unidentified) has been soaked is sometimes given to an infant to drink to make it stop crying. It is also used by adults as love magic or to make one strong and suppress hunger in the morning or while working in the garden.

By about one year of age the child may be handed small pieces of boiled yam, banana, sweet potato, taro, or manioc or a small piece of sago and coconut-meat flatbread during the family meals in the morning and evening or on demand during the day. Breastfeeding continues on demand.

Breastfeeding is said to stop a woman's menses for one year. When they start again it indicates that she is strong enough to garden on a regular basis.

Children aged one to three may be left in the hamlet with older sisters or brothers while the mother goes to the garden or to collect forest or marine foods. An unmarried sister or niece of either parent or a grandmother may also watch the toddler, but child caretakers are a common sight. Girls are more often drafted as babysitters, but pre-adolescent brothers also take their turn. Older children are generally patient and indulgent, leading their tiny charges around the hamlet and taking them on short expeditions to the shore to fish or the forest to look for wild fruit or nuts. The child caretakers stoke the fire, peel and boil vegetables or make a sago flatbread, which young children often beg for, and solicitously feed the toddler, then induce it to nap while they play quietly nearby.

Children begin to chew areca nut, betel pepper and powdered lime by about four years of age (Lepowsky, 1982).

Food Taboos

The range of foods from which adults select their diet is extremely wide despite the increasing local emphasis on horticulture. But the foods which may be given to children under the age of about three years are limited by a set of food taboos. According to Vanatinai custom, children below weaning age may not be given any animal protein foods, fruits, greens, storebought foods, or any other food categorized as *love*, which means both "sweet" and "greasy". Otherwise it is said that the child would fall sick and die. However, coconut cream is *love*, and young children are given vegetables boiled in coconut cream and the broth in which the vegetables have been boiled.

These taboos are said to be ancient and part of *taubwaragha*, "the way of the ancestors." Nevertheless a few people believe that the taboo on animal protein need only be observed until a child is six months of age, before which it is unlikely to be eating much supplemental food anyway. One middle-aged man who holds this view said that a traditional first supplemental food given at three months of age was grasshoppers which had been roasted over the fire and that nowadays the flesh — but not the fat — of a possum (*wodoya*) and the liquid in which it is boiled are given to a three month old infant as a first supplemental food. But in most families the taboo on animal protein foods is rigidly adhered to, for as one mother said, "We must take good care of our children, for we are living very far from hospitals." Further research is necessary to compare actual food intakes with statements about the proper diet for infants and young children and the foods which are proscribed to them.

Other taboos limit the consumption of certain foods for older children. Particularly greasy foods such as wild pig fat, possum fat and crab roe are believed to be detrimental to children's health and are forbidden until adolescence. Certain areas of the fringing reef and certain streams are said to be *silava*, which means "sacred" or "taboo". Adults and children alike should not eat fish or shellfish from these areas on pain of serious illness. Young children should not even be allowed in their vicinity, or they might fall sick. Many of these areas lie directly below settlements, and this taboo may well be congruent with Western notions of sanitation. It is noteworthy that the danger to children and the special danger of shellfish from these areas are stressed.

Few other taboos limit adult diet on Vanatinai. People eat their own and their fathers' totem birds and animals, unlike on Misima Island to the northwest, where the father's totem bird is forbidden. Vanatinai people do not usually eat their fathers' totem fish. Fish should not be eaten or touched before yams are planted, or the yams will disappear. Dog is never eaten on Vanatinai. It is eaten by Misima adults but forbidden to Misima children, again because it is too greasy. The dog on Vanatinai is economically valuable for hunting pig, which is plentiful on that island, while there are very few wild pigs on densely settled Misima. This may explain why dog is eaten on the latter island and its neighbors but not on Vanatinai. The Vanatinai people eat crocodile, while their immediate neighbors to the northwest, the culturally and linguistically distinct people of the East Calvados Islands, are repelled by this culinary habit. People on Vanatinai (and elsewhere in the Louisiade Archipelago) never eat or molest snakes or bats. Sometimes snakes may be spirits (such as Rodio, the creator spirit), place spirits (silava) or human sorcerers in the form of a snake. Bats are the familiars of sorcerers. Marsupial fruit bats and flying foxes, however, are considered delicacies. Octopus are plentiful in the lagoon, but they are not eaten. Silava may take the form of an octopus.

Van Der Hoeven (1958) emphasized the "magic-religious" aspect of the "protein taboo" for young children on the northwest coast of New Guinea, and concluded that because protein foods are scarce and highly valued, "a gift of proteins (or its negative equivalent, the protein taboo) is reserved for higher Powers." On Vanatinai as well, the protein taboo for children under weaning age is part of a covenant with the ancestors. In exchange for supernatural protection from death or illness, the infant and its parents do not eat highly valued protein foods for certain periods defined by custom. Violations of the taboo would anger the ancestor spirits, who control the health and prosperity of their descendants. The following section suggests why a protein taboo for young children may have come to be part of a covenant with the ancestors on Vanatinai.

CHILDHOOD FOOD TABOOS AS A CULTURAL ADAPTATION

Consideration of the evidence available on nutritional status and diet of young children in the Louisiade Archipelago suggests that the presence of traditional food taboos prohibiting children under the age of about three from eating animal protein foods in the Vanatinai region contributes to the higher reported rates of child malnutrition and particularly the rates in children aged one to three years, which are the highest in Milne Bay Province. Similar taboos against the feeding of animal protein foods to young children were formerly followed in the Misima language area, which includes the Deboyne-Renard (Panaeati) and West Calvados census divisions, but these beliefs have nearly disappeared due to pressure against them from health workers, missionaries, and government personnel, according to local informants. The Misima area has lower reported child malnutrition rates. The taboos are still very strong in the more remote Vanatinai-East Calvados area. Existing nutritional surveys of Rossel Island children have given contradictory data, but Rossel Island beliefs on proper diet for young children are quite different. (See Lepowsky, 1979.) More research on nutritional status and food intake of young children in the various parts of the Louisiade Archipelago is necessary in order to draw firm conclusions on the relationship between food taboos and child nutrition.

Meanwhile, the question remains as to why these childhood food taboos have arisen and why they have been maintained on Vanatinai. Similar taboos against the consumption of animal protein foods by young children, with the associated warning that otherwise the child may sicken or die, have been reported from many parts of the world, including West Africa (Hendrickse, 1966; Ogbeide, 1974), East Africa (Gerlach, 1969), India (Jelliffe, 1957), Malaysia (Wolff, 1965; Bolton, 1972), and New Guinea (Van Der Hoeven, 1958). Cassidy (1980) referred to taboos on giving animal protein foods to young children as part of her argument that toddler malnutrition from this and other causes is a form of culturally sanctioned benign neglect which may be deleterious to the individual but beneficial to the group as a form of population control, natural selection and adaptation to prevalent conditions of food scarcity. She and most other writers concerned with such childhood food taboos have assumed that appropriate intervention or re-education are ethical and necessary. Bolton (1972), writing about similar childhood food taboos among various groups of aboriginal Malays, saw an element of greed in restricting choice foods to adults and particularly to adult men. Cassidy argued at length that people who enforce these child food taboos really love their children and are not practicing malign neglect. But do the restrictions on giving animal protein to young children on pain of their sickness or death which have arisen in various parts of the world in any way benefit the individual child?

The Vanatinai animal protein taboo for young children has a number of consequences. The child is dependent on breastmilk for protein. Continued lactation is known to inhibit conception. Traditional child spacing beliefs on Vanatinai hold that the next child should be born only when the first "is already playing with his friends," and in theory the postpartum sex taboo should last as long as the mother lactates, or the child will sicken. Long-term lactation may also cause the mother's percentage of body fat to decline and depress fecundity (Frisch, 1975; Frisch and McArthur, 1974; Hamson, Boyce and Platt, 1975), protecting the well-being of the nursling by inhibiting conception of a younger sibling who would displace it from the breast, although the relationship of body composition to fecundity is a controversial subject (Johnston *et al.*, 1975; Scott and Johnston, 1982).

Breastmilk contains antibodies to a variety of infectious agents (Jelliffe and Jelliffe, 1978; Ogra, Fishaut and Theodore, 1980; Carlsson *et al.*, 1980). The child may also benefit from high milk consumption by reducing the amount of possibly contaminated water which it drinks and thus the threat of diarrhea. Springs and streams from which water is drawn are sometimes fouled by pigs and dogs, and a few water sources are downhill from settlements.

Fatty foods which are prohibited to a young child by Vanatinai custom are not easily digested by children in the first years of life (Barness, 1979; Fomon, 1974) and might cause diarrhea, a life-threatening condition in the very young because of the rapid dehydration it may induce and the inhibition of nutrient absorption accompanying it.

Malaria

Plasmodium falciparum and *P. vivax* malaria are endemic on Vanatinai. The island was described as mesoendemic by the Public Health Department Malaria Service in 1973 after epidemic foci in the

Louisiade Archipelago had received special attention in the DDT residual spraying program (Parkinson, 1974). Its vectors are *Anopheles punctulatus*, *A. farauti*, and *A. koliensis*, while *A. farauti* is the only vector found on the other, smaller islands of the archipelago (Spencer, Spencer and Venters, 1974). These three anopheline species comprise the *A. punctulatus* group which, "(g)iven optimal conditions ... are as dangerous a group of vectors as *A. gambiae* and *A. funestus* in tropical Africa" (Peters, 1960, p. 242). *A punctulatus* is most likely to bite humans indoors, and its biting activity peaks around midnight. It prefers transient sunlit pools as breeding areas. *A. farauti* and *A. koliensis* are more adaptable, breeding both in natural and artificial bodies of water. Both favor disturbed areas. *A. farauti* tolerates brackish water and is common in tidal creeks and along the coastal shelf. It bites primarily after dusk and will also feed on pigs and dogs (Van Dijk and Parkinson, 1974; Spencer, Spencer and Venters, 1974).

Malaria is a parasitic disease which disproportionately affects the very young, although infants up to six months of age are protected by fetal hemoglobin and the antibodies in breastmilk (Maegraith, 1967; Pasvol *et al.*, 1976). Survivors gain a degree of partial immunity which increases as the individual reaches adulthood (Molineaux and Gramiccia, 1980; Nardin *et al.*, 1979). *P. falciparum* malaria is a major cause of the death of young children on Vanatinai and elsewhere in lowland Papua New Guinea. The malaria death rates in the Vanatinai region are unknown, but in some parts of Papua New Guinea, "20 per cent of infant deaths have been attributed to malaria" (Black, 1972, p. 682).

Vanatinai adults frequently suffer severe malaria attacks as well. Even if it does not kill directly, malaria suppresses the immune response, leaving the individual more susceptible to infectious and parasitic diseases (Bell, 1978). It also destroys red blood cells, suppresses appetite, and induces high fevers which burn off the host's caloric reserves and lead to negative nitrogen balance (McGregor, 1982). Thus endemic malaria may increase malnutrition rates in a population.

Malaria and Diet

The majority of studies of the interaction of malnutrition and infection indicate that they act synergistically, with malnutrition lowering an individual's resistance to infectious diseases. In 81% of the case studies and experiments reviewed by Scrimshaw, Taylor and Gordon (1968), malnutrition and infection acted in this way. But their survey of malaria studies showed that over half the time the interaction of overall malnutrition or deficiencies of particular nutrients in animals or humans with malaria was antagonistic: the more poorly nourished individual seemed to be more resistant to malaria or did not manifest as severe symptoms. The reviewers conclude that the *Plasmodium* in such cases may be more susceptible than the animal or human host to nutrient deprivation and may either "starve" or be unable to multiply in the host's red blood cells. Later writers have suggested that if the host red blood cells, because of nutrient deprivation, cannot maintain normal function and membrane integrity and cannot divide normally, the *Plasmodium* will not be able to thrive (Eaton *et al.*, 1976). In recent reviews of the relationship between malaria and malnutrition, Beisel (1982, p. 749) stated that, "Malaria may be considered an illustration in humans of an antagonistic interaction between severe malnutrition and a parasitic disease," and McGregor (1982, p. 798) noted that although malaria may have an adverse impact on host nutrition by contributing to low birth weight, protein-energy mal-nutrition, and the pathogenesis of anemia, there is "little convincing evidence that malnutritional states in humans materially enhance the severity or lethality of plasmodial infections."

Researchers have reported a relationship between decreased susceptibility to malaria and host deficiencies of a wide array of nutrients, including protein, iron, Vitamins A, C, and E, P-amino-benzoate and the B vitamins thiamine, riboflavin, pyridoxine and pantothenic acid (Scrimshaw, Taylor and Gordon, 1968). Animal studies have shown that host folate deficiency correlates with reduced severity of plasmodial infection (Katz, 1982). A milk diet has also been shown to suppress malaria infection (Maegraith, Deegan and Jones, 1952; Bray and Garnham, 1953), and human infants on all-milk diets are known to be resistant to malaria (Maegraith, 1967). Nomadic African children on a milk diet fed grain after a famine had a higher incidence of cerebral malaria (Murray *et al.*, 1978a). Edington (1967) and Hendrickse *et al.*, (1971) suggest that under-nutrition protects children against cerebral malaria based on evidence from West Africa. Hendrickse (1967) concludes that in Nigeria lethal infections and serious morbidity from *P. falciparum* are less frequent in malnourished children.

Recent research in Africa has indicated that increasing the dietary intake of iron of anemic patients in Tanzania (Byles and D'Sa, 1970; Masawe, Muindi and Swai, 1974) or of Somali nomads accustomed to an all-milk diet (Murray *et al.*, 1975; Murray *et al.*, 1978a) precipitated attacks of previously asymptomatic malaria. The Murrays (1978b), also reported activation of prexisting brucellosis and tuberculosis. Based upon their observations of Somali nomads during the Sahelian famine (1978b), they concluded that the famine actually suppressed malaria and other disease symptoms and that refeeding with relief foods activated these diseases, a phenomenon first reported by Ramakrishnan during the Bengal famine of 1943 (cited in Hendrickse, 1967, and Murray, Murray and Murray, 1980a; see also Murray *et al.*, 1977a; Murray *et al.*, 1978c; Murray, Murray and Murray, 1980b; Sharma *et al.*, 1979). The Murrays concluded that, "within certain limits undernutrition in humans and animals appears to decrease susceptibility to infection with viruses, malaria, and some bacteria, probably those preferring an intracellular environment" (1977b, p. 482).

In those who survive malaria attacks the *Plasmodium* may deplete the host's reserves of key nutrients during the course of the disease but then be unable to live and reproduce optimally, leading to a diminution in host symptoms.

DISCUSSION

I suggest that the food taboos on Vanatinai which restrict young children's intake of animal protein foods may be a long-standing cultural adaptation to an environment of endemic malaria. The reported prevalence of mild and moderate undernutrition which may result from this practice may be a trade-off for increased resistance to potentially lethal and debilitating malaria attacks.

The Misima language area islands, which have lower reported child malnutrition rates than the Vanatinai-East Calvados region where the taboos are still observed, are less swampy, better drained and support fewer malaria vectors than Vanatinai (Spencer, Spencer and Venters, 1974, p. 23).

The Vanatinai food taboos may further benefit young children by emphasizing the importance of breastmilk in their diet, which confers some immunity to a variety of diseases and which may itself help suppress malaria. Breastfeeding may reduce consumption of potentially contaminated water. Prolonged nursing inhibits conception, protecting the nursling's health by preventing its displacement from the breast by a newborn. The prohibited fatty protein foods may also induce diarrhea in the very young. The taboo on animal protein minimizes heme intake and results in a diet low in iron, which may suppress malaria symptons.

Adaptation to a low-protein diet is widely found in Papua New Guinean adult populations, which often defy conventional nutritional standards by functioning well on presumably protein-deficient diets (Ferro-Luzzi, Norgan and Durnin, 1975, 1978; Macpherson, 1963; Norgan, Ferro-Luzzi and Durnin, 1974; Oomen and Corden, 1970). Western-derived standards of adequate protein intake may therefore not be applicable to Papua New Guinean populations. Young children fed a low-protein diet may grow up to be shorter but may then require a smaller amount of protein to remain healthy than do Western adults (Stini, 1971, 1979). This is an important consideration even on coastal Vanatinai, where animal protein foods are only eaten occasionally.

It is possible that in lowland Papua New Guinea, the evolution of populations adapted to low-protein diets came about not only because of the relative scarcity of animal protein foods away from the coast but because such a diet offers more protection against malaria and perhaps other parasites. Childhood food taboos on animal protein consumption may have evolved concurrently with a biological adaptation to low-protein diets for the same reason. Similar taboos may have evolved as a cultural adaptation in other malaria-endemic areas of the world.

Cultural adaptation to environment is an ongoing process. The physical environment itself changes over time. On Vanatinai shifts in rainfall patterns and changes in sea level may have occurred during the unknown period of time the island has been settled. The area of mangrove and sago swamp on the island may have been larger or smaller in the past. It is impossible to tell when the *Anopheles punctulatus* group or *Plasmodium falciparum* and *P. vivax* were introduced, when malaria became endemic among the island's human population, or whether the island's first settlers brought malaria with them. Therefore it is impossible to reconstruct the environmental conditions which prevailed when the childhood food taboos on animal protein consumption first arose, to determine whether the taboos were adopted before or after the first settlers reached the island, or to document a correlation between the taboos and the threat of malaria.

As population density increased on Vanatinai over thousands of years, more primary rainforest was cut and burned for garden land, creating more grasslands, standing water, and disturbed areas where *Anopheles* mosquitoes could live. The government edict of 1943 forcing settlement in larger villages in closer proximity to the coast and swamps rather than small dispersed hamlets at 800-1000 foot elevations on the steep interior ridgetops probably contributed to malaria endemicity among the population, since *A. farauti*, the vector most dominant after residual spraying, tolerates brackish water and is often found in tidal creeks and along coastal shelves (Spencer, Spencer and Venters, 1974). The greater number of visitors to the island since World War II and of islanders migrating elsewhere in search of work and then returning home has probably contributed to an increase in malaria endemicity.

Increased contact with the rest of Papua New Guinea also may have contributed to a greater prevalence of influenza and its frequent complication, pneumonia. I observed in 1978-9 that influenza sweeps the island several times per year and that the pneumonia which follows kills a few people of all ages. Childhood food taboos may be adaptive in reducing the severity of malaria attacks among children aged about six months to three years and thus the mortality rate from malaria, but the suboptimal nutritional status to which they probably contribute may place children of the same age at greater risk of dying from pneumonia, which is currently a major killer on the island and throughout Papua New Guinea.

Presumably the childhood food taboos arose during a time when Vanatinai diet, or the diet of the first settlers in their original home-lands, was oriented more heavily toward collection of wild foods rather than the cultivation of a range of starchy tubers. The restriction of animal protein foods might have a different nutritional impact upon a child whose diet regularly included such items as the wild nuts and legumes which grow abundantly on the island and whose lactating mother regularly ate small game, wild nuts, legumes, greens and other wild foods. It is therefore possible that the childhood food taboos are less adaptive now in ensuring the viability of children by giving partial protection against malaria attacks than they were one hundred years ago due to a change in local diet and subsistence strategies and an increased prevalence of other infectious diseases which are synergistic with malnutrition.

Cultural adaptations to endemic malaria have been noted in other parts of the world. Houses in the hills of Vietnam are built on piles,

D

with pigsties underneath. Malaria vectors feed on pigs which are on their flight level, while the smoke from cooking fires in the house keeps them out of the human living area (May, 1958).[†] Brown (1981) saw a variety of practices in Sardinia as cultural adaptations to malaria, including the hilltop settlement pattern, summer residence shifts for the wealthy, inverse transhumance, restrictions on mobility of pregnant women, and folk medical theories of malaria etiology which further reduce exposure and susceptibility. Durham (1982) suggested that the widespread West African wet-season taboo on the consumption of yams, which contain cyanogenic glycosides that may inhibit the sickling of hemoglobin, is a "cultural mediation" of a biological adaptation to malaria. Etkin and Ross (1983, p. 254-5) concluded that among the Hausa of northern Nigeria, medicinal plants used by local practitioners to suppress malaria symptoms may be therapeutic, increasing levels of intracellular oxidation, and found that they were also "important dietary elements at the end of the rainy season," the peak risk period for plasmodial infections. They labelled medicinal and dietary use of these plants "biocultural adaptations to disease."

Numerous authors have offered evidence that malaria and low host intakes of a variety of nutrients are antagonistic. If there is in fact a relationship between Vanatinai food taboos on feeding animal protein foods to young children and malaria prevalence and mortality among this population, government or health personnel who wish to intervene to improve the health of the islanders should be aware of the possible consequences of their actions and take precautionary measures. For example, if further research indicates that it would be beneficial overall to child health, a nutrition education campaign encouraging the giving of fish and other local protein foods to children under three could be accompanied by education about the connection between standing water, mosquitoes and malaria, discussion of the favorite breeding sites and biting times of the *A. punctulatus* group and suggestions on minimizing exposure, especially that of young children. An intensive program to drain swampy areas adjacent to settlements, to reduce the amount of standing water in the hamlets, and to cut tall grass in hamlet clearings could be carried out at the same time.

[†]On Vanatinai houses are also built on poles, and the pigs and dogs which *A. farauti* feed on scavenge underneath. Some islanders cook in detached ground-level kitchens. People build smokey fires of green wood at night on the ground under their black-palm slatted floors if mosquitoes are too irritating.

Cultural change is normal in human societies, and present conditions on Vanatinai may not represent an adaptation to environment prevailing over centuries but the impact of a series of recent historical events upon a constantly changing island society and environment. An adaptive relationship between childhood dietary proscriptions and malaria would not rule out the possibility of intervention. The hypothesis emphasizes the need to investigate all of the interrelations among dietary practices, health and environment before attempting to change the customary behavior of a population. The particular interrelations which it suggests may be significant in many other parts of the tropical world.

ACKNOWLEDGEMENTS

Financial support for this research was received from the following institutions and is gratefully acknowledged: United States National Science Foundation, Chancellor's Patent Fund and Department of Anthropology of the University of California, Berkeley. The Institute of Applied Social and Economic Research of Papua New Guinea funded my 1981 transportation to and from Port Moresby. A National Institutes of Health, National Institute of Child Health and Human Development Public Health Service Fellowship has supported me during the period in which this paper was written.

I wish to thank the late Dr. Colin Lewis, Milne Bay Provincial Health Officer in 1978 and 1979, Dr. Festus Pawa, Milne Bay Provincial Health Officer, 1981 to present, and the staff of the Milne Bay Provincial Health Department for their assistance and their encouragement of this research. An earlier draft of this paper was presented at the Symposium on Infant Care and Feeding in Oceania in Hilton Head, South Carolina in March, 1982. Helpful comments on earlier drafts were made by Dr. Sheldon Margen, Dr. Frederick Dunn, participants in the Symposium on Infant Care and Feeding in Oceania, and three anonymous reviewers.

The data in this paper could not have been collected without the interest, cooperation and moral support of the people of Vanatinai and the Louisiade Archipelago. I give them my deepest thanks and hope that this research may eventually benefit them and their children.

REFERENCES

Bailey, K.V. (1965). Quantity and composition of breastmilk in some New Guinean populations. *J. Trop. Pediatr.* **11**, 35-49.
Barness, L. (1979). *Developmental Nutrition.* Ross Laboratories, Columbus, Ohio.
Beisel, W. (1982). Synergism and antagonism of parasitic diseases and malnutrition. *Rev. Inf. Dis.* **4**(4), 746-750.
Bell, R.G. (1978). Undernutrition, infection and immunity: The role of parasites. *PNG Med. J.* **21**(1), 43-55.

Black, R.H. (1972). Malaria. In P. Ryan (Ed.), *Encyclopaedia of Papua and New Guinea.* Melbourne University Press, Melbourne.

Boediman, D., D. Ismail, S. Iman, Ismangoen and S.D. Ismadi (1979). Composition of breast milk beyond one year. *J. Trop. Ped. Env. Ch. Hlth.* **25**(4), 107-110.

Bolton, J.M. (1972). Food taboos among the Orang Asli in West Malaysia: A potential nutritional hazard. *Amer. J. Clin. Nutr.* **25**, 789-799.

Bray, R.S. and P.C.C. Garnham (1953). Effect of milk diet in *P. cynomolgi* infections in monkeys. *Brit. Med. J.* **1**, 1200-1202.

Brown, P.J. (1981). Cultural adaptations to endemic malaria in Sardinia. *Med. Anthropol.* **5**(3), 313-339.

Byles, A.B. and A. D'Sa (1970). Reduction of reaction due to iron dextran infusion using chloroquin. *Brit. Med. J.* **3**, 625-627.

Carlsson, B., J. Cruz, L. Mellander, and L. Hanson (1980). The mechanism of immunity provided by breast-feeding. In S. Freier and A. Edelman (Eds.), *Human Milk: Its Biological and Social Value.* Excerpta Medica, Amsterdam, pp. 122-131.

Cassidy, C. (1980). Benign neglect and toddler malnutrition. In L. Greene and F. Johnston (Eds.), *Social and Biological Predictors of Nutritional Status, Physical Growth, and Neurological Development.* Academic Press, NY, pp. 109-139.

Durham, W. (1982). Interactions of genetic and cultural evolution: Models and examples. *Human Ecol.* **10**(3), 289-323.

Eaton, J.W., J.R. Eckman, E. Berger and H. Jacob. (1976). Suppression of malaria infection by oxidant-sensitive host erythrocytes. *Nature* **264**, 758-760.

Edington, G.M. (1967). Pathology of malaria in West Africa. *Brit. Med. J.* **1**, 715-718.

Etkin, N. and P. Ross (1983). Malaria, medicine and meals: Plant use among the Hausa and its impact on disease. In L. Romanucci-Ross, D. Moerman, and L. Tancredi (Eds.), *The Anthropology of Medicine: From Culture to Method.* Praeger, New York, pp. 231-259.

Ferro-Luzzi, A., N.G. Norgan and J.V.G.A. Durnin (1975). Food intake, its relationship to body weight and age, and its apparent nutritional adequacy in New Guinean children. *Amer. J. Clin. Nutr.* **28**, 1443-1453.

Ferro-Luzzi, A., N.G. Norgan and J.V.G.A. Durnin (1978). The nutritional status of some New Guinean children as assessed by anthropometric, biochemical and other indices. *Ecol. Food Nutr.* **7**(2), 115-128.

Fomon, S. (1974). *Infant Nutrition.* Second edition. W.B. Saunders Co., Philadelphia.

Foster, G. and B. Anderson (1978). *Medical Anthropology.* Wiley and Sons, New York.

Frisch, R.E. (1975). Critical weights, a critical body composition, menarche, and the maintenance of menstrual cycles. In E.S. Watts, F.E. Johnston and G.W. Lasker (Eds.), *Biosocial Interrelations in Population Adaptation.* Mouton, The Hague, pp. 319-352.

Frisch, R.E. and J.W. McArthur (1974). Menstrual cycles: Fatness as a determinant of minimum weight for height necessary for their maintenance or onset. *Science* **185**, 949-951.

Gerlach, L. (1969). Socio-cultural factors affecting the diet of the Northeast Coastal Bantu. In L.R. Lynch (Ed.), *The Cross-Cultural Approach to Health Behavior.* Farleigh-Dickenson University Press, Rutherford, NJ. pp. 383-395.

Harrison, G.A., A.J. Boyce and C.M. Platt (1975). Body composition changes during lactation in a New Guinea population. *Ann. Human Biol.* **2**(4), 395-398.

Hendrickse, R.G. (1966). Some observations on the social background to malnutrition in tropical Africa. *African Affairs* **65**, 341-349.

Hendrickse, R.G. (1967). Interactions of nutrition and infection: Experience in Nigeria. In G. Wolstenholme and M. O'Connor (Eds.), *Nutrition and Infection.* CIBA Foundation Study Group No. 31. Little, Brown, Boston, pp. 98-111.

Hendrickse, R.G., A.H. Hasan, L.O. Olumide and A. Akinkunmi. (1971). Malaria in early childhood. An investigation of five hundred seriously ill children in whom a clinical diagnosis of malaria was made on admission to the Children's Emergency Room at University College Hospital, Ibadan. *Ann. Trop. Med. Parasitol.* **65**, 1-20.

Huxley, T.H. (1935). T.H. Huxley's Diary of the Voyage of H.M.S. Rattlesnake. Julian Huxley (Ed.). Chatto and Windus, London.

Jeffreys, M.D.W. (1971). Pre-Columbian maize in Asia. In C. Riley (Ed.), *Man Across the Sea: Problems of Pre-Columbian Contacts,* University of Texas Press, Austin, pp. 376-400.

Jelliffe, D. (1957). Social culture and nutrition: Cultural blocks and protein malnutrition in early childhood in rural West Bengal. *Pediatrics* **20**, 128-138.

Jelliffe, D.B. and E.F.P. Jelliffe (1978). *Human Milk in the Modern World: Psychosocial, Nutritional, and Economic Significance.* Oxford University Press, Oxford.

Johnston, F., A. Roche, L. Schell and H.N. Wettenhall. (1975). Critical weight at menarche: Critique of a hypothesis. *Amer. J. Dis. Child.* **129**, 19-23.

Katz, M. (1982). Discussion: Malaria and malnutrition. *Rev. Inf. Dis.* **4**(4), 805.

Leonard, D. (1980). *Report on Food and Nutrition in Milne Bay Province.* Milne Bay Development Study. Milne Bay Provincial Government, Alotau, Papua New Guinea.

Lepowsky, M. (1979). *A Preliminary Report on Cultural Factors Affecting Health and Nutrition: Sudest Island and the Louisiade Archipelago, Papua New Guinea.* Report submitted to the Provincial Health Officer, Milne Bay Province. Mimeographed and distributed by the National Planning Office, Port Moresby, Papua New Guinea.

Lepowsky, M. (1981). *Fruit of the Motherland: Gender and Exchange on Vanatinai, Papua New Guinea.* Unpublished doctoral dissertation, University of California, Berkeley. *Dissertation Abstracts International* **42**(12), Part 1, 5174A. University Microfilms International No. DA821200 8.

Lepowsky, M. (1982). A comparison of alcohol and betelnut use on Vanatinai (Sudest Island). In M. Marshall (Ed.), *Through a Glass Darkly: Beer and Modernization in Papua New Guinea.* Monograph No. 18. Papua New Guinea Institute of Applied Social and Economic Research, Boroko, Papua New Guinea, pp. 325-342.

Lepowsky, M. (1983). Sudest Island and the Louisiade Archipelago in Massim exchange. In J. Leach and E. Leach (Eds.), *The Kula: New Perspectives on Massim Exchange.* Cambridge University Press, Cambridge, pp. 467-501.

Lepowsky, M. (1985). Gender, aging and dying in an egalitarian society. In D. Counts and D. Counts (Eds.), *Aging, Gender and Dying: Transforming Categories in Oceania.* ASAO Monograph 10. University Press of America, Washington, D.C. In press.

Lewis, C. and D. Henton (1979). *Nutrition Report: Child Nutrition in Milne Bay.* Milne Bay Provincial Health Department, Alotau, Papua New Guinea.

Macgillivray, J. (1852). *Narrative of H.M.S. Rattlesnake ... Including Discoveries and Surveys in New Guinea, the Louisiade Archipelago, Etc. ...* Two Volumes. T. and W. Boone, London.

Macpherson, R.K. (1963). Physiological adaptation, fitness and nutrition in the peoples of the Australian and New Guinea regions. In P.T. Baker and J.S. Weiner (Eds.), *The Biology of Adaptability*. Clarendon Press, Oxford, pp. 431-468.

Maegraith, B., T. Deegan and E.S. Jones (1952). Suppression of malaria (*P. berghei*) by milk. *Brit. Med. J.* **2**, 1382-1384.

Maegraith, B. (1967). Interaction of nutrition and infection. In G. Wolstenholme and M. O'Connor (Eds.), *Nutrition and Infection*. CIBA Foundation Study Group No. 31. Little, Brown and Company, Boston, pp. 41-58.

Masawe, A., J. Muindi, and G. Swai (1974). Infections in iron deficiency and other types of anaemia in the tropics. *Lancet* **2**, 314-317.

May, J. (1958). *The Ecology of Human Disease*. MD Publications, New York.

McGregor, I. (1982). Malaria: Nutritional implications. *Rev. Inf. Dis.* **4**(4), 798-804.

Molineaux, L. and G. Gramiccia (1980). *The Garki Project: Research on the Epidemiology and Control of Malaria in the Sudan Savanna of West Africa*. World Health Organization, Geneva.

Murray, M.J., A.B. Murray, N.J. Murray and M.B. Murray (1975). Refeeding, malaria and hyperferremia. *Lancet* **1**: 653-654.

Murray, M.J., A.B. Murray, M.B. Murray and C.J. Murray (1977a). Parotid enlargement, forehead edema, and suppression of malaria as nutritional consequences of ascariasis. *Amer. J. Clin. Nutr.* **30**, 2117-2121.

Murray, M.J. and A.B. Murray (1977b). Suppression of infection by famine and its activation by refeeding — a paradox. *Persp. Biol. Med.* **20**, 471-483.

Murray, M.J., A.B. Murray, M.B. Murray and C.J. Murray (1978a). The adverse effect of iron repletion on the course of certain infections. *Brit. Med. J.* **2**(6145), 1113-1115.

Murray, M.J., A.B. Murray, N.J. Murray, and M.B. Murray (1978b). Diet and cerebral malaria: The effect of famine and refeeding. *Amer. J. Clin. Nutr.* **31**(1), 57-61.

Murray, M.J., A.B. Murray, M.B. Murray and C.J. Murray (1978c). The biological suppression of malaria: An ecological and nutritional interrelationship of a host and two parasites. *Amer. J. Clin. Nutr.* **31**, 1363-1366.

Murray, M.J., A.B. Murray and N.J. Murray (1980a). The ecological interdependence of diet and disease in tribal societies. *Yale J. Biol. Med.* **53**, 295-306.

Murray, M.J., A.B.Murray and C.J. Murray (1980b). An ecological interdependence of diet and disease? A study of infections in one tribe consuming two different diets. *Amer. J. Clin. Nutr.* **33**(3), 697-701.

Nardin, E.H., R. Nussenzweig, I. McGregor and J. Bryan (1979). Antibodies to sporozoites: Their frequent occurrence in individuals living in areas of hyperendemic malaria. *Science* **206**, 597-599.

National Planning Office (1978). *National Nutrition Survey*. National Planning Office, Port Moresby, Papua New Guinea.

Norgan, N.G., A. Ferro-Luzzi and J.V.G.A. Durnin (1974). The energy and nutrient intake of 204 New Guinean adults. *Phil. Trans. R. Soc. Lond. B.* **268**, 309-348.

Ogbeide, O. (1974). Nutritional hazards of food taboos and preferences in Mid-West Nigeria. *Amer. J. Clin. Nutr.* **27**(2), 213-216.

Ogra, P., M. Fishaut and C. Theodore (1980). Immunology of breast milk: Maternal-neonatal interactions. In S. Freier and A. Eidelman (Eds.), *Human Milk: Its Biological and Social Value*. Excerpta Medica, Amsterdam. pp. 115-121.

Oomen, H.A.P.C. and M.W. Corden (1970). *Metabolic Studies in New Guineans*. Technical Paper No. 163. South Pacific Commision, Noumea, New Caledonia.

Osifo, B. and A. Onifade (1980). Effect of folate supplementation and malaria on the folate content of human milk. *Nutr. Metab.* **24**, 176-181.

Parkinson, A.D. (1974). Malaria in Papua New Guinea 1973. *PNG Med. J.* **17**(1), 8-16.

Pasvol, G., D.J. Weatherall, R.J.M. Wilson, D.H. Smith and H.M. Grilles (1976). Fetal haemoglobin and malaria. *Lancet*, **1**(7972), 1269-1272.

Peters, W. (1960). Studies on the epidemiology of malaria in New Guinea. *Trans. R. Soc. Trop. Med. Hyg.* **54**, 242-260.

Robson, J.R.K. and G.R. Wadsworth (1977). The health and nutritional status of primitive populations. *Ecol. Food Nutr.* **6**, 187-202.

Scott, E. and F. Johnston (1982). Critical fat, menarche, and the maintenance of menstrual cycles: A critical review. *J. Adol. Health Care*, **2**, 249-260.

Scrimshaw, N.S., C.E. Taylor and J.E. Gordon (1968). *Interactions of Nutrition and Infection.* WHO Monograph No. 57, World Health Organization, Geneva.

Sharma, O.P., R.P. Shukla, C. Singh, and A.B. Sen (1979). Suppression of malaria infection by starvation — a biochemical study. *Indian J. Med. Res.* **69** (February), 251-254.

Spencer, T., M. Spencer and D. Venters (1974). Malaria vectors in Papua New Guinea. *PNG Med. J.* **17**(1), 22-30.

Stini, W. (1971). Evolutionary implications of changing nutritional patterns in human populations. *Amer. Anthropol.* **73**, 1019-1030.

Stini, W. (1979). Adaptive strategies of human populations under nutritional stress. In W. Stini (Ed.), *Physiological and Morphological Adaptation and Evolution.* Mouton Publishing Company, New York, pp. 387-407.

Van Der Hoeven, J.A. (1958). Taboos for pregnant women, lactating mothers and infants on the north coast of Netherlands New Guinea. *Trop. Geog. Med.* **10**, 71-76.

VanDijk, W.J., and A.D. Parkinson (1974). Epidemiology of malaria in Papua New Guinea. *PNG Med. J.* **17**(1), 17-21.

White, D.L. (1893). Descriptive Account, by David L. White, Esquire, of the Customs, etc., of the Natives of Sudest Island. *British New Guinea Annual Reports.* Appendix U, pp. 73-76.

Wolff, R.J. (1965). Meanings of food. *Trop. Geog. Med.* **17**, 45-51.

CHAPTER 5

INFANT FEEDING AND HEALTH CARE IN KADUWAGA VILLAGE, THE TROBRIAND ISLANDS

SUSAN P. MONTAGUE

INTRODUCTION

Infant and child nutrition in Papua New Guinea has received a good deal of attention but still remains problematical. The *Report of the Nutrition Monitoring Group* (Nutrition Monitoring Group, 1980) on its 1979-80 Milne Bay Province survey of infant and child weight-by-age reports that

> During the first year of life the total number under 80% weight was 36%. During the second year of life this increases to 67%. There is a corresponding increase in those under 60% weight for age from 6% to 10%. The percent of those under 60% weight for age does not remain at this level but drops again. However the total percent under 80% weight for age remains fairly steady. (1980, p. 3).

The authors conclude:

> Weight for age results show a distinct change between the first and second years of life. This changed pattern remains relatively constant after 12 months of age. This indicates health and nutrition problems in the weaning age group (6 months to 24 months) from which children do not recover. (1980, p. 4).

The surveying team collected interview information on diet and feeding practices, and the authors cautiously suggest:

> The dietary information collected could explain the patterns of growth seen. After 12 months 35% of children were not receiving any breast milk. After 18 months 69% of children were not receiving any breast milk. These children then would be totally dependent on the food given. The typical diet reported was composed mainly of bulky starch staple food such as sweet potato, taro, yam, etc. Green leaves were fairly commonly given but there was very little use made of protein foods or high energy food. Frequency of food consumption was also low.

> Such a diet is bulky so that it is difficult for the child to ingest enough food to meet its energy and other nutrient needs. In addition the child has only two or three opportunities per day to eat. Even small amounts of breast milk would provide a useful supplement of energy and some protein.
>
> Increasing the fat content of the diet and the frequency of food consumption would permit children to consume more nutrients, especially more energy. (1980, p. 10).

In a sense this picture does not seem problematical. Low weights-for-age correlate with dietary interview data sufficiently well to suggest that, in Milne Bay Province, malnutrition does constitute a significant health threat to infants and children. Yet the authors of the Monitoring Group's report are cautious about asserting this point as fact because interview material may contain omissions that necessarily bias the data towards a picture of dietary inadequacy.

This paper offers additional material designed to augment the interview and weight data presented in the Nutritional Monitoring Group's (1980) report. The purpose is to assist with an accurate diagnosis of the local infant and child nutritional situation in the Milne Bay Province of Papua New Guinea.

METHODOLOGY

The data consist of field observations gathered during three visits to Kaduwaga Village, Kaileuna Island, the Trobriand Islands. The first visit lasted 14 months in the years 1971 and 1972. The second lasted two months in 1980, and the third lasted two months in 1981. All three visits were spent living in the village and the latter two in residence in the household of the village chief, Katubai Kariguai. During all three visits infants and children were observed both being fed and eating on their own, adults were talked with about how they view the diets of their offspring and information was gathered on their ideas about what could and should be done to improve their children's health. The data are largely cultural in that the focus is on people's ideas about what constitutes a healthy child, an adequate diet, proper feeding practices, and so forth. Information was not systematically collected on dietary variation within the village, nor was an attempt made to examine, either ideologically or in practice, the remedial efforts designed to correct a situation wherein the standard dietary formula was not growing a healthy child ("healthy" by local standards). It should be pointed out, however, that these omissions were not entirely fortuitous. Observations of infant and child feeding patterns largely coincided with the things that adults said about them, and most Kaduwagan children were locally perceived to be growing properly. The only exception, on both counts, was an infant twin girl whose mother had died in childbirth. Her sister had been put to a wetnurse, but none could be found for her. She was being fed on powdered milk, as per instructions from the hospital, and was perceived to be malnourished.

DATA

Before describing the cultural information collected, it is necessary to offer a brief description of Kaduwaga Village.

Kaduwaga Village

Kaduwaga is a large village by Trobriand standards with a stable population of around 300 residents. Located on the western shore of Kaileuna Island, it fronts on the ocean and also has a shallow coral shelf extending out for about a third of a mile. Residents have access both to deep and shallow marine foodstuffs, like to eat fish at least once a day and usually do so. In addition, women often collect shell-fish to add variety to meals. Although pigs are available, and highly regarded as a food, pig meat is rarely eaten.

Kaduwaga possesses extensive garden lands. Except in years of extreme drought, it is not difficult for villagers to grow varied crops of vegetables. Kaduwagans, like other Trobrianders, prize abundance in their most highly valued staple foods — yams and taro — and try routinely to grow more of these than is needed for annual consumption. Yams function as currency as well as comestibles. Kaduwagans also grow sweet potato as a dietary staple, along with various types of greens, squashes and manioc. The latter are not eaten on a daily basis but are spread out over the week to add variety to meals. Villagers also grow bananas, which are eaten much as are squashes and greens. Coconuts are abundant. They are eaten ripe, their juice is drunk green and their milk is involved in virtually all cooking.

In addition, Kaduwagans have dietary resources from the ring of bush that encircles the land side of the village. Bush foods are eaten selectively in good times and heavily in times of crop failure due to drought. Like other Papua New Guineans (John Woichom, personal communication), Kaduwagans see bush tracts as necessary to survival and wonder at Western perceptions of bush as extra land available for development exploitation.

Water at Kaduwaga comes from underground streams that surface by the shore's edge. It is sweet and uncontaminated. Next to the village is a mangrove swamp. It is the only aspect of the local landscape that presents a clear disease hazard to local residents. Malaria is transmitted by mosquitos that breed there. Villagers, however, indicate that the health threat posed by malaria is not as great as it is perceived by Western health officials. Their houses are sprayed yearly with DDT by government patrols, and the local health officer is equipped with malaria suppressant pills. Villagers resent the house spraying and have repeatedly requested that it be stopped. They do not routinely take the pills and say that attacks are sufficiently infrequent that it is preferable occasionally to be sick. I saw no evidence that infants and children were frequently ill with malaria.

Kaduwagan Attitudes and Practices About Infant and Child Feeding

Malinowski (1929) commented that Trobriand Islanders are very clean with regard to person, food and waste disposition. This definitely is true for modern Kaduwaga villagers. They bathe in fresh water at least once a day and often twice or three times a day. They also bathe their infants frequently and nag their school age children to bathe. Water for infant baths is heated, as sometimes is water for child and adult baths.

Foodstuffs are carefully cleansed in fresh water before cooking. After cooking they are carefully covered to avoid contamination. Cooked foods kept for more than one day are recooked each day to retard spoilage and to ensure their wholesomeness. Food utensils are washed thoroughly after each meal, and each person is served on his/her own plate with his/her own individual fork and cup.

Villagers defecate and urinate into the ocean or bush. The government has encouraged them to build latrines out over the ocean, but villagers say that it is not necessary. Pigs are allowed to wander freely about the village, and dogs are kept to eat their feces. Pigs are fed cooking scraps and are thought, therefore, not to need to eat their own feces.

It is important to note that Kaduwagan hygienic practices are not

linked to Western ideas about contagion and disease. Instead they are
linked to a traditional set of ideas about the nature of human bodies
and of life processes.

Health and health care is a topic of consuming interest to
Kaduwaga adults. They hold that each person's body is created,
grown and maintained through the coordinated labor of other people.
At no time does a Kaduwagan act as a bodily self-sufficient being.
Adulthood is defined by actively contributing to the system of working to grow others upon which one's own bodily growth and maintenance is contingent. Infants, children and adults alike live and grow
because adults invest labor in creating and growing them. The investment is high; the work is dangerous, arduous and time consuming.
Not surprisingly, therefore, Kaduwaga parents always are on the alert
for any sign of health impairment in their offspring. And they always
are on the alert to ensure that the care given to their children is first-
rate care. Many non-Trobriand, and some Western-trained
Trobriand, residents in the islands complain that local parents are
sloppy or unconcerned with regards to infant and child care, but a
visit to a Trobriand household reveals that the opposite is the case.
Parents are almost excessively anxious about their children, because,
in addition to their own heavy investment in growing their offspring,
they feel that children lack mental responsibility and lack strong
bodies. Kaduwagans fear that their children will do things injurious to
their own health, and that their children quickly will succumb should
their health seriously be impaired.

Kaduwagans recognize three sources of poor health. One is
improper diet, another is sorcery, and a third is injury. In this paper
the focus will be on diet.

Diet

Kaduwagans hold that all human bodies are grown out of *kanua*. The
term "kanua" can be translated as meaning human-body-substance-
building food, or men's food, and it consists of yams and taro. Kanua
is the foodstuff that grows bodies that are distinctively human bodies.
It also is the foodstuff that keeps human body substance alive. It is
called men's food because men possess the environment-controlling
magic that creates the conditions necessary for its growth. It consists
of yams and taro, objects that are grown in the ground. Human
beings are characterized as beings of a certain degree of bodily

solidity, suitable to residence on the solid ground. Their characteristic and environmentally suitable degree of bodily solidity is produced through the steady consumption of hard foodstuffs grown in the solid earth. Although today yams are considered to be the ideal type of kanua, vocabulary usage indicates that in the not too distant past taro was the ideal type of kanua. This fits with Lepowsky's (personal communication) suggestion that yams spread through various parts of Milne Bay in relatively recent times. The spread of yams into the Trobriand Islands, however, has not curtailed taro production. Taro and yams are grown together in the same garden plots, and Kaduwagans eat as much taro as they do yams.

Kaduwagans do not classify manioc as kanua. They grow it and they eat it, but they say that it does not build human bodies and that it will not keep human body substance alive. It is classed instead, as *kawenu.* "Kawenu" can be translated as meaning human-body-substance-individuating food, or as women's food, and it consists of wild or exotic vegetal food. Kawenu is food that alters the density of human bodies. Different lineages of Trobriand people consist of people grown out of different patterns of kawenu consumption. Kawenu is women's food because it is gathered by women rather than being grown through men's magical environment-controlling efforts. This is not to say, however, that all kawenu is gathered in the bush. Women often plant it in order to have a supply located in a convenient spot. Kawenu is either food that can grow wild or that is exotic in origin (perceived as not coming from Milne Bay sources).

With respect to food production, men are responsible for providing kanua, the food that grows human body substance; and women are responsible for providing kawenu, the food that individuates human body substance. It should be noted, however, that all adults are expected to work in the kanua gardens. Women are responsible for rendering all foodstuffs into edible form and for feeding other people, primarily the members of their own nuclear families. Women render foodstuffs edible in two ways, through cooking and through transformation in their own bodies. The latter process creates both the animate menstrual blood that constitutes the body substance of their offspring and the mother's milk that will feed their offspring until such time as the infants are able to consume cooked foods.

Pregnancy is thought to consist of a process wherein mother's menstrual blood both increases in quantity and congeals in form to create the body of a new infant child. A newly born infant is thought both to be small (made up of a minimal amount of body substance) and to be comparatively unsolid. Nursing mothers are anxious to increase both the size and solidity of their infants' bodies. Kanua, the food that grows human body substance and produces the characteristic degree of solidity appropriate to human body substance, lies at the root of their efforts in this direction. Ideally, it would be nice if newly born infants could do like adults and consume kanua in cooked form. But they cannot. Instead they must consume it in liquid form, mother's milk. Mother's milk is called *nunu*, and the term for infant suckling also is nunu.

Very few Kaduwaga mothers have trouble nursing their infant offspring. In fact, when asked about this, women initially have trouble understanding. They indicate that milk failure is rare and is no problem. The only women said to be unable to nurse adequately are women with severe health problems, and these women are thought generally unfit for pregnancy and infant care.

There is very little bottlefeeding at Kaduwaga Village. Women do not feel that bottle milk, made out of Western kawenu, is likely to grow good solid bodies for their infants. As proof, in 1971 they cited the example of the twin girls mentioned above. They noticed that, at two years of age, the twin fed by the wetnurse was running around playing, while her sister fed on powdered milk was still unable to walk. As they said, "Her little legs just cannot hold her up."

Kaduwagans deny practicing infanticide, and there is no reason to doubt the truth of their denial. But they also mark the social start of the infant's life as the moment when it first is fed. Kaduwaga women were surprised that Western women do not have the right to kill an unwanted child at birth. They commented that the child is not a social being yet, only a product manufactured by a woman inside her own body. Living people are human beings who have begun to participate in social interaction, such as the taking and giving of food. Once that process is initiated, killing becomes murder.

Traditionally, new infants were born inside a house. They and their mothers remained in seclusion inside the house for several months until their mothers' skins turned white (until they lost their tans and their skin color matched that of their infants). During the period of seclusion, men were banned from the house and mothers were attended by other women. Today some children are born inside village houses and some are born at the Losuia hospital on Kiriwina Island in the Trobriands. Most Kaduwaga women who go to the hospital to give birth perceive themselves to be at exceptionally high risk. One went because she was in her late thirties and already had borne eight children. She said, "The more children you have and the older you get, the greater chance that you will run into trouble." After giving birth, she agreed to be sent to Alotau, the Milne Bay Provincial capital, to have her tubes tied. Another went because she had had a difficult birth with a previous child and was afraid of dying in childbirth should difficulties arise again. She, too, noted that she was getting older, and commented that birth dangers rise with increased age.

Today there is less seclusion of mother and infant. Women who go to the hospital are unlikely to seclude themselves because the child already has been out in public on the journey home. And women who give birth in the village are unlikely to seek seclusion for as long a period as they once did because health patrol officers advise against it. Mothers are told that fresh air and sunshine is good for their infants. My impression, however, is that seclusion has advantages as well as disadvantages. While it does keep the infant inside a smokey darkened house for around six months, it also provides a controlled physical environment that focuses the new mother's entire concern on her infant child. The seclusion house offers an entirely infant-centered world during the period when the threat of infant mortality is high. The fact that men are wholly banned from the house ensures that a child will be at least 15 months old before its mother gives birth again.

In 1971 I was told that a mother should not have sexual inter-course while she was nursing, and I also was told that a woman should nurse until her child could walk well, for example, walk one way to the gardens. Sexual intercourse was said to harm mother's milk, but, in point of fact, the only criticism I heard of a woman who actually became pregnant while nursing was that she was so keen on sex that she was going to have another child too soon. In the years between 1971 and 1980, birth control devices became available (intrauterine devices and pills), and today mothers are less concerned about the potentially pernicious implications of engaging in sexual intercourse while nursing.

More concern is expressed over keeping the temperature of mother's milk suitably warm. One of the important aspects of infant seclusion is external and internal temperature control. Mother and child sit on a platform over a hearth, and fires are built to ensure that, externally, the infant is continually surrounded with warm air. Mothers who do not seclude themselves wrap their infants to protect against external temperature fluctuations. They also worry about internal temperature fluctuations induced by cold foods and are concerned that their milk remain constantly warm. At night nursing women wear clothes on top of their heads because, they say, warmth leaves the body primarily through the head and their own bodily heat loss will cool the milk they feed to their infants.

Infants are fed on demand. Village mothers do not appear to be concerned about emptying the breast at each feeding or balancing feedings so that each breast is used in turn. My hostess said that she uses whichever breast is most comfortable for her and her infant. Children are cuddled and snuggled and talked and sung to while nursing.

Weaning usually poses no particular problem, either to mother or child. Toddlers repeatedly are told about how grown up it is to drink from a cup and are praised as being adult-like whenever they try to do it. When a mother perceives that her walking toddler is competent to handle a cup, she absents herself for a few days, leaving the child in the care of other adults. These cannot nurse it, and they praise it for being so grown up as to be able to drink entirely from a cup. My hostess commented that when a mother returns, the child, confident and proud of its accomplishment, usually will spurn the breast as babyish.

Mothers appear to believe that their normal diet produces their own bodily substance and routinely produces their menstrual blood; special diets for pregnancy do not appear to be necessary. They may, however, restrict the dietary intake to ensure that their infants will not be overly large at birth. But if they do this, they also carefully monitor themselves to ensure that their diet is sufficiently large to keep their own bodies sturdy and energetic. Monitoring consists of paying close attention to how much physical work they are able to perform during their pregnancies.

After giving birth, Kaduwaga mothers alter their diets in order to create the best milk for their infants. They restrict the intake of foods thought to contribute to the creation of milk that is difficult for the newly born to digest. Prominent among the discarded foodstuffs are fish and cooked leftovers. These are dead foodstuffs, foodstuffs whose smell during cooking connotes decay. Mothers feel that their newly born infants have weak digestive systems at best, and they therefore restrict their own postpartum dietary intake to more digestible, "living" foods.

There is concern among health professionals that such postpartum dietary restrictions contribute to infant malnutrition. (See *The Report of the Nutritional Monitoring Group*, 1980, p. 9.) Again, however, interpretive caution is warranted. Kaduwagan mothers eat fish up until the time they give birth, and they begin eating fish again as soon as they perceive their infants are able to tolerate the consumption of small amounts of soup. The dietary restriction only lasts a few weeks at most, and over that time span it is likely that the mother's nutritional reserves are adequate for her infant.

In addition to feeding their infants soup, mothers soon offer masticated foods and/or mashed foods.

From the point of view of health care experts, it is at the point of dietary transition that infant nutrition becomes highly problematical. In Milne Bay, the Nutritional Monitoring Group reports that the transition from mother milk to solid foods involves decreasing consumption of mother's milk, infrequent feedings (two or three times a day) and predominantly starch staple food (1980, p. 10). At Kaduwaga, however, this is not the characteristic feeding pattern. Infants continue to nurse on demand until they are weaned. They may not need to nurse as much when eating solid foods, but the milk is available to them. Parents sometimes *say* that the bulk of meals, served thrice daily, consists of starch staples, but this is when they describe the basic ideal diet. Observations and more detailed conversations suggest that infants and children do not conform to the basic dietary ideal. They prefer eating fish to either yams or taro, and they also prefer to snack throughout their waking hours rather than to eat three solid meals. Just as American parents urge their children to eat their vegetables, so Kaduwagan parents urge their children to eat their kanua. And just as American parents warn their children that snacks will destroy their appetites for dinner, so Kaduwagan parents do likewise. Parents recognize the discrepancies and cope with them by largely allowing infants and children to eat what and when they want while adjuring them to eat properly and insisting that they do eat at least some kanua daily.

If Kaduwaga infants and small children eat a sufficient quantity and variety of foods to constitute an adequate diet during the period of transition from nursing to solid food consumption, why does their weight fail to increase to the degree considered normal on an international scale? Part of the reason is local cultural expectations about proper body configuration. There is no such thing as a plump Kaduwaga child or adult. The proper body is one that is sturdily muscled but lacking in visible fat deposits. All people should eat enough to produce and maintain sturdy musculature. But they should not eat *more* than that. Eating beyond this point is considered selfish gluttony because it is a waste of capital. Kaduwagans tend to over-look the correlation of muscle and exercise because, in their world, everyone routinely exercises a lot. The thin Kaduwagan who eats more naturally will add muscle, not fat, to his or her body.

Mothers of nurslings view their infants' plump fat bodies with pride. But as the infants grow, mothers shift their focus to their off-springs' sturdy legs and energetic running. Infant fat is a good sign and indicative of a healthy baby. But it also connotes a lack of musculature in a basically soft, squishy body. As the infant grows more solid, muscles develop, and muscular development and overall energy become the criteria by which the body is judged well-fed and healthy or poorly fed and healthy or poorly fed and unhealthy. The Nutrition Monitoring Group reports that in many areas of Milne Bay, malnutrition is recognized neither as a disease nor as a common health hazard for infants and children. (1980, p. 9). This is true, too, at Kaduwaga Village. Adults perceive that, in the ordinary course of development, children put on the proper amount of musculature and are physically active. By local standards, children could not do this if they were poorly fed. Again, the malnourished twin was recognized as such because her body (specifically her legs) was not taking on the proper degree of sturdy muscle and she could not run about.

It is interesting, with respect to musculature, that Kaduwagans assess their children's legs, and not arms, when rating development. They note that upper arm development occurs at puberty. Locally, upper arm development is thought to be correlated with increased manual labor. Children use their legs a lot, running around all day long, but they do not use their arms a lot in such a manner as to pro-mote arm strength during their play activities. As older children begin to garden and to take on other adult work activities, their upper arms, shoulders and torsos begin to fill out with muscle, just as their legs began to fill out when they grew towards walking.

Kaduwagan Views About Improving Infant and Child Health

From the Kaduwagan perspective, the major problem with infant and child health is not nutritional. Instead, it is that the government does not provide adequate emergency health care. Illness and injury are thought to represent the most serious infant and child health hazards.

Indeed, they are thought to represent the most serious adult health hazards. The local hospital is located on another island, and there is no way that Kaduwagans can transport a seriously ill or injured person to it. Nor is there any way that they can get sophisticated medical personnel and supplies out from it to the village quickly to cope with the ill or injured person's needs. Acquisition of a motor boat would help this situation but, despite years of lobbying by villagers, no motor boat has become available.

The problem of access to emergency medical care pertains not only to geographically remote spots in the Trobriands like Kaduwaga Village. In 1980 an expatriate man residing on the same island as the hospital died from hepatitis before transport was arranged to get him to proper treatment facilities. In part his death was caused by the government's decision to de-emphasize the local hospital's role as a center for the care of serious illness and injury. He had to be flown to Alotau, about 130 miles away on the mainland of Papua New Guinea, for treatment and was mortally ill by the time he arrived.

Kaduwagans say that it is nice that medical patrols come to their village to instruct people in better health practices. But they also say that such patrols are of limited utility, and they wonder why some of the money spent on patrols cannot instead be spent to improve local emergency health care.

DISCUSSION

The purpose of this paper is to help the Nutrition Monitoring Group and other interested health care professionals to arrive at a more accurate diagnosis of local infant and child nutritional conditions. Obviously, there remains a serious problem. If the Nutritional Group's survey perhaps suffers from a lack of informational depth, this study definitely suffers from a lack of broad geographic coverage. One of the advantages, however, of cultural studies is that, within a recognizable cultural region, many ideas and attitudes are found spread across the landscape. Bearing this point in mind, the juxtaposition of the studies of the Nutrition Group and this study does offer some food for thought.

The Kaduwagan observations and conversations reported here confirm the already suspected fact that interview survey data on infant and child feeding and nutrition patterns in Milne Bay Province contain significant dietary omissions. It also confirms that these omissions tend to bias the data towards the presence of infant and child malnutrition. Equally important, however, my observations and conversations provide some information about the nature of the omissions. Kaduwagans, in initial and brief interviews, report their infants' and children's feeding and diet in terms of locally ideal adult patterns. Proper adults eat two to three meals a day and they consume kanua (yams and taro) as their staple foods. (It should be noted that, even for adults, this should not be taken to mean that they do not eat a good many other things as well. Rather, it is a verbal indicator of proper dietary orientation.) The fact that, throughout the region, the survey team found a constant stress upon yams and taro consumption and on eating two to three meals a day, suggests that parents throughout the region probably do share Kaduwagan perceptions about proper dietary reporting. Insofar as this is likely, it might be worthwhile for a future interview team to try couching their questions in terms of how parents cope with getting their recalcitrant children to eat the right foods at the right times. If interviewers acknowledge that children everywhere naturally want to eat all the wrong things, parents might be less defensive and more willing to discuss actual dietary intake.

My observations and conversations also suggest that local perceptions of the properly developing body have a lot to do with the recorded local drop in weights as compared to the international standard during the second year of life. Kaduwagans prize plump babies but not plump toddlers and children. They prefer that children and adults be on the thin side so far as is consonant with sturdy muscle development and high energy levels. If this preference is shared by their regional neighbors, weight *per se* is a dubious index for making any inferences about the presence of widespread malnutrition.

In addition, my observations and conversations suggest that local perceptions of dietary adequacy probably are reasonably accurate. Kaduwagans possess and eat abundant quantities of fish, tubers, squashes, greens, bananas and coconuts. With the exception of dairy products, their diet seems little different nutritionally from the diets of people in areas of the world where the presence of malnutrition is not suspected. While it is not clear that all of the residents of Milne Bay Province eat as well as do Kaduwagans, it is worth considering that others of them who say that malnutrition is not a local problem also may know more or less what they are talking about. This is a valuable avenue for future researchers to explore because, if it is the case that they do know, then scarce and expensive nutritional resources could be channeled more appropriately to people genuinely in need of them.

Finally, my conversations and observations suggest that research on infant and child nutrition needs always to be considered in relation to the broader issue of general infant and child health care. While it is easy to see that, from the perspective of centralized health care planners, the broader issue is best dealt with by breaking it down into specific topics, it also is easy to see that, from the perspective of local people, some of those topics are of greater significance than are others. To Kaduwagans, for example, the topic of emergency health care looms as far more significant than does the topic of infant and child nutrition. While it is recognized that nutritional and other health survey teams always have more than enough work to do, it might prove worthwhile for them to inquire about local perceptions of significant topics, especially in instances where they are told that their particular topic is not of local concern. Such a procedure at least offers the hope of alerting planners in related areas to needs that they could fulfill.

ACKNOWLEDGEMENT

My three trips to Kaduwaga Village were funded, in order, by the National Institute of Mental Health, the Dean's Fund of Northern Illinois University, and the National Endowment for the Humanities. I wish to express my gratitude to all of them.

REFERENCES

Nutrition Monitoring Group (1980). *Report of the Nutrition Monitoring Group.* Provincial Health Office, Division of Health, Alotau, Milne Bay Province, Papua New Guinea.
Malinowski, B. (1929). *The Sexual Life of Savages.* Harcourt Brace and World, New York. pp. 444-446.

CHAPTER 6

SOCIAL, ECONOMIC AND ECOLOGICAL PARAMETERS OF INFANT FEEDING IN USINO, PAPUA NEW GUINEA

LESLIE CONTON

INTRODUCTION

Nutritionists, public health workers, social scientists and others who have experience in developing countries are concerned about the crisis in infant feeding practices signalled by the worldwide decline in human lactation since 1930 (Jelliffe, 1976; Latham, 1977; Mata, 1978). The nutritional, immunological, behavioral and economic benefits of breastfeeding and the adverse effects of early weaning and bottlefeeding in transitional societies are well documented elsewhere (Jelliffe and Jelliffe, 1971; Jelliffe and Jelliffe, 1978; Porter and O'Connor, 1975). Observations in developing countries indicate a variety of reasons for premature weaning, but much remains unknown. The trend toward earlier weaning is assumed to be associated with forces of modernization, westernization, urbanization, industrialization, commercialization and medicalization, or some combination of these (Hull, unpublished observations, 1982),[†] especially when women are employed in the wage labor force (Harrell, 1981; Jimenez and Newton, 1979; Mata, 1978; Nerlove, 1974; Raphael, 1979; Van Esterik and Greiner, 1981). The complex of variables associated with social change has not been adequately categorized nor systematically researched, however. Latham (1977, p. 200) observes that there is a poor correlation between these "determinants" and a decline in breastfeeding; forces of modernization do not always result in a decline in lactation. The socio-cultural

---

[†]Hull, V. (1982). Breastfeeding and Fertility: The Sociocultural Context. Paper prepared for the WHO/NAS Workshop on Breastfeeding and Fertility Regulation: Current Knowledge and Policy Implications, Geneva, Switzerland.

consequences of these "izations" cannot be understood outside their cultural contexts. Instead, different patterns of infant feeding largely reflect differences in ideology and politics. Furthermore, the cultural blinders worn by some Western investigators, accustomed to a social environment unsupportive of breastfeeding, obscure or skew the meaning of relative terms such as "breastfeeding on demand," "frequent" or "indulgent" breastfeeding. As Raphael (cited in Hull, 1982, p. 4) comments: "Missionaries or male anthropologists, our main reporters on how others live, have been loathe to look at — even more to see — a bare-breasted female with a baby attached."

While more biomedical research is necessary to understand the anti-infectious properties of breastmilk, longitudinal field investigation of the cultural contexts — social, economic, religious, symbolic, demographic and environmental — in which infant feeding traditionally[†] occurs is also critical to the development of appropriate field methodologies, techniques and operational systems designed to maintain or promote healthful infant feeding practices in modernizing regions. Much of our knowledge about infant feeding practices comes from cultures undergoing rapid development, where a crisis already exists. There is a need for detailed ethnographic description of infant feeding patterns in regions minimally altered by modernizing forces in order to delineate cultural baselines, in regard to infant feeding, against which the impact of development may be identified and measured. We need to examine the cultural contexts in which breastfeeding is maintained in the face of modernization and westernization. Can we identify the specific factors, or combination of variables, that promote healthful infant feeding practices and the mechanisms by which traditional feeding patterns are altered? We know little about the individual decision-making processes involved in feeding infants and in accepting or rejecting changes in feeding methods. Neither do we understand sufficiently the social environment in which mothers care for and interact with their infants or the conceptual bases for infant feeding practices. In order to promote successfully infant nutrition and health, the design of appropriate health care programs must be informed by social scientists who have an adequate knowledge of the local situation, in concert with biomedical researchers, educators, communication specialists and

---

[†]In this paper, *traditional* refers to cultures and practices not significantly altered by modernization or westernization. It is a convenient term but does not imply stasis or homogeneity in practices.

administrators. Furthermore, the indigenous actors themselves should play a major role in such decision-making, if health programs are to be responsive to their expressed needs.

This case study identifies some of the social, economic and ecological variables that define the decision-making context of infant feeding in the Usino area, Upper Ramu District, Madang Province, Papua New Guinea. These practices are embedded in complexes of cultural beliefs, customs and behaviors. Because infant feeding practices are directly and effectively linked to infant survival, they are subject to strong evolutionary pressures of cultural selection. There is, therefore, a likelihood that these cultural complexes have adaptive significance not immediately apparent to the outsider. Often there evolves a wisdom in the system neither explicitly recognized nor consciously enacted by the insiders. This paper urges a holistic and longitudinal understanding of these practices in cultural context, prior to modernization, because the effects of the "izations" can only be understood relative to a baseline. This paper provides such a baseline in an area only marginally altered by economic development.

METHODOLOGY

Infant feeding practices were documented during two ethnographic field sessions, from June, 1974, until July, 1975, and July through August, 1981. In 1974-75, in order to collect balanced and accurate data in this sex-segregated society, the author worked primarily with female informants and focused on male-female relationships, while a male colleague studied local political leadership with male informants. This initial team approach provided the ethnographic foundation for the 1981 study. In 1981, under the auspices of the Institute for Applied Social and Economic Research, Papua New Guinea, investigations were conducted on reproductive decision-making and the value of children in the same region and previously collected data on infant feeding practices were supplemented.

The data in both field sessions were derived from customary anthropological field techniques of participant observation and informal intensive and extensive interviewing. Conversations were conducted in Neo-Melanesian (Pidgin) in all nine hamlets of the culture group.[†] All inhabitants were bilingual in Neo-Melanesian and

---

[†]The term, Usino people, designates the lowland speakers of the Usino language.

the Usino language. In order to place these localized data in a wider cultural context, government officials on the provincial and district levels, health workers, teachers, clergy and local leaders familiar with the Usino people were interviewed.

The entire Usino population in 1981 consisted of approximately 395 persons, of which 158 were adults (82 females, 76 males). During both field sessions, all persons over the age of 15 were observed in interaction with infants and were engaged in conversation about childrearing, but 13 men (17 %) and 28 women (34 %), 26 % of the total adult population, were primary respondents.

Some anthropological field procedures differ in several important ways from standardized interview or questionnaire techniques employed in other disciplines (Back and Stycos, 1959). In participant observation research, the researcher lives fulltime with the people under study, under the same conditions, engaged in description of and conversation with informants from morning until night (and sometimes through the night). Ideally, the anthropologist gradually establishes relationships of trust and respect with informants prior to discussion of sensitive topics. This enables her to frame questions in an understandable way; to address delicate questions in a culturally acceptable manner; to allay suspicion, resistance or hostility that a foreign interviewer might otherwise encounter; and to minimize distortion in reporting and interpretation of the cultural setting. By establishing a role which is understandable and non-threatening, the ethnographer collects data not only on ideal behavior (what people say they do) but also from observation of actual behavior. This enables the researcher to gauge variability in beliefs and in behavior.

Random-sampling procedures are neither always practicable nor necessarily desirable. To minimize the real possibility of fabrication and distortion, primary informants are selected partly on the basis of proven reliability and partly for the diversity they represent. Consistency checks within the same interview, between successive interviews and between respondents, as well as longitudinal observation also help identify the loci and extent of unreliability. Primary informants tend to be people competent and willing to respond to the researcher's questions. Standardized written or oral questionnaires, like random sampling, are not particularly useful in small-scale pre-literate societies where news travels faster than the interviewer and where most respondents are forewarned of the nature of the interview, or even prompted as to the answers. Often more accurate data are elicited by seizing an appropriate moment to pose a question in a concrete way rather than in a hypothetical way, by rewording questions, by dropping topics or introducing new ones, by flexibility and innovation in interactional style, or by *not* asking questions and instead watching unobtrusively. Occasionally, these methods compromise our statistical precision in favor of overall accuracy in meaning and content, resulting in improved reliability.

Intensive community-based research as well as large-scale comparative studies and implementation of varied approaches are required to expand our empirical, theoretical, and policy-relevant knowledge of infant feeding practices. In-depth research into individual attitudes and decision-making processes with regard to infant feeding, inquiry into the cultural meanings ascribed to parameters that shape these practices, and direct observation of behavior as it occurs in cultural context augment survey and biomedical research by outsiders and provide a crucial source of information on the forces affecting parental actions. A sound knowledge of local conditions, married with innovative and sensitive questions, can elicit valuable information rarely obtained by other techniques. Small community-based studies can highlight the needs of individuals, and in keeping with growing commitment to localized primary health care programs, can form the foundation for improved service and more responsive health care delivery.

INFANT FEEDING IN USINO

The Modern Context

Usino, the name commonly given to the lowland region centered east of the Ramu River and six and one-half km southwest of the Usino Patrol Post, is a steamy forested locale of rich biotic resources and low population density, about 2.5 persons per sq.km. At the time of first study, the 250 Usino people resided in three centralized villages. Several villagers were employed at the government station, the center of the Ramu Subdistrict between 1967 and 1981. By 1981 Walium, located on the new Lae-Madang Highway, succeeded Usino as Upper Ramu District headquarters. As a result, all government personnel moved to Walium, and the Usino airstrip and health center closed. With the demotion of Usino Government Station back to a patrol post, Usino people were cut off from their primary source of cash income. In addition, as a result of an influenza epidemic, most Usino villagers abandoned their former settlements along the single dry-season track in favor of small isolated hamlets, scattered in the bush between the Ramu River and the Usino Patrol Post. Since 1975 the population has increased from 250 to 350 persons, owing in large part to an unusually sizeable cohort of newly-reproductive women, and in part to the return of all wage laborers and their families.[†]

The Usino subsistence base has changed little in the past four generations. Usino people remain horticulturalists engaged in supplementary gathering, hunting and fishing. Most Usino adults lack formal education and access to wage labor at present. Attempts at commercial production of coffee, rice and peanuts have been uniformly unsuccessful, and cattle projects have engendered few profits. Prior to the establishment of Usino Patrol Post and airstrip in 1967, access to the port town of Madang entailed a four-day walk. Following this, few villagers have availed themelves of air transport to Madang. Since 1974 a feeder road into Usino Patrol Post from the Lae-Madang Highway facilitates travel to Madang, Lae and the Highlands, but Usino villagers do not frequent Madang, let alone the more distant areas. The cultural changes of the last eight years have been largely recidivistic, in response to increased relative isolation resulting from government withdrawal from the Usino region and from economic development in other parts of the Madang Province. With no perceived opportunities in the cash sector, a sense of relative deprivation and pessimism pervades Usino.

In the period from 1974 to 1981, the number of trade stores at the patrol post has grown from one to two, but there is still only one poorly-stocked trade store, with an unpredictable schedule, in the vicinity of the Usino hamlets. Although Usino people covet western goods, and imported wares such as pots, dishes, lanterns, clothes, chairs (and even a radio-cassette) have found their way to Usino, cultural emphasis is still on such items of traditional wealth as clay saucepans, wooden bowls, spears and pigs. The maintenance and repair of imported items is not of concern to either sex. In the future, modernization and economic development may well alter women's

---

[†]The 1981 study included one village of 45 persons not studied in 1974-75, hence the 1981 population figure of 395. Since these villagers are neither living on nor utilizing traditional Usino land, they do not figure in the calculation of population density.

workload, which may in turn foster a change in infant feeding practices (see Nardi, ch. 16), but neither women's household and productive tasks nor men's roles and tasks have changed significantly so far.

In 1980 a Lutheran catechist and bush-material chapel were installed in Usino Village, the largest (100 persons) and most central village. Church activities, limited to a Sunday morning service, involve about 30 adults and their children. In contrast to the observations of Counts (ch. 9) and Nardi (ch. 16), there are no new social activities competing for women's time nor are there economic incentives for earlier weaning or supplemental feeding. This may prove adaptive for Usino infants dependent on breastmilk.

The total number of Usino children attending primary school at Usino Patrol Post increased from two to about twelve in the past seven years, but loss of children's labor has not appreciably increased parents' workload because only children who are economically expendable are permitted to attend school.

Schooling itself, however, may have an eventual impact on infant feeding practices. Two young women who completed primary school are now marriagable. As a result of their education, they show increased receptivity to Western values and concepts of health. Infant feeding practices in Usino may change if their non-schooled agemates follow their lead in this behavioral pattern as they have in others. For instance, unlike other women, both women are embarrassed to expose their breasts. Traditionally breasts of young women were considered beautiful, objects of pride and erotic significance. In 1974-75 women only donned blouses for special events and when travelling to the patrol post. By 1981, many women wore blouses in the villages, removing them for garden work. The oldest women continued to expose their breasts. Usino women recognize a significant change in dress in the past eight years: "Before, we did not hide our breasts. Later, when skirts came in, we wore them, and after some people went to Madang and told us we should be ashamed to show our breasts, we put on blouses. We don't like to hide our breasts; they are our decorations, our beauty. Young girls like to show them off; they attract men." Westernizing influences such as norms of modesty, reinforced by changing modes of dress, have changed behavior in Usino, but not attitudes related to breastfeeding. Public breastfeeding continues to be viewed as a natural process.

Initiation of Breastfeeding

Usino women do not usually discuss their knowledge of infant feeding practices. Most adult skills and responsibilities are learned by observation and practice, rather than by verbal instruction. Unlike transitional societies where mothers become socially isolated and must be taught how to breastfeed (Mata, 1978), at Usino breast-feeding is deemed a natural, if inevitable, event, and knowledge of feeding behavior is thought to be innate. Consequently, my investigation of these practices met with initial amazement ("Why does she ask us all this? All women know how to breastfeed!"), followed by mirth at the thought that I did not know how to feed infants. Unlike the situation reported by Barlow (ch. 8), advice on this topic is rarely sought or offered, and most aspects of infant feeding are not the subject of conscious examination, discussion or comparison in Usino. Therefore, some feeding practices and their underlying conceptual bases appear idiosyncratic rather than shared.

Breastfeeding is the primary form of infant feeding until weaning between one and two years of age. Besides breastmilk, supplemental fluids given to nurslings (via cup, bowl or bamboo tube) are water, vegetable broth (mainly for sick children), and sugar cane juice sucked from the peeled stalk. There is no bottlefeeding in the region. Were bottles easily available, few Usino women could afford them. Parents with sick children sometimes purchase canned or powdered milk from the patrol post stores. But such purchases are infrequent because, in addition to parental lack of cash, the stores are infrequently stocked and are a two to four-hour walk from the hamlets. Walium Health Center sporadically supplies infant formula to malnourished babies, but few Usino mothers will bother with the time consuming trip. They cite lack of money as the cause for their reluctance. Dry biscuits, soda pop, tinned meat and fish, and white rice occasionally show up in Usino, but food from trade stores is not a regular part of the diet of Usino children or adults.

After the mother gives birth, she sleeps, eats a large meal with soup (believed to be lactogenic) and applies hot water or hot banana stalk compresses to her breasts to stimulate milk flow and to ease the pain of engorged breasts. The mother initially expresses the colostrum (referred to as "yellow," "dirty" or "sick" milk) and waits for the "true, white" milk to appear. All Usino women but one consider colostrum, as they define it, to be contaminated. Some say that it will poison the child because it is associated with pregnancy. Some believe that it is milk that remained when the mother weaned her previous child and resumed sexual relations. All women agree that semen entering the milk turns it "yellow" or "black", and such milk will sicken a nursing child. This is the conceptual basis for postpartum abstinence. Sexual relations endanger the health of nurslings, so a conscientious mother will not resume sexual relations prior to sevrage.[†]

Breastfeeding commences on the morning *following* birth; a whole day may elapse before the infant is first fed. The infant is thus deprived of some nutrition in its first hours as well as some immunological benefits of colostrum (Gyorgy, 1967; Jelliffe and Jelliffe, 1975; Mata, 1978; Mata and Wyatt, 1971). This could be an adaptive mechanism that permits weak or physically distressed neonates to die of natural causes before mother-child bonding has occurred. Intentional infanticide is not publicly condoned, but informants state that an unmarried woman, a woman who has too many children or too many of one sex, or the mother of twins might resort to infanticide. There are no obviously retarded, deformed or otherwise abnormal infants in Usino, suggesting that either infanticide or benign neglect may be operative. These decisions are made by mothers immediately following parturition. Such practices are facilitated by the ritual isolation of the mother-child pair during and following birth. Infant mortality is a fact of life for Usino women and sometimes results from rational economic decision-making on the part of women attempting to balance family size with subsistence demands, perceived economic opportunity and health status (Conton, unpublished observations, 1983).[‡]

The Usino belief that true milk appears on the morning following birth may allow the survivor of these first critical hours to derive nutritional and immunological benefit from remaining colostrum and early breastmilk, culturally designated as "true, white" milk. That no other supplements are given to the neonate during the first few hours

---

[†]The term *sevrage* is used in the literature to denote the complete cessation of breastfeeding (Hull, unpublished observations, 1982), while *weaning* refers to the process of establishing dependence on food other than breastmilk, which may be a gradual process. *Beikost* refers to supplemental solids given to the infant prior to sevrage.

[‡]Conton, L. (1983). Reproductive decision-making in the Upper Ramu District, Papua New Guinea: Cognitive aspects of adaptive problem-solving. Paper presented to American Anthropological Association, Symposium on Adaptive Problem Solving: Methods and Examples. Washington, D.C.

obviates the possibility of infection from introduced foods. I was unable to observe what was meant by yellow or true milk, but the timing of initial breastfeeding suggests that Usino definitions of colostrum differ from those of other Melanesian societies as well as Western medicine (Gegeo and Watson-Gegeo, ch. 13; Oomen and Malcolm, 1958). In some Melanesian societies, similar beliefs in the polluting qualities of colostrum, but different definitions of it, result in the withholding of nourishment for the first few days of life. Were Usino women to adopt similar (or western) definitions of colostrum or mature breastmilk, they would probably postpone breastfeeding for several days, also. The existing delay in initiation of breastfeeding in Usino is not accompanied by problems in establishing lactation.

At present, the mother and child remain in ritual seclusion for one week. They are isolated from all men but are visited frequently by the mother's agemates and kinswomen, the father's kinswomen, and other village women who cook for her, converse with her, minister to her needs, or keep her silent company. There are no specific obligations in this regard, but couples who hope to adopt the child later may use this opportunity to curry favor with the natural parents by providing special support. The mother is surrounded by women with whom she shares the experience of motherhood, and with whom she is familiar. This supportive environment reflects the solidarity of Usino women occasioned by preferential intra-community marriage. Postmarital residence within the Usino community gives a woman continued economic and emotional support from her natal family and from her female agemates, with whom she has lifelong contact. This female support system facilitates transition to motherhood, providing a secure and relatively unstressful environment in which the new mother receives assistance from more experienced women. Physical, emotional and social support during the onset of breastfeeding enhances successful lactation and survival of the infant (Raphael, 1973; Wilson, 1975). Because in Usino the main tensions in interpersonal relations are between the sexes, prohibition of male contact during this critical period may further promote mother-child bonding and successful initiation of lactation. Usino marriage and postmarital residence patterns strongly affect the social environment of early lactation, as well as later infant feeding and care.

A few generalizations can be made regarding the scheduling of breastfeeding in Usino. Infants are fed on demand as signalled by the child's restlessness or fussing. This type of infant behavior also reminds the mother to transfer the baby to a full breast. The breast frequently is offered to soothe or distract an upset child and is withdrawn when the child is calmed. Not all feedings result in a significant intake of milk, but over the course of a day, a child will suckle many times. Children do not nurse on any time schedule, except that a few mothers never breastfeed upon retiring for the night. Even so, they sleep with their infants rather than with their husbands during lactation. If night feeds are less suppressant of ovulation than are daytime feedings, owing to the normally higher levels of prolactin during sleep (Sassin *et al.*, 1972), perhaps curtailment of nocturnal feedings does not significantly shorten lactational amenorrhea and thereby increase fecundity.

Usually a mother will not regularly leave a nursling in the care of others in order to attend to other work tasks until the child is eating supplementary semisolids and is crawling. As a result, infants accompany mothers everywhere in netbags — to the gardens (where women sling them on their backs while they work or suspend them from tree branches or command other children to watch them), to other villages, or to the patrol post. Trips to unfamiliar areas endanger babies, who are vulnerable to attack by bush spirits. Therefore, Usino mothers with infants restrict their travel to the Usino region and always travel in daylight in the company of others. The mother and child pair is rarely separated during the first few months, but a few industrious mothers and women without spouses leave their infants in the care of older children, their sisters, their parents or in-laws while they garden for a few hours. These mothers say that their babies adjust quickly to such schedules and are given sugar cane juice or water if they cry. If that fails to satisfy, the child-minder will recall the mother from her work or bring the child to her. It is not generally considered responsible to leave a dependent infant untended for more than a few minutes. A woman leaves her sleeping baby hanging in a netbag in her house while she visits within the village, assured that others will call her should the child awaken. Sometimes, however, a frustrated mother, interpreting her infant's incessant cries as signs of hunger, rebels at its persistent demands and wanders out of earshot. Her husband or older female relatives chastise her, if not physically punish her, for such irresponsible behavior. Babies are not left long in the care of their fathers because crying babies are said to infuriate men, who will "throw the babies away." Infant crying for any length of time is thought to induce illness or soul loss. This belief increases the likelihood that infants will be breastfed frequently.

E

Breastfeeding and Women's Work

All women return to garden work about a month after delivery, carrying their infants with them. A mother waits this long for fear that pigs will destroy the garden fences and root up crops if they "smell her wind" (a reference to the aura of pollution that surrounds her after childbirth). Adherence to this pollution belief indirectly enables her to focus her attention on the newborn and to rest during this first month.

Women's labor is somewhat circumscribed by regular and frequent breastfeeding. Mothers complain at the difficulty in completing their regular tasks when burdened by a dependent child. They see the nursling as a physical restraint on the mobility and independence they value, but they perceive no alternative to breastfeeding. Despite their complaints, women's work is generally compatible with breastfeeding and childcare, owing to classificatory kinship and to patterns of marriage and postmarital residence. Because a woman lives near almost all her relatives, consanguines as well as affines, she has household help and many potential childminders to choose from. The kinship system insures that each child has several "mothers" who can be expected to share responsibility for it. While the infant requires the mother's attention for regular breastfeeding, small tasks, such as collecting firewood and fire-tending, washing dishes or clothes, hauling water or preparing produce and meals, devolve on older offspring or on related children. Childminders' responsibilities increase as their charges grow less dependent on mother's milk. Child caretakers accompany mothers to the gardens and tend the infant while the mother works. One of their main tasks is to carry infants, thus enabling the mother to carry heavy loads of produce, meat, portable valuables or personal items. Because of this support network, infant feeding requirements impose few restrictions on mothers' horticultural activities or participation in important activities, such as inter-village visiting and trading, work parties, ceremonial exchanges and feasting.

Once a child accepts supplemental food, the mother may leave it for several hours or all day, balancing the needs of her child, the food requirements of her family, and her other economic obligations. An ambulatory child who breastfeeds only in the mornings and evenings poses no real restriction on the mother's activities.

Although both sexes regard childrearing as women's major contribution to society, Usino women also play key economic roles within the community. Marriage and postmarital residence patterns foster economic partnerships and mutual economic assistance between females. Usino women potentially can be economically independent of their husbands in the extradomestic distribution of valued goods (Conton, 1977). In-group marriage patterns, then, not only promote female economic and emotional solidarity, but make available to women a large number of potential helpers to ease their workloads, enabling them to meet both their childcare responsibilities and their voluntary extra-domestic obligations.

Adoption and fosterage are additional means by which women can recruit household help, or conversely, bring family size into balance with available labor and subsistence resources. Parents prefer their own children, but if natural family sex-age ratios are unsatisfactory, they can recruit children of the desired age and sex when required, or relieve themselves of children through these same institutions. Adoption rarely occurs before sevrage, but one overburdened couple who consistently broke postpartum sex taboos, promised their newborn to another couple with one son who were experiencing difficulty in conceiving more children. The adoptive mother, who had not lactated in four years, assured me that if she applied hot water to her breasts, and a ritual specialist administered the appropriate spell, she would soon be able the nurse the child herself. (I recorded two other similar cases of induced lactation.) In another case of prospective adoption, no fewer than three women told me that a mother, recently delivered of twin boys, had promised each woman one of the sons to adopt.

The most serious inhibitor of women's extradomestic economic participation in Usino is not lactation requirements, but rather a woman's perception of her own poor health, or that of family members. Because her husband and children contribute to the household economy, their illness results in an increase in a woman's workload. In addition, her horticultural and other economic activities are circumscribed while she cares for ill family members, and more so if she seeks treatment for them at the Usino Patrol Post or Walium Health Center. Many Usino mothers are only marginally healthy at best, and some of their complaints about dependent infants stem as much from their own fatigue as from their desire to be physically autonomous. A mother preoccupied with her own physical distress demonstrates little interest in the well-being of her children.

Food Restrictions During Lactation

Almost all lactating mothers follow food taboos in order to protect the health of their nursing offspring. But there are no specific universal restrictions, nor is there a common body of knowledge shared by all mothers. Three younger women in their twenties, who are the most amenable to Western childcare practices as promulgated by the Maternal and Child Health (MCH) sisters, claim to follow no food taboos while lactating. They also claim that their babies eat everything, but close observation revealed that none of them fed their infant taro or meat until they were nearly a year old. Taro, like meat, is considered too "strong" for infants; some say it causes bloated stomachs. I rarely observed premastication of meat for babies.

As in other parts of Oceania, many mother-child food restrictions, though idiosyncratically followed, are based on the belief that through breastmilk the child will acquire the negative characteristics of the food in its natural state. For instance, to eat crocodile or fish that have scales will give ringworm (*Tinea imbricata*) to the infant. To eat wallaby will give the child crooked legs, and its ability to stand up will be impaired. If the mother or child eats crayfish, the child will fall back on its buttocks, "like a crayfish." Opossum meat will likewise cause the child to sit down all the time. Some mothers eat no meat for a month following delivery to avoid a swollen stomach. For the same reason, some avoid tobacco. Others refuse pandanus grease in order to spare the infant illness. Bandicoot and cassowary meat are said to cause colds, and sago to cause swelling. No woman follows all these taboos; it is a matter of personal preference, each woman adhering to one or two such restrictions. A few middle-aged women said they observed fewer taboos with each successive child.

Aside from food restrictions (which mostly involve animal protein) and postpartum sex taboos, there are no strictures on lactating women. Women with young children exercise caution in several contexts to insure the health of their children, but these precautions are not directly related to lactation. As mentioned, women complain about centering their lives around nursing children because they would prefer to be less sedentary. This desire for restored mobility is reflected in the rationale for many food taboos. The restricted foods are said to delay the child's motor development. The sooner a child can walk, the better, according to Usino parents. Mothers view their limited mobility not as a rule, law or taboo, but rather as an inconvenience inherent to motherhood.

The origin of cultural traits such as food taboos is essentially untraceable and may have occurred under conditions different than the present. Present protein intake may not be a reliable indicator of protein intake in the past. Usino people rationalize and maintain food restrictions by the beliefs described. They may manipulate these rationalizations with changing circumstances, and they may rationalize actions in terms other than those truly operant. Native rationality need not be verbalized, conscious or even elicitable to be operationally insightful and adaptive (Boehm, 1982). Present-day emic rationalizations for infant feeding practices may bear little relation to the conditions that spawned them, but they may be practically efficacious. Lepowsky (ch. 4) argues that restrictions on the child's consumption of animal protein, for example, may be adaptive because they increase the dependence of infants on breast-milk. She cites the following benefits of breastfeeding, well-documented elsewhere. Frequent suckling may prolong lactational amenorrhea (Anderson, 1983) and protect the health of the nursling in several ways. The anti-infectious properties of breastmilk protect the infant against intestinal viruses, bacteria and diarrheal disease. By reducing the amount of contaminated water ingested and by replacing fatty foods, not easily digested by infants, breastfeeding decreases the likelihood of acute diarrhea, which quickly leads to infant death because of rapid dehydration, impaired absorption of nutrients and metabolic loss of essential nutrients (Brown, 1978). Unlike other forms of animal protein, breastmilk better enables infants to absorb zinc and to retain water (Mata, 1978), thereby decreasing the risk of enteric infection. If adult diets are low in protein, a mother may be too debilitated to nourish both an infant and a fetus, and if weaned too early, the older child will be susceptible to nutritional deficiency, dysentery and intestinal parasites (Abernethy, 1979; Jelliffe, 1968). Additional arguments for the adaptability of prolonged breastfeeding are that it may promote optimum parent-child interaction that has a positive influence on maternal attitudes toward prevention and control of infectious disease (Mata, 1978), and that it is the cheapest means of feeding a child in the first six months, especially considering the cost of treatment of infectious disease (Mata, 1978).

Ecological interdependence between diet and disease is suggested by studies of *pig bel*, a serious and often fatal necrotizing enteritis of children in the New Guinea Highlands and elsewhere (Lawrence and Walker, 1976). Although the probable responsible factors, the B toxin of *Clostridium welchii* type C and poor nutrition, are present in lowland regions, *pig bel* is rarely encountered, probably due to the variety of foods available. The antienzyme content of sweet potatoes, the highlands staple, prohibits the destruction of the B toxin found in meat, and the two are usually eaten together. Children dependent on breastmilk rather than on animal protein or tubers, in regions with reliance on one staple, may more successfully avoid the disease. Perhaps milk diets without animal protein supplements can improve the individual fitness of infants in these ecological settings (Lepowsky, ch. 4), but more documentation of the interrelationships between nutritional deficiency and disease is required before we can assert that attempts to improve the nutritional status of infants in tropical lowland regions by encouraging early animal protein supplements may disrupt delicately balanced, culturally-maintained biological systems that protect humans from some serious indigenous diseases. However, some cultural practices or courses of action, such as limiting protein intake for lactating mothers and infants, may appear arbitrary or even maladaptive because their connection to survival and reproduction has not been carefully studied. (See Lepowsky, ch. 4.) Outside attempts to alter specific indigenous practices, such as infant feeding, without due consideration of the decision-making context and ecological setting producing those practices, may reflect the cultural imperatives of change agents rather than the irrationality of indigenous cultural systems.

Weaning

Despite their expressed ambivalence regarding dependent children, most Usino women believe that a child should wean itself gradually and not be forced. They say that with time, children wean themselves. As they learn to walk and play, they naturally forget about suckling and seek the breast only for solace when upset. Occasionally, however, mothers apply ginger or chili pepper to their breasts to speed the process. A few women wean their children by leaving them at home when they travel. I observed three cases of prolonged breast-feeding. In two cases, children who were about five years old and who lacked younger siblings continued to suckle. In another, a boy of four or five was allowed to breastfeed while his newborn sibling slept. He had been weaned previously, according to tradition, but once "he saw his younger sister drink, he cried for milk again."

The ideal expressed by all informants, both male and female, is that sexual relations should not resume before the child is weaned and eating solid food (sweet potato, taro, cassava, plantain, meat). According to most informants, the child's limbs must be strong before it is weaned. Strong limbs are evidenced by unaided walking. A few informants interpret strength as vigorous crawling instead. For reasons stated above, a mother endangers the health of her infant if she engages in sexual intercourse before her child is weaned. Contaminated milk causes the child to develop slowly, to have tiny buttocks and spindly legs "like a chicken," to remain small, frail, emaciated and unable to stand. Minimal acceptable birth spacing is two years at present. This suggests that, at the earliest, sexual relations resume about one year after the birth of a child, subsequent to weaning.

Informants state that pre-contact postpartum taboos lasted for three or four years because men, concerned with maintaining their strength and fighting prowess, would not risk these to sleep with their wives. Dependent children were also more of a liability then, because they were vulnerable during raids and because men were too preoccupied with raiding to plant gardens large enough to support them. Earlier sevrage and the concomitant decrease in the length of the postpartum taboos stemmed directly from voluntary acceptance of pacification in the 1920s and 1930s. It appears that closer child-spacing was the intended result, but whether or not this reflects a decision-making process based on some form of cost-benefit analysis is unclear (Conton, 1981; 1983). These changes occurred *prior* to the introduction of a cash economy or economic development in the region. Since then, there has been no appreciable change in the age of weaning despite intermittent participation in the cash sector and access to wage labor.

Despite the expressed ideal that weaning should occur prior to the resumption of sexual relations, several women stated that children sometimes weaned themselves when their mothers again became pregnant; children were said to dislike the unpleasant taste of the breastmilk. Others said that the fetus required the breastmilk to grow. Informants did not recognize any inherent contradiction between the universally held ideal and this practice, nor did they offer any explanation or apology for continued breastfeeding concomitant with sexual relations. (See Counts, ch. 9.) Many lactating women in pre-industrial societies conceive without menstruating (Harrell, 1981). Likewise, many Usino women do not recognize a nine-month gestation period. Thus, they might unwittingly continue nursing during early pregnancy. This explanation of self-weaning, however, still does not explain the apparent breach of moral code prohibiting such behavior.

Usino women do not agree in theory or in practice about supplementary feeding. According to them, the timing of the introduction of semisolid and solid food varies individually, according to the inclinations of both the mothers and children. Most mothers introduce mashed papaya, sweet potato, pumpkin and banana, and broth between six months and one year, if they experience no problems with lactation. The type of food introduced depends largely upon seasonal availability and convenience. According to older informants, traditional practice favored withholding solid food until the baby was walking, but three informants gave supplemental food soon after birth because their babies "were not satisfied with breastmilk and cried for food." Today solid food is given whenever the child demands it, once it has teeth. Biting of the breasts signals the child's readiness for insects as a supplement.

Maternal and Child Health (MCH) staff, who usually visit the patrol post for monthly clinics, have convinced about five young mothers to introduce semisolids at the age of four to six months. As in other parts of Papua New Guinea, the nursing sisters encourage breastfeeding by example, taking frequent breaks from their work to nurse their own infants (Marshall, ch. 2). Some staff members, however, demonstrate minimal concern for or understanding of rural mothers and their problems, treating them in a condescending, sometimes rude, manner. Obrist (unpublished observations, 1983)[†] makes similar observations in the Eastern Sepik Province. Educated wives of teachers, police, clergy and traders stationed at the patrol post also attend the same clinics; they make it a point to insult the uneducated bush women of Usino. Usino women enjoy the all-day social gathering afforded by the baby clinics, and many desire western medical care for their infants. In July and August, 1981, approximately 27% of the eligible Usino mothers attended each clinic. Humiliated by outsiders, however, most Usino women resist the advice of the nursing sisters, and instead, look toward tradition for answers and security with regard to infant care.

[†]Obrist, B. (1983). The study of infant feeding: Suggestions for further research. Paper presented at the twelfth annual meeting of the Association for Social Anthropology in Oceania, New Harmony, IN.

Other social and ecological factors combine to make prolonged breastfeeding and late introduction of beikost appear to be adaptive patterns. Usino babies are regularly bathed in the nearby stream in which all villagers bathe daily. This major source of drinking water is frequently defiled by pig, cattle and human feces. Dishes and bowls are sometimes rinsed in the water between meals. For this reason, the use of powdered milk or infant formula, which require water for mixing, could lead to acute infant diarrheal disease. Prolonged and regular lactation decreases infant ingestion of polluted water from local sources. Also, babies spend some of their time on ground contaminated by betel spittle and feces of pigs, chickens, dogs and other infants. The rest of the time they are in close physical contact with a multitude of relatives, a number of whom are suffering from communicable disease. Neither situation is hygienic for infants whose hands are frequently in their mouths, but infants dependent on breastmilk for nutrition rather than on formula or food passed through many hands may have a better chance to remain healthy than those weaned or given beikost early.

Insufficient Lactation

Usino mothers identify insufficient lactation by the same symptoms as reported by Gussler and Briesemeister (1980). (See also Greiner, Van Esterik and Latham, 1981). It appears to be a common problem, resulting from mothers' illness or emotional distress, such as occurs following a death in the extended family. According to the regional MCH nursing sisters, the medical team from the Madang College of Health Sciences stationed at the Usino Patrol Post, and the district nutritionist (Claire Marshall, personal communication, 1981), Usino people suffer primarily from respiratory tract infection, especially pneumonia; gastroenteritis and intestinal infections; skin disease, especially ringworm; malnutrition;[†] parasitic infestations and tropical ulcers. Anemia, leprosy, filariasis and tuberculosis are not unknown. Although most women identify liquids such as broth, water, tea, sugar cane juice and plenty of food as galactogogues, serious illness in

---

[†]Considerable debate among medical researchers in Papua New Guinea centers on the issue of malnutrition and how to recognize it (Peter Heywood, personal communication, 1983). None of the health specialists consulted had calculated actual rates of malnutrition for the Usino area.

Usino is culturally expressed by social withdrawal and an unwilling-ness to eat or drink. This has obvious negative consequences for the health of lactating mothers and nursing infants. Unhealthy mothers frequently complain that their suckling infants are ill, hungry or weak. That illness is linked clearly with decreased breastmilk production probably reflects not only the mother's negligible calorie intake, but also decreased suckling frequency. Even though mothers substitute other liquids for breastmilk when they are ill, it is not surprising that infants remain hungry and indisposed. This is due in part to the ailing mother's disinterest in others, including her dependent baby, and in part to the infant's ingestion of unclean water. The 1981 mortality rate for all children under five years of age was 28 %, slightly less than the 1975 rate of 30 %.[†]

Older informants agreed that wetnursing was condoned in situations of maternal illness, insufficient lactation or maternal death. Reports of infant death following insufficient lactation, however, suggest that wetnursing has not been standard practice for some time. Wetnursing was never observed during this study, and interviews elicited contradictory data about this practice. Only one of my primary informants claimed to have actually done it, but several others said they *would* do it to aid their sisters or friends. Conversely, though, some of these women said that under the same circumstances they would not consider giving their own infants to other women to nurse. Two women mentioned "germs" in this connection, a signal that some rudimentary adaptation of Western medical knowledge was informing their responses. Disavowal of the existence of the practice by a few younger women reflects outside influence.

CONCLUSIONS

It seems likely that when a cash economy is established in Usino, the present self-sufficient economic system and traditional value system will rapidly give way to systems that emphasize the acquisition of cash. An increased interest in and dependence on the purchase of

---

[†] Infant mortality figures are based on pregnancy histories collected in 1975 and 1981 from 91 women, nine of whom were deceased by 1981, leaving 79 parous females and three non-parous females in 1981. Under-reportage of births is a common error in studies using retrospective fertility data, but this was minimized by cross-checking on separate occasions with the respondent herself and with other female relatives.

imported commodities will probably encourage earlier weaning or mixed feeding strategies if Usino women, materially-oriented by tradition, continue their heritage of full participation in economic enterprise, or if there is a breakdown in traditional marriage and residence patterns that foster female solidarity. Also, if men come to predominate in the wage labor economy, women's workload will likely increase. Earlier introduction of supplementary foods and earlier weaning without improvements in overall hygiene could result in increased infant and child mortality. Breastfeeding presently is a major buffer between life and death for many infants. Economic opportunity *might* presage a change in infant feeding practices. However, unless the overall health of Usino people improves, their own perception of their marginal health status will continue to inhibit initiation and sustenance of modern cash enterprises above and beyond traditional subsistence and economic obligations (Conton, unpublished observations, 1983). By their own evaluation, health recovery is a prerequisite for economic development in Usino. Women's self-defined poor health inhibits their ability to care for their children, so improvements in adult health would also benefit infants.

This case study and others like it, which delineate the cultural context of infant feeding practices, are a preliminary step toward a holistic appraisal of how forces of modernization and westernization influence infant and child health. Successful delivery of appropriate modern health technology and care to rural non-modernized populations, such as in Usino, relies upon awareness of the complexity of interrelationships between economics, social structure and physical environment that structure infant care and feeding patterns. A focus on the socio-economic environment of mother-child interaction may enable health personnel to develop programs more responsive to needs of mothers and infants. More important, it may help national policy-makers to become more cognizant of, and concerned about, the impact of their economic development programs on the health of their people.

ACKNOWLEDGEMENTS

I thank Kathleen Barlow, Peter Heywood, Carol Jenkins, Leslie Marshall, Susan Montague, Brigit Obrist and Alison Orr-Ewing for their critical comments on an earlier draft. They are in no way responsible for any shortcomings of the final product. For introducing me to the medical literature that proposes an interdependence between diet and malaria, and for suggesting the adaptive nature of infant food (protein) taboos, my thanks also to Maria Lepowsky. I also acknowledge the assistance of Elliott Gehr in numerical tabulation, and especially, the men and women of Usino who so patiently answered my questions and accustomed themselves to my presence.

REFERENCES

Abernethy, V. (1979). *Population Pressure and Cultural Adjustment.* Human Sciences Press, New York.

Anderson, P. (1983). The reproductive role of the human breast. *Cur. Anthropol.* **24**, 24-45.

Back, K.W. and J.M. Stycos (1959). *The Survey Under Unusual Conditions: Methodological Facets of the Jamaica Human Fertility Investigation.* Monograph No. 1, Society for Applied Anthropology, Ithaca, NY.

Boehm, C. (1982). A fresh outlook on cultural selection. *Amer. Anthropol.* **84**, 105-125.

Brown, R. (1978). Weaning foods in developing countries. *Amer. J. Clin. Nutr.* **31**, 2066-2072.

Conton, L. (1977). *Women's Roles in a Man's World: Appearance and Reality in a Lowland New Guinea Village.* Unpublished doctoral dissertation, University of Oregon, Eugene, OR. *Dissertation Abstracts International* **38**, A3583.

Conton, L. (1981). *Reproductive Decision Making and the Value of Children: Usino, Upper Ramu District, Madang Province.* Report to the Population Programme of the Office of Environment and Conservation. Papua New Guinea Institute for Applied Social and Economic Research, Boroko.

Greiner, T., P. Van Esterik, and M.C. Latham (1981). The insufficient milk syndrome: An alternative explanation. *Med. Anthropol.* **5**, 233-260.

Gussler, J. and L.H. Briesemeister (1980). The insufficient milk syndrome: A biocultural explanation. *Med. Anthropol.* **4**, 145-174.

Gyorgy, P. (1967). *Human Milk and Resistance to Infection.* CIBA Foundation Study Group No. 31. Little, Brown and Co., Boston.

Harrell, B. (1981). Lactation and menstruation in cultural perspective. *Amer. Anthropol.* **83**, 796-823.

Jelliffe, D.B. (1968). *Infant Nutrition in the Sub-tropics and Tropics.* World Health Organization, Geneva.

Jelliffe, D.B. (1976). World trends in infant feeding. *Amer. J. Clin. Nutr.* **29**, 1227-1237.

Jelliffe, D.B. and E.F.P. Jelliffe (1971). The uniqueness of human milk. Symposium. *Amer. J. Clin. Nutr.* **24**, 968-969.

Jelliffe, D.B. and E.F.P. Jelliffe (1975). Human milk, nutrition and the world resource crisis. *Science* **188**, 557-560.

Jelliffe, D.B. and E.F.P. Jelliffe (1978). *Human Milk in the Modern World: Psychological, Nutritional and Economic Significance.* Oxford University Press, Oxford.

Jimenez, M.H. and N. Newton (1979). Activity and work during pregnancy and the postpartum period: A cross-cultural study of 202 societies. *Amer. J. Obstet. Gynecol.* **135**, 171-176.

Latham, M.C. (1977). Infant feeding in national and international perspective: An examination of the decline in human lactation and the modern crisis in infant and young child feeding practices. *Ann. NY Acad. Sci.* **300**, 197-209.

Lawrence, G. and P.D. Walker (1976). Pathogenesis of *Enteritis necroticans* in Papua New Guinea. *Lancet* **1**, 125-126.

Mata, L. (1978). Breastfeeding: Main promoter of infant health. *Amer. J. Clin. Nutr.* **31**, 2058-2065.

Mata, L. and R.G. Wyatt (1971). The uniqueness of human milk: Host resistance to infection. *Amer. J. Clin. Nutr.* **24**, 976-986.

Nerlove, S.B. (1974). Women's workload and infant feeding practices: A relationship with demographic implications. *Ethnol.*, **13**, 207-214.

Oomen, H.A. and S.H. Malcolm (1958). *Nutrition and the Papuan Child: A Study in Human Welfare.* South Pacific Commission Technical Paper No. 118. Noumea.

Porter, R. and M. O'Conner (1975). *Parent-Infant Interaction.* CIBA Foundation Symposium No. 33. Elsevier/Excerpta Medica, Amsterdam.

Raphael, D. (1973). *The Tender Gift: Breastfeeding.* Prentice-Hall, Englewood Cliffs, NJ.

Raphael, D. (1979). *Breastfeeding and Food Policy in a Hungry World.* Academic Press, New York.

Sassin, J.F., A.G. Frentz, E.D. Weitzman and S. Kapen (1972). Human prolactin: 24-hour pattern of increased release during sleep. *Science* **177**, 1205-1207.

Van Esterik, P. and T. Greiner (1981). Breastfeeding and women's work: Constraints and opportunities. In E. Baer and B. Winikoff (Eds.), *Breastfeeding: Program, Policy and Research Issues.* Special issue of *Stud. Fam. Plann.* **12**, 184-197.

Wilson, S.F. (1975). Matri-patrilocality and the birth of the first child. In D. Raphael (Ed.), *Being Female: Reproduction, Power and Change.* Mouton, The Hague, pp. 73-86.

CHAPTER 7

INFANT CARE AND FEEDING PRACTICES AND THE BEGINNINGS OF SOCIALIZATION AMONG THE MAISIN OF PAPUA NEW GUINEA[†]

ANNE MARIE TIETJEN

INTRODUCTION

Parents in all societies raise their children in a manner that is generally congruent with the demands of their physical environment and their economic system, and with the system of of beliefs that has developed (Benedict, 1938; Mead and Newton, 1967; LeVine, 1974). The existence of cultural continuities between adult beliefs and childrearing practices is the basis for socialization. Even the care of infants reflects aspects of the ideology of the culture, as Caudill and Frost (1973) have shown.

In this paper it will be argued that infant care and feeding practices represent the beginnings of socialization. It will be shown that continuities exist between adult systems of belief and infant care and feeding practices among the Maisin people of Papua New Guinea, and that the particularly Maisin pattern of infant care and feeding practices serves as a means for beginning to integrate infants into Maisin society.

The period of infancy presents a special problem in the study of socialization because infants are in the early stages of developing the capacity to use language, on which much socialization is thought to depend. Infancy is generally recognized as the period in which the affectional bonds necessary for the success of the socialization process are established. Recent research has shown, too, that infants have

†The research on which this paper is based is supported by grants from the Spencer Foundation and from the Social Sciences and Humanities Research Council of Canada.

many capacities that enable them to learn from their environments from birth on (for example, Bower, 1974) and to be active participants in social interaction with their caregivers (Bell, 1968; Freedman, 1974). Piaget's work (1952) has shown that infants learn about their environments through direct actions, using their developing sensorimotor skills before they are able to use language. These findings raise the questions of whether thematic content of socialization may begin to be communicated in early infancy and, if so, to what degree and by what processes?

Current approaches to socialization emphasize the importance of social interaction between children and their caregivers as the means by which childhood socialization occurs (Zigler and Child, 1973). Processes involved in socialization include direct experience, observation and verbal instruction. Harkness and Super (1982) have suggested that children abstract the rules of culture from the regularities of the physical, social and psychological parameters of their environments just as they abstract the rules of grammar from the regularities of the speech environment.

The care and feeding of infants provide many opportunities for intense social interaction. In these interactions infants may be introduced, through sensorimotor and affective processes, to actions which are informed by important cultural beliefs. This may be seen as the beginning of socialization.

METHODOLOGY

The data reported here were gathered during two series of interviews with women in Uiaku and Ganjiga villages. In the first set of interviews, 20 women with children elementary school age and younger were asked a few very general questions about their infant care and feeding practices along with other questions about childrearing. The answers provided some very rich ethnographic material, but this approach could not give an indication of the distribution of knowledge or beliefs or practices, since new questions were raised by the responses of each successive woman in this first sample. In order to get a more accurate picture of this distribution, ten grandmothers and ten currently nursing mothers were selected for a second set of interviews. These women were asked a standardized set of detailed questions based on information obtained in the first interviews. In addition, systematic observations of maternal and infant behavior during feeding were done with 21 mother-infant pairs.

THE SETTING

The Maisin villages of Uiaku and Ganjiga are situated on opposite banks of the mouth of a broad shallow river that empties into Collingwood Bay, Oro Province. Uiaku and Ganjiga were first contacted by Europeans in 1890, and an Anglican mission outstation and school taught by Solomon Islanders were established there in 1902. Most men and a few of the younger women in the villages have spent some time working or going to school in urban centers. Despite early contact and relatively good education, however, Uiaku and Ganjiga have remained isolated and in an economic backwater. They are linked to the rest of the country by an airstrip at Wanigela, 12 miles to the north, accessible only by boat or by walking along the beach (and crossing eight rivers and a mangrove swamp). There is a cooperative trade store in Ganjiga that operates sporadically, and three other very small operations run even more sporadically from individuals' houses.

The Maisin are hardworking subsistence gardeners who also gather wild foods, hunt and fish. What little cash income they have comes primarily from remittances from relatives employed in the towns; from the sale of *tapa* (bark cloth) made by nearly all of the village women; or to a very limited extent, from the sale of copra. The absence of regular shipping in Collingwood Bay has prevented any real development of the area.

Because of the villages' relative isolation and lack of development, many traditional beliefs and customs have survived and co-exist with the teachings of the Anglican church and other Western introductions. The existence of a mission clinic at Wanigela and a travelling baby clinic that comes to the villages bi-monthly has improved infant health (primarily through immunizations and other drugs) without substantially altering Maisin infant care and feeding practices. Practices described here are still currently observed by most Maisin women in the villages unless otherwise indicated. The few differences between the older and younger women will be reported where relevant.

Maisin infant care and feeding practices reflect adult systems of belief regarding pregnancy, birth and early infancy as periods of vulnerability from which the child gradually emerges; the powers inherent in foods; the importance of interdependence with other people; and the social significance of food. Beliefs concerning pregnancy, birth and infancy as periods of vulnerability and beliefs about the powers inherent in foods seem to have developed in response to environmental dangers to mothers and infants. Beliefs about the importance of interdependence with others and about the social significance of food contain both the substance and the means of socialization. They help to define the kinds of opportunities for social interaction that will exist within the culture and the meanings of those interactions.

Maisin ideas about vulnerability and about the powers of foods are verbally communicated and are manifest in interactions among adults rather than in direct interaction between caregivers and children. Their transmission requires a certain level of language and cognitive skills on the part of the child. Beliefs in interdependence and in the social significance of food, on the other hand, are demonstrated, often nonverbally, in interactions in which the child takes part from an early age. Aspects of these two beliefs, then, may be communicated to the child during infancy.

PREGNANCY, BIRTH AND INFANCY AS PERIODS OF VULNERABILITY

The Maisin value children highly. They recognize the dangers inherent in pregnancy, birth and infancy, and have developed means of dealing with them. Many beliefs and practices acknowledge these periods as ones in which the developing child is both especially vulnerable and especially malleable, gradually emerging from vulnerability as it gains physical strength and takes its place in the social system.

Pregnant women are thought to be highly susceptible to spirits and sorcery which, they believe, is practiced on them by rejected former suitors. In order to prevent miscarriage, stillbirths and congenital deformities, they must take special precautions. Traditionally, women did not talk about their pregnancies with anyone but their husbands. Young women now report that they converse about their pregnancies with other women making the all-day trip to the weekly pregnancy clinic at Wanigela, but otherwise maintain silence about their state.

Pregnant women do not go to their gardens alone, as spirits and sorcerers are likely to do their work when the woman is unprotected. Women are also careful to cover their breasts and abdomens when going outside at night during pregnancy, because spirits are believed to be especially active at night. The wearing of blouses, a modern introduction, is sufficient to eliminate this risk.

Some complications of delivery, such as the child being born with the umbilical cord around its neck, are thought to be caused by the pregnant woman encountering webs or web-like objects. In order to guard against this, pregnant women do not wear necklaces or make string bags and must avoid walking into spider webs. When going along a path toward the garden, they must either be preceded by another person or hold a large leaf in front of them to break any webs that might block the path. This belief is still held by the younger women, although some disregard the prohibitions, viewing the risks as not very serious.

The regularity with which women attend the weekly prenatal clinic at Wanigela also reflects the strength of the Maisin belief that special care must be taken of pregnant women. Most women attend at least monthly throughout their pregnancy and many attend every week. Women give birth in the clinic if complications in the delivery are expected. Four of the twenty mothers interviewed had delivered at least one of their children at the clinic at Wanigela (or another clinic if they had been living away from the village at the time). Two others, both middle-aged mothers, had given birth on canoes on the way to Wanigela.

The events surrounding birth also reflect Maisin beliefs in the vulnerability of mother and child. Birth takes place in the home of the child's parents, usually with the child's father and one or more experienced female relatives attending. After the birth both mother and child are washed. The "rubbish" that is in the baby's mouth at birth is cleaned out with a finger, and the baby is given water to drink. The baby's tongue is scraped with a clam shell in order to ensure that it will learn to speak Maisin properly, and its limbs are "straightened" with hands warmed over a fire. Sticks from this fire must not be taken to start fires at other houses (ordinarily a common practice) to guard against sorcery or harmful actions by spirits toward the child. Sticks for this fire should be put on neatly and not broken, in order to prevent the baby from developing a rash on its neck, armpits and groin. Just after birth the mother sits near the fire, heats a broken piece of a clay pot in the fire, wraps it in tapa and holds it against her abdomen to encourage the blood to flow.

The Maisin take care during the early days and weeks after birth to ensure that the baby's spirit, which seems to be somewhat easily distracted, becomes firmly attached to its body. Following delivery the umbilical cord and placenta are wrapped in tapa cloth and taken into the bush, where they are either buried deep in the ground or placed in the branches of a tree in order to prevent their being destroyed by animals. If the placenta is eaten by fish or animals, the Maisin believe the baby's spirit will not become properly attached and the child will suffer from madness. A few of the older women reported that the baby's spirit was believed to reside in the placenta before birth. The person who takes the placenta to the bush must return to the house by the same route in order to ensure that the spirit of the child will not get lost in the bush. If the spirit becomes lost, the baby will cry inconsolably until its spirit finds it. The Maisin also believe that returning to the house by the same route after disposing of the placenta ensures that the woman will have more children.

The mother's first food after the birth is cooked for her. After her bleeding has stopped, however, she must cook for herself in separate pots and must not eat from any other pot during her confinement. This practice was consistently explained as a means of protecting the baby from a variety of unspecified illnesses.

Except in the case of firstborn children, whose confinement with their mother may last up to one month, mother and child remain in the house for several days, at least until the stump of the umbilical cord dries and falls off. At dawn on the day that they will emerge, the mother stands just outside the house holding the baby and tosses it in the air several times, exhorting it to climb betelnut and coconut trees and hunt many pigs if the child is a boy, and to fetch water and chop wood if it is a girl. This action may be repeated many times over the first year of the baby's life. Some women reported that this tossing shows the baby how it feels to move, thus preparing it for the hard work ahead. This explanation indicates an awareness on the part of Maisin mothers that their actions during care and feeding of their infants are the beginnings of socialization.

Within the first week after birth, if the mother and child are well, the baby is taken to the family garden in a string bag. A piece of tapa (pillows are usually used now) is placed in the bottom of the bag and another (often a *lap lap* or other piece of cloth now) is folded around the bag to protect the baby from the sun. On the way to the garden, the mother leaves a token at any crossroads or fork in the path. For a son the token is small straight sticks crossed, representing spears. For a daughter the mother leaves a small ring made of grass to represent the larger grass rings women formerly used to balance round-bottomed clay pots of water on their heads, or a small replica of the strap used by women to carry large bundles of firewood on their backs. These signs ensure that on the return trip the baby's spirit will find its way home.

In the garden the mother plants a long stick in the ground and from it suspends the string bag containing the baby. On leaving, she throws the stick into the bush, where spirits will not be able to find it and use it to harm the baby. The mother calls aloud, "Come, we'll go home now!" to encourage the baby's spirit to come back to the village with the mother and baby. The spirit must be called until the baby begins to smile and recognize people, which the Maisin interpret as a sign that the baby's body and spirit are united.

During lactation the baby's health is protected by the mother's observance of extensive food proscriptions, which will be discussed in a later section of this paper, and by the father's observance of a custom dictating that he must not touch or step over the mother's or child's clothing or sleeping mat. Intercourse is prohibited until after the child has been weaned. Children whose parents have intercourse while the mother is still lactating will, the Maisin believe, have large heads and stomachs, and small hips and legs. In fact, children are rarely spaced less than two years apart.

Infants and toddlers are carefully guarded from physical dangers including the sun, falls, knives, sandflies and mosquitoes. Spirits become somewhat less threatening as the child grows; but as the child's understanding of language emerges, many parents try to keep their children from going into dangerous places by frightening them with spirits, particularly one known as Babau.

Most children are weaned between the ages of 18 and 24 months, although there is no fixed chronological age for weaning. Such external circumstances as a food shortage may prolong the nursing period. The health and vigor of the child, the ability to eat the same foods as adults do, to speak and to run are taken as signs that the child has emerged from the most vulnerable period of its life and can function physically independently of its mother.

THE POWERS INHERENT IN FOODS

Closely related to Maisin beliefs about pregnancy, birth and infancy as periods of vulnerability are beliefs about the powers inherent in foods. There are beliefs about the powers of specific foods that apply to the general population; but perhaps the greatest number of specific beliefs in this area apply to lactating mothers.

There are no food prescriptions or proscriptions for pregnant women. Some women report eating less during pregnancy than they ordinarily do, some report eating more and some have specific food cravings which are indulged if possible. Herbs mixed with water are sometimes taken to prevent miscarriage. *Ibeka* or hibiscus leaves, which become slippery when boiled, may also be used during labor to speed delivery or to help expel the placenta if necessary.

Beginning at birth the mother follows strict regulations about her food intake. Many of these suggest notions of sympathetic magic, and others suggest attempts to find nutritional solutions to common health problems.

The first food eaten by mothers after giving birth is usually the *warubi* banana (*Musa* spp.), a relatively soft and sweet variety, cooked in water. The resulting soup, which is said to be just like breastmilk, is also fed to the mother and used to wash her breasts before feeding the infant. Maisin women believe that this helps make the milk come quickly. Two other varieties of bananas, both relatively soft and sweet, may be substituted for the warubi if it is not available. Traditionally, the only other food the mother eats while waiting for her baby's cord to fall off is boiled taro, or sago in times of food shortage. Five of the ten young women interviewed reported that they had observed these restrictions for the first week. Three others included greens in their diet at this point as a result of the clinic's promotion of greens. One young woman ate some European foods during her confinement, and another maintained that any food except pig was permissible during confinement.

Before Maisin mothers give the breast to their infants, they expel the colostrum. They believe that it has been inside the mother during pregnancy and is, therefore, old and likely to make the baby sick. All of the women interviewed denied that it had semen in it, despite their contention that semen can enter and spoil the milk of a lactating mother.

After the period of confinement is over, a mother can begin to eat other foods. A type of white worm found at the base of mangrove trees is believed to be especially good for nursing mothers. The worms are cooked in water and eaten along with the resulting soup, said to closely resemble breastmilk. A particular kind of shellfish boiled twice in coconut milk until very tender is also considered to be good for lactation. Sweet potatoes and papayas and any European foods may be eaten after confinement.

On many other foods, however, there are traditional restrictions.

No baked foods may be eaten because "there is no juice in them for the baby." The women believe that eating baked foods or dry coconut will make the baby's skin dry and will give the baby a big spleen. Only small, tender pigs and fish may be eaten, and then only the most tender parts in order to prevent an enlarged spleen in the infant. Some women explained that pig and fish were not to be eaten because the baby would develop a taste for them and cry for them when they were unavailable. Several species of fish are prohibited because they are too tough, and one type, a flat fish with spots which lives in salt water, is believed to produce scabies in the child of a breastfeeding mother. Eating crabs is thought to inhibit the production of milk. Mangoes and pumpkin are not eaten because of the belief that mothers who ingest them during lactation will have babies with yellow skin. Guavas and some other fruits are believed to produce diarrhea. The child of a nursing mother who touches cassowary meat with her hands will, the Maisin believe, have boils on its head. Cassowary meat eaten with a fork, however, is safe. All of the grandmothers interviewed reported observing these restrictions at least until their children were crawling. All but one of the young mothers reported observing most of them for three months or more. Several women in both the older and younger groups expressed the idea that they no longer had to be so careful about what they ate because now they had doctors to make the children well if they got sick.

Children begin to eat semi-solid foods when they get their first teeth. Ripe bananas and cooked potato are usually the first foods given. Taro, rice, biscuits, and other foods are usually introduced when the baby has four teeth on top and four on the bottom. Most children do not eat pig and fish until they have been weaned.

Insufficient milk is reportedly not a serious problem for Maisin women. None of the women interviewed admitted to having had the problem themselves, but all said that if it did occur it could be corrected by eating more food, especially taro. In the only serious case that came to my attention the infant was spoonfed powdered milk mixed with water and was thriving.

INTERDEPENDENCE WITH OTHERS

Maisin social organization reflects the importance of close interdependence among people at several levels. First, while there is no

Maisin word for the nuclear family, this is the unit that takes primary responsibility for childrearing. Close relationships are also usually maintained with kin on both the mother's and the father's side of the family, with multiple mothers, fathers, grandparents, etc. variously involved in a child's life. Closest contact is usually with the child's clan members, since residence is patrilocal, and children are taught to identify with their clan. Children quickly learn to address grandparents, aunts, uncles, siblings and cousins by the appropriate kinship terms.

European influence has probably increased contact and interdependence among clans. The church, Village Council, school, Women's Club and Youth Club all have community projects that cut across the clan lines.

Adults related by blood or marriage help one another in their gardens, in building houses, borrowing goods and money, exchanging food and childminding. Maisin parents teach their children to help anyone who needs it and give them responsibilities within the family and clan at an early age. Children are also taught that, after a successful fishing expedition or coconut climb, they should give some of the fruits of their efforts to anyone they see on their way back to the village.

The child's interdependence with people is recognized from the moment the mother knows she is pregnant. The Maisin expect the child's father to take a great deal of responsibility for ensuring the safe delivery of the child. They consider frequent intercourse to be necessary for pregnancy to occur but harmful to the growth of the fetus after pregnancy has been established. The husband must also care properly for his wife by making sure she does not have too much heavy work to do, seeing that she is not left alone, and walking before her on the way to the garden to keep her from walking into spider webs. His failure to provide adequate care may result in his being blamed for any congenital physical or mental abnormalities that may occur in the child. A Down's syndrome boy was pointed out to me as an example of a child whose father had not looked after his wife properly during her pregnancy. Maisin frequently interpret such abnormalities as the result of vengeance sorcery performed by the angered relatives of the wife as a punishment of her husband or, less frequently, as the revenge of a rejected suitor against the woman. Children whose mothers are unmarried or whose fathers die before they are born are believed to be "difficult to teach," not only because of the father's absence after birth but also because of the lack of the father's protective influence on the mother during pregnancy.

Fathers usually attend their children's birth and may be the mother's only attendant if she has given birth several times previously. If others are present, the father's main function is to hold and soothe his wife. By all accounts, the father's participation in the birth is a traditional custom of the Maisin, and one that is regarded as very important. Fathers may touch the baby from birth and, thereafter, take an active role in child care. This role increases in scope after about the third month, when some mothers begin to leave their babies at home while they go to the garden for as long as half a day at a time. The father is often the caregiver at such times throughout infancy. When a mother is separated from her child for this long during early infancy, she may leave green coconut juice for it to drink or she may squeeze milk from her breasts onto the baby's hands so that it will smell the milk and suck its fingers when it wakes.

Some of the child's kin begin to take part in the child's life while the mother is still pregnant. The woman's mother, sister and in-laws often help with heavy work during this time. The birth is usually attended by one or more female relatives experienced in midwifery, as well as the father. These assistants massage the woman's back during labor, assist with the delivery, wash the mother and child, cook food for the new mother and help her with her chores for the first few days after delivery. It is usually a grandmother or another female relative, and not the mother, who straightens the baby's limbs with hands warmed over the fire just after birth.

The child announces its own birth to neighbors with its cries. Those who hear may call to the parents to ask the child's sex, and close relatives may come to the house to see the baby shortly after birth. Placing the tokens at intersections along the garden paths on the baby's first trip to the garden also serves to announce the birth and the child's sex to the village.

During the mother's week of confinement in the house with her infant, the Maisin observe a custom referred to as *jon*. This custom dictates that all of the usual inhabitants of the house where the infant is born must sleep in that house rather than elsewhere. If household members sleep abroad during this time, the child will have small sores on its body, according to Maisin belief. At another level, the custom reflects the solidarity of those closest to the child in an action intended to protect the child from harm.

If a woman dies in childbirth or while her infant is still suckling, the child may be entrusted to the care of another lactating woman, usually a close relative of either the baby's mother or father. It is not unknown for a woman who has not previously lactated to be able to breastfeed an orphaned baby through massage and application of hot water and herbs to her breast. Traditionally, if such arrangements could not be made, however, motherless babies were fed green coconut juice, but expected to die.

Maisin women rarely breastfeed children other than their own if the child's mother is living. When this does occur it is always between close kin. Three of the young mothers and four of the grandmothers interviewed said they had fed someone else's child at least once, though none admitted to having allowed her child to be fed by another woman. Many women believe that if their child is breastfed by another woman or allowed to suck on the breast of a non-lactating woman it will become mute. Several women also pointed out that other women's milk may contain semen, which would harm the baby. Like the Kaliai women (Counts, ch. 9), Maisin women believe that semen can enter the breastmilk, although they did not specify the mechanism by which this occurs.

For the first three months of its life, the baby is almost constantly in the company of its mother. It is breastfed on demand and sleeps next to the mother on her sleeping mat at night. The mother takes the child almost everywhere during the early months, until it "can see people and smile." At this point, usually about three months, older children (usually at least school age) may begin to hold the child and stay with it for brief periods of time. The role of older children in caregiving steadily increases. It is not uncommon to see several school-age girls playing active games while holding baby siblings on their hips. If there are no older sisters, older brothers help their mothers in the care of their younger siblings. In polygynous households, wives may share the care of one another's children to a limited extent.

Weaning represents a significant step in the course of the child's integration into the group. It is accomplished by sending the child to sleep with someone other than the mother. Sometimes this is the father, in which case the child remains within the same house, but children are often sent to stay with grandparents for a few days to mark the end of weaning and the beginning of a new phase in their integration into the social group. Children who have been adopted into another household go there to live at the time they are being weaned.

Maisin infants experience interdependency from an early age.

Their needs influence the actions of others, and others' needs influence how and by whom infants will be cared for. This interdependence grows in complexity as the child develops. The infants' daily interactions with parents, siblings, clans-people and other kin have begun communicating this idea.

THE SOCIAL SIGNIFICANCE OF FOOD

As in most cultures, food plays a key role in the social life of Maisin adults. The giving and receiving of cooked food is the concrete manifestation of all Maisin social bonds. The word for "to look after" another person or group (*kaifi*) implies that food is given to that person or group. Kin exchange cooked food on a daily basis (often delivered by toddlers), even in times of famine. Failure to provide enough food for close relatives may sometimes result in serious social rifts and is a common explanation for the use of sorcery.

Food is an essential ingredient in all life transition rituals. A pot of cooked food sent from the parents of a young man to the parents of a young woman is the first step in making a marriage proposal. The groom's relatives provide further gifts of food to the bride's family at the time of the wedding and at the birth of the couple's firstborn child. When the firstborn reaches marriageable age, an initiation ceremony is held at which the firstborn's father's kin provide food and other gifts for the guests. At funerals and at ceremonies marking the termination of mourning, the giving and receiving of food is also a central part of the proceedings. According to Maisin etiquette, guests at large formal exchanges may not leave until their host feels that he has given them enough food. During times of food shortage, life transition ceremonies are postponed, sometimes for a year or more, until the gardens can produce enough food to celebrate the event properly. Food is also used as payment for services. For example, if a man is building a house or canoe, the women of his clan will cook food to give to the men who assist him.

Just as adult social bonds are expressed in terms of provision of food in adulthood, the child's first social bond is established through the provision of food. Maisin mothers feed their babies on demand and offer the breast as a source of comfort when the baby cries. Mothers frequently talk to their babies during feeding and play informal social games with them as the baby sucks. Structured observations of mother-infant interaction during feeding, using a measure devised by Barnard (1978) to assess mothers' sensitivity to their infants' cues, social-emotional growth fostering and cognitive growth fostering, indicate that Maisin mothers regard feeding as a social situation and use it to provide social stimulation as well as nourishment to their infants.

As soon as the baby is able to sit up and eat semi-solid foods, it sits with other members of its household at mealtimes, usually on its mother's lap, and shares the mother's portion with her. Household members usually eat two meals a day together.

With weaning comes change in the child's social status and relationships with family members. In recognition of these changes and as consolation or reward, parents make a special effort to provide plenty of good food, especially taro, pig and fish, for the child during this time. Provision of good food continues to be a primary means used by Maisin parents to reward their children throughout childhood.

CONCLUSION

The ethnographic data presented here show that cultural continuities exist between some of the beliefs and practices most important to Maisin adults and the manner in which Maisin infants are fed and cared for. They also show that, in addition to meeting the infant's physiological needs, infant care and feeding provide opportunities for interaction between infants and caregivers in situations that can serve as the beginnings of socialization. In these interactions the infant is introduced, through sensorimotor and affective processes, to actions associated with some of the important ideas of the culture. As the child's cognitive and social capacities develop, he or she comes to know and understand increasingly complex aspects of the culture, including the beliefs that inform the actions.

Ethnographic research can reveal the goals of socialization and the kinds of opportunities that are provided for socialization within cultures during the course of child development. Much more psychological research is needed in this area to clarify the processes of socialization and to show precisely what is understood about cultural beliefs and practices by children at various points in development.

REFERENCES

Barnard, K. (1978). *Nursing Child Assessment Training Scales.* University of Washington School of Nursing, Seattle.

Bell, R.Q. (1968). A reinterpretation of the direction of effects in studies of socialization. *Psychol. Rev.* **75**, 81-85.

Benedict, R. (1938). Continuities and discontinuities in cultural conditioning. *Psychiatry* **1**, 161-167.

Bower, T.G.R. (1974). *Development in Infancy.* W.H. Freeman and Company, San Francisco.

Caudill, W.A. and L. Frost (1973). A comparison of maternal care and infant behavior in Japanese-American, American, and Japanese families. In W.P. Lebra (Ed.), *Youth, Socialization and Mental Health. Mental Health Research in Asia and the Pacific.* The University Press of Hawaii, Honolulu, vol. 3, pp. 3-15.

Freedman, D.G. (1974). *Human Infancy: An Evolutionary Perspective.* John Wiley and Sons, New York.

Harkness, S. and C. Super (1982). The development of affect in infancy and early childhood. In D. Wagner and H. Stevenson (Eds.), *Cultural Perspectives on Child Development.* W.H. Freeman and Company, San Francisco, pp. 1-19.

LeVine, R.A. (1974). Parental goals: A cross-cultural view. In H.J. Leichter (Ed.), *The Family as Educator.* Columbia University Teachers College Press, New York, pp. 52-65.

Mead, M. and N. Newton (1967). Cultural patterning of perinatal behavior. In S.A. Richardson and A.F. Guttmacher (Eds.), *Childbearing: Its Social and Psychological Aspects.* Williams and Wilkins, Baltimore, pp. 142-244.

Piaget, J. (1952). *The Origins of Intelligence in Children.* (M. Cook, Trans.) International Universities Press, New York.

Zigler, E. and I.L. Child (1973). *Socialization and Personality Development.* Addison-Wesley Publishers, Reading, MA.

CHAPTER 8

THE SOCIAL CONTEXT OF INFANT FEEDING IN THE MURIK LAKES OF PAPUA NEW GUINEA[†]

KATHLEEN BARLOW

INTRODUCTION

Patterns of infant feeding and growth in modernizing societies do not exist in isolation from the cultural and socio-economic environments in which they occur. They result from beliefs and values which operate within particular ecological environments and socio-economic situations (Barry, Bacon and Child, 1959; LeVine, 1974; Whiting, 1975; Whiting and Child, 1953). By investigating traditional practices and contexts for infant care and feeding, their effectiveness, and the meanings that underlie them, the impact of changes in many arenas (social, political and economic) on the health and well-being of infants and mothers can be understood better.

This discussion concerns the social context of infant care and feeding in the Murik Lakes of Papua New Guinea, a fishing and trading society located near the mouth of the Sepik River. The Murik have well-formulated ideas of how pregnant women, nursing mothers, and infants should behave and be cared for by family and community. Many features common to Melanesian societies are present: support of close kin for mother and infant; high social and symbolic significance attached to feeding and food substances; belief that infancy is a period of exceptional vulnerability; and specific arrangements for and

[†]This research was supported by a dissertation research grant from the University of California, San Diego, the Wenner-Gren Foundation for Anthropological Research and by the Institute for Intercultural Studies. Research affiliation in Papua New Guinea was with the Education Research Unit of the University of Papua New Guinea.

valuation on spacing children two or more years apart. (See Doan[†] and Obrist[‡], unpublished observations, 1983; Tietjen, ch. 7.) Among the Murik, feeding a mother and her infant has social significance far beyond the necessity of providing adequate food. Claims to membership in social groups are acted out in this arena, introducing elements of competition and contention that sometimes improve and sometimes strain caretaking arrangements. Imbalances in family composition, mobility and consequent labor shortages also present problems in providing for mothers and infants. To some extent these problems are by-products of the ecological and socio-economic conditions of Murik life, and to some extent they involve recent changes resulting from participation in the cash economy.

This paper describes the general conditions of Murik life and the ideal norms which they espouse for cooperation and support of pregnant and nursing mothers and infants. A case in which an adoption claim was made on the newborn child illustrates the symbolic importance of feeding as a way of negotiating membership in social groups and obligations among kin. This case also demonstrates the influence of family composition, labor shortages, mission and school influences and outmigration on infant feeding and caretaking.

METHODOLOGY

This research is part of a larger study of how Murik children become socially competent adults. The data were gathered during 17 months' field work between February, 1981, and August, 1982, while resident in a Murik village. The data on ideal practices, norms and expectations about infant care were collected from focused interviews with parents and grandparents, and from comments made in the course of other conversations and activities about how things ought to be, once were or might have been. Observations of on-going events provided contrastive data about actual situations. Informants were asked about their reasons for particular actions and their points of view on the activities. I was particularly fortunate to be adopted into a family which had several young childbearing women and to be included as a

[†]Doan, H.M. (1983). Infant care and breastfeeding in selected villages in Papua New Guinea: "Traditional" and recent practices. Paper presented at the twelfth annual meeting of the Association for Social Anthropology in Oceania, New Harmony, IN.

[‡]Obrist, B. (1983). The study of infant feeding: Suggestions for further research. Paper presented at the twelfth annual meeting of the Association for Social Anthropology in Oceania, New Harmony, IN.

listener when they expressed their personal opinions about what was happening. These conversations helped to differentiate expected or normative actions from accommodations to the situation at hand.[†]

MURIK SOCIAL LIFE

Ecology, Family Structure, and Social Change

The Murik are a small population (about 1400) of fishermen and traders living in a precarious ecological situation. The villages are built on narrow sandspits which separate the open ocean from vast inland mangrove lakes near the mouth of the Sepik River. Travel among the villages is by canoe, except for one village which can be reached by road from Angoram, a small town on the Sepik River. The abundant waters of the ocean and lakes provide their primary and fiercely guarded seafood resource. Land, fresh water, staple starch (sago) and other vegetable foods are often in short supply. The Murik routinely go to houses built deep in the mangroves to fish intensively and accumulate a surplus for trade or market. Trading activities take adult workers away from the village, sometimes for long periods. These temporary absences result in labor shortages for those who stay at home. The sexual division of labor is, therefore, not a strict one. Men and women share many tasks in order to meet the daily demands of subsistence and childrearing. This problem is sometimes more acute because of prolonged visits to town and out-migration for work and school.

Family composition is flexible and a matter of dispute. The core social unit is the sibling group. There is competition to recruit children into the sibling group of each of their parents. An important means to advance a claim is to feed the child. Social acts of parenting are considered more important than biological parentage. Marriage is brittle, and most people marry several times in the course of their adult life, whether due to divorce or death of one spouse. Families are very often composed of children of different parents and include adopted children. Individuals commonly maintain multiple ties to actual, step and adoptive parents, and an individual's loyalty, affection and work may be claimed by several sets of kin.

[†]The personal names have been changed. Traditional names have been replaced by other traditional names, not necessarily belonging to the same family. Mission names have been replaced with other Christian names.

F

The negotiation of claims over membership in different groups begins even before a child's birth. The key element in establishing a claim is feeding the pregnant woman, and later the mother and infant. Mother's and father's siblings, grandparents and potential adopters proclaim their loyalties and intentions by contributing or failing to contribute to feeding a mother and infant.

Western influence has come in several forms. The Catholic mission, which has been in the area for over 60 years, is now quite tolerant of traditional ways, but non-responsive to the Muriks' entrepreneurial approach to Christianity. The Seventh Day Adventist Mission runs a school in the central village — where we worked — and is adamantly opposed to traditional religious and ritual practices. Most people in this one village send their children to the mission school, though they do not practice the SDA religion. The school occupies much of the children's time, reducing their availability for caretaking of younger siblings, helping with household chores and fishing.

The Murik also participate in the cash economy. A few families in each village own outboard motors. They earn cash income by taking smoked fish and fresh shellfish to Wewak, the provincial capital, some 75 miles to the northwest on the coast. Although the distance to town is not great, bad weather and unreliable outboards make travel to town for market, schools and medical facilities difficult.

There are three small Murik settlements in Wewak which are also perched on thin strips of beach and resemble the lake villages. The town Murik are sometimes employed in business. From time to time they return to the village to fish intensively and bring smoked fish to the market. Others sell carvings and baskets to tourists. Most have children in school.

The Significance of Food and Feeding

The Murik diet is very monotonous, but rich in protein. The daily fare is two meals of fish and sago. There are occasional shortages of staple foods when many people have only one small meal or nothing for a day or two. (Children are then fed first, but not as much as they would like.)

Fruits, nuts, vegetables and meats — occasionally available through trade, small-scale gardening efforts, hunting and foraging — are highly prized. A few items, mainly rice, oil and flour, are purchased to fill in when trade and fishing are hindered by weather and social circumstances or to add to the special foods at a ritual meal.

Obligations to give food extend far beyond the basic requirements of providing for the immediate family. Every ritual occasion, from introducing a newborn child to the community to ending the mourning period, involves an exchange of food. Arguments are settled by sharing a meal. Work is compensated by feeding the workers. Ceremonies and performances of songs and dances require presentations of food to the performers. Generosity and personal prestige are measured in terms of the ability and willingness to feed others. The meaning of food exchange is that the food-giver is superior and abundant, like a mother, while the receiver is inferior and dependent, like a child.

Special arrangements to feed an expectant mother or a mother and newborn child go on for months at a time and compete with the other occasions on which it is appropriate to give food. Members of the extended family and community evaluate a family's interest in a new child by the care given to mother and child before and after birth. Conflicting adoption claims may arise when the family appears unwilling (in a case of suspected illegitimacy or antagonism between the spouses' families) or unable to care for them adequately during this period of dependence. The child's new family membership is established by feeding both mother and child.

IDEOLOGY OF PREGNANCY, BIRTH, AND INFANCY

The image of a nursing mother with a plump, healthy baby at her breast embodies the Murik ideal of love, nurturance, generosity and bliss. Ideas about the care of pregnant women, new mothers and babies are well-formulated. The mother and child, before and after birth, are cherished, indulged and, of course, well fed. The following description of the ideal circumstances for pregnancy, childbirth and the seclusion following childbirth is a composite of descriptions told by men and women, young and old, and of comments made when actual circumstances fell short of the ideal.

Pregnancy

Ideally, children are born at least two or more years apart. Parents should abstain from sexual intercourse until the child is walking and talking. There are several reasons for this prohibition. Some senior people say that it is to safeguard the husband's strength and health from the woman's polluting influence. Others say that the mother's milk will be spoilt and the nursing child will become sick. Child-bearing women point out that it gives the mother time to recover her strength between births and to give her other young children adequate attention and training. Women whose children are four and five years apart are praised for giving superior care to each child during its early years.

A pregnant woman is particularly vulnerable to dangerous spirits and enemies of one of the parents or grandparents and, therefore, should not discuss her condition except with her mother and older sisters. She must stay close to home and be circumspect in all behavior in order to avoid drawing the attention of some person or spirit with evil intentions who may make her miscarry or cause the child to be deformed or stillborn.

The expectant mother should eat only the best quality food — fresh, unblemished and well-formed. Certain foods must be avoided altogether since they may affect the child's appearance and/or personality.[†] The mother may eat as much or as little as she likes. Others in the family must make an extra effort to provide her with whatever food she wants. Her younger siblings find and husk dry coconuts, which pregnant women crave.[‡]

[†]The foods prohibited to pregnant women have a sympathetic relationship to the personality and appearance of the child. Most of the foods which endanger the baby's health are uncommon, strange in appearance by Murik standards or very seldom available. Animals from the bush are suspect. Flying fox is prohibited for fear the child will have shrivelled buttocks or become a person who steals by night. Small nocturnal marsupials may not be eaten lest the child have big, bulging eyes and be greedy for other people's food. Certain kinds of fish with grotesque shapes are avoided for fear the child will be similarly disfigured. Other fish are the animal vehicles of malicious spirits who will kill the child *in utero* by stealing its soul substance. The mother may not eat anything twisted or deformed or the child may have a twisted limb. *Laulau*, or the Malay apple (*Yambosa gomata*), a seasonal bell-shaped fruit, is thought to cause big ears and a tendency to eavesdrop on other people's affairs. In addition to these restrictions, the pregnant woman should not witness burials, planting, knots or lashings being tied as these activities may delay the birth and cause the child to suffocate or be imprisoned in the womb.

[‡]Pregnant women are said to both crave coconut meat and eat it in large quantities. This may be because coconuts are high in calcium, which is otherwise lacking in a diet of fish and sago (Townsend, personal communication, 1982).

As the pregnancy progresses, the woman stays at home, relieved of work by her husband, her own mother and siblings and her husband's mother and siblings, in that order. For the last few months before the birth, she rests, sleeps and eats. Her mother and sisters take over her household duties. Her husband supplies whatever food she wants to eat, including any special cravings for a particular kind of fish or shellfish, etc. He is eager to do these tasks out of devotion to his wife and enthusiasm over his new child.

Birth and Early Infancy

Shortly before delivery, the pregnant woman goes to the birth house located a short distance outside of the village. The birth house, a specially built house away from the village, is built by the husband as a further sign that he cares for his wife and their new child. Several men and women described such a house. It should be large and airy, with two hearths so that visitors may gather for conversation and cooking without disturbing the mother and child. It should have a verandah so they may sit outside on warm afternoons. The mother's husband and her own family should make sure that water, firewood, mosquito nets, bedding and a lamp are available for use. Older women referred to the large, comfortable houses which their husbands had built for them. Young women described the ones their husbands would have built if time and resources had been available.

An experienced woman gives birth by herself and does not cry out in pain. A first-time mother is assisted by her mother, aunt or a midwife who coaches her and ties the cord. The mother disposes of the placenta by burying it and washes the baby and herself. A "strong" woman carries out as much of the birth procedure as she can. She spares any other woman from becoming contaminated with the pollution of birth and thereby passing it along to any men in her own household when she returns home.[†]

[†]As in most of Melanesia, elaborate ideas about the polluting effects of menstruating and postpartum women on male strength and health (and sometimes on other women as well) require that women separate themselves at these times. However, among the Murik these ideas are not strictly enforced. It is believed that men who become polluted by women will have shortness of breath and arthritic knees in their old age — both very common afflictions. Thus, most older men seem to be suffering the consequences of earlier carelessness on the part of both men and women.

The woman and baby remain in seclusion for several months until the child is both fat and strong. Her younger sister or sisters stay in the birth house to act as personal maids and messengers, providing her with a prolonged period of rest and relaxation. Her mother and sisters cook the best food that is available and deliver it twice a day. Other women come to visit, bringing presents of food such as green coconuts, bananas, nuts and special greens or fruit. These gifts are meant to strengthen the mother and child and to celebrate them.

At first the new mother's only task is to satisfy the child's every need. She keeps the baby clean, dry, and warm or cool enough. She has a small fire to warm the baby's bath water and, after bathing it, warms her hand over the fire and gently straightens its arms, legs, toes and fingers. She guards the baby from mosquito bites and makes sure that it sleeps with a gentle breeze wafting across its face.

The mother feeds the baby on demand, paying careful attention to how well and how much it drinks, always assuming that more is better. As soon as the umbilical cord falls off (in about seven to ten days), the mother begins feeding the child sago pudding dipped in water or fish broth. The Murik consider this food "the bones of the ancestors," and eating it is an essential part of becoming Murik. The mother feeds this to the baby before she nurses it, pinching off tiny bites which slide down easily until the child refuses to swallow any more. In a few weeks the baby fills out with a pleasing layer of rolling fat. Murik women then feel the child has the strength to withstand an illness, such as a malaria attack, and are reassured of its stamina.

As the mother's strength returns, she begins weaving baskets for the women who are helping to feed and care for her. She is indebted to them for their work and repays it by weaving many baskets which they will later sell or trade.

When the baby is "strong," the mother and child leave the birth house for her parents' house. There they are taken care of by her husband, mother and siblings for several more months.

The overall result of these arrangements for mother and child is a period of nurturance for both of them. As the mother lavishes care and attention on her new infant, her kin lavish care and attention upon her. The hoped for result is a healthy, strong and charming baby who will become a productive, hard-working and generous member of society.

Adoption

Adoption is quite frequent (between 20% and 25%) and is a means of adjusting the size and sex ratio of a family's labor force. There are well-defined expectations about the role of future adoptive parents during the months before and after birth. A couple who wishes to adopt a child (usually their brother or sister) works for the natural mother during her pregnancy and postpartum seclusion. By supplying her with fish, sago, coconuts and firewood, they demonstrate their willingness to do the work of raising a child. They also relieve the husband and mother's family of working to support a child who will not be theirs.

After the child is born, the adoptive mother takes the child and encourages it to suck until her milk comes in so that she may relieve the natural mother of all work for the child and be able to claim that she has fed the child herself from its earliest days.[†]

When the natural mother and the child emerge from seclusion, the adoptive parents make a small feast for them to pay for the child (now about K20,[‡] formerly pigs' tusks or dogs' teeth ornaments). This ceremony is called "cutting the breast" and precludes all future claims on the child by the natural parents. No one may tell an adopted child about his or her natural parents until the child is grown and has children of his or her own. A grown child who is told of his or her adoption will deny its natural parents, saying "I have no other parents than the ones who raised me."

These ideals about pregnancy, birth, early infancy and adoption are the standards by which actual performance is evaluated. The more closely they are approximated, the less likely others are to question the membership of the new child in its family, and the less likely they are to propose to support alternative claims.

[†] Informants assured me that induced lactation is possible, but they could not remember any specific instances. They remembered one case in which a newborn who was one of twins was given to a woman whose child had died in childbirth only the day before. This child was nursed by its adoptive mother from the very start. In one other instance, an adopted baby whose mother died in childbirth was nursed by its adoptive mother's sister. She was weaning her second child (about 2 ½ years old) and drank lots of water and coconut juice to increase her milk supply for the new baby. Other women said they had raised adopted children from the age of a few weeks on sugar cane boiled in water and sago pudding.

[‡] K = kina. One kina = U.S. $ 1.14.

CASE

Given the conditions of social life and the standards for ideal care, it is interesting to examine the ways in which people handle the differences between what is desirable and what is possible in actual situations. The following case describes how one family managed the care and feeding of a woman and her new baby, from pregnancy through postpartum seclusion. The case is typical in the way the idioms of work and food are used to negotiate claims on the new child. It is atypical in that this family was faced with more severe problems than most.

Kuma and Aupai had six children under the age of 14. Both parents had grown up in the village, but Kuma's parents had divorced and left the village. Aupai's natural parents still lived in the village, and he retained some ties with them. His adoptive parents had moved to Wewak, leaving three teenage children in the village to attend the mission school. Aupai and the three younger step-siblings were adopted from different families. Aupai and Kuma's own house had fallen into disrepair, and they all lived in the house belonging to the adoptive parents in Wewak.

Aupai and Kuma took responsibility for feeding everyone, and the three teenagers, though in school, helped from time to time. Kuma was unable to make real demands on the two girls, Moinso and Aran, because they were affines and, therefore, had to be treated with respect. Their own oldest son, who was old enough to provide real help, was out of the village attending school. They not only did without his labor between vacations, but worked hard to earn the money for his school fees, room and board.

Fortunately, Aupai had a gill net which enabled him to catch fish in quantity, and he had several adult brothers in his natal family with whom he cooperatively made trips to trade for sago. He also owned many coconut trees, planted for him by his natural and adoptive fathers. Nevertheless, he and Kuma worked hard to support eleven people. There was constant arguing over food at their house.

When Kuma became pregnant again, it was clear to many people that they would be hard-pressed to support another child. Her last child, a boy, had been adopted by her older sister. Several people mentioned that they might be interested in adopting the baby. However, the couple had only one daughter, and they were definite about wanting to keep a female child. The interested parties waited to find out the sex of the child before offering food or help.

Kuma worked hard throughout her pregnancy and was chopping firewood only a few hours before the child was born. Aupai had no time to repair the rickety shack that passed as a birth house. He had difficulty finding time to repair their own delapidated house so that when Kuma left the birth house she would not have to return to his adoptive father's house. (Both of them were aware that the older man would probably refuse to return to the house if Kuma came there from the birth house.)

When it was time for the birth, Kuma went to the teacher's wife at the mission and asked her to deliver the baby. Then she waited in her house until the contractions were close together and went surreptitiously to the birth house. The teacher's wife was watching for her and followed, delivering the baby about 15 minutes later. She called to one of Kuma's neighbors who was walking within sight of the birth house and asked her to bring a bucket of water for washing and a cloth to wrap the baby. (The woman later suggested that since she had done part of the work at the birth, Kuma should consider giving the child to her childless younger sister. This claim was weak because the work was trivial and never followed up with food.)

Later, Kuma's own children brought her some clothes, bedding, a mosquito net, food and fresh water to drink. Aupai himself cooked for her. Moinso and Aran moved into the birth house to enjoy special status as assistants. They resisted Kuma's efforts to send them home to help Aupai cook and watch the younger children. As a result, three or four of the other six children were usually in the birth house with Kuma and the baby. They made their first appearance there while she was still having contractions following the birth.

Aupai called on his natural mother for temporary help, and she sent some sago bread and fish. In a few days, his adoptive mother arrived from Wewak and took over the cooking and some of the childminding. She was also gone for hours each day fishing. Kuma remained in the birth house only until the baby's umbilical cord fell off, then went to the house which Aupai had repaired and resumed the duties of cooking for the family.

Others were critical of her for endangering the men of the community by returning to the village so soon after giving birth. They did not offer to help take care of her family so that she could remain in seclusion. A few days later, Kuma and Aupai had a fight over another woman. When he chased her with a stick and she ran across the village, the older men and women were horrified and screamed at both of them. For several days, they talked of little else, and pointed to their large family of closely spaced children as a disgrace.

The Adoption Claim

The new baby was a boy, small and thin, but otherwise healthy. Kuma sighing said, "If I hadn't had to work so hard he would have been big and fat like the other children." Several people sent food the first evening and it was rumored they might ask to adopt the child. Matthew and Delilah only had two children, the younger of whom was about six years old. They asked for the baby on the basis of Delilah's father and Aupai's natural mother being siblings.

After a few days, when the senior men in each family and the mother had agreed, Delilah began to bring meals to the birth house. Observers inquired disapprovingly to what extent Delilah and Matthew had done the hard work they should have to acquire the child. Kuma and Aupai were clearly hard-pressed to care for seven small children and unlikely to change their minds about the adoption.

Kuma began to tell women visitors that Delilah and Matthew would take the baby as soon as they could get formula and bottles. They did not try to do this. (They would have had trouble getting a prescription to do this, and they could not afford the expense.) Kuma continued to nurse the baby and complained that they wanted the child but did not want to do the work.

Delilah tried nursing him to get her milk to come in and was unsuccessful. She provided the baby's cooked food and assumed responsibility for watching him while he slept in the afternoons. When the baby was about two months old, Delilah became ill, and Kuma again took full responsibility for the child. She was clearly becoming very attached to this baby and was proud of his rapid growth. All of the older siblings were well aware that he was actually their brother but was being adopted. They went with their mother on regular afternoon visits to the adoptive parents' house. In a few weeks when she began to recover, Delilah again cooked for the baby, and her husband and two children played with him.

DISCUSSION

Kuma and Aupai's situation was difficult and complicated. Some elements of the idealized situation were present, but in other ways the case diverges widely. The supportive kin group recognized some obligations but not others. In addition, the positive and negative effects of recent change brought about by the cash economy, missions and schools are revealed.

Family Composition

One source of difficulties was the make-up of the family. Ideally there should have been three sets of willing caretakers — a devoted husband, Kuma's parents and siblings, and her husband's parents and siblings. They would also have had children spaced far enough apart that the older ones could help with minding younger ones and could do some of the domestic work. The ideal also assumes a balanced sex ratio among children, so they would not have had to postpone the decision to adopt the child until after it was born. This delayed decision-making about adoption is not uncommon, but it did deprive them of much-needed help during Kuma's pregnancy.

Absent Personnel

The Murik are well accustomed to substituting personnel when circumstances require it. Adoption is basically one means of making permanent adjustments in a family's labor supply. In caring for mothers and infants, there is a hierarchy of resort to ensure a sufficient support network. The first rank caretakers are the husband and the mother's mother and sisters. If they are unable to work or are temporarily absent, the husband's family is expected to and usually does provide help. What is unusual about Kuma and Aupai's case is the compound problem that both her parents, her only sister and his adoptive parents no longer lived in the Murik villages and were of little help. Kuma occasionally talked during her pregnancy of going to stay with her father or mother until after the birth; but getting to Madang, a coastal town to the south, was an expense they could not afford. The second rank caretakers (Aupai's adoptive parents) were also living in town and helped only by allowing the family to live in their house in the village.

In the absence of these people, Kuma and Aupai received some help from his natal family — parents and brothers — in return for work which he did with and for them. They mostly relied on their own resources, which meant that Kuma's pregnancy and postpartum seclusion were not periods of rest and relaxation. Often women who have had one or two successful births do work throughout most of their pregnancy but are relieved of most work in the last six to eight weeks. They also have a few weeks in the birth house to recover and care for the infant without the demands of minding all of their children.

Adoption

Aupai's status as an adopted son meant that both his adoptive and natural parents were inclined to let the other ones perform the grand-parental role. There was a history of arguments over him. Both sets of parents felt short-changed because Aupai maintained ties with the other set, though neither would relinquish their claim on him. They expected him to help support them in their old age. This is a common problem for adopted children who sometimes benefit by receiving food, coconut palms, land and occasional help from two sources, but are in turn the object of claims and conflict. Aupai's new son would no doubt find himself in a similar situation.

The three teenage siblings were only recently adopted from a different family than Aupai. They did not feel obligated to help raise Kuma and Aupai's youngsters. Their relationship to Kuma as affines provided a certain distance. She could not make requests of them, and the two girls used this status to advantage. The older boy spent hours spear-fishing after school to help provide enough food. These siblings, though old enough to work, were more a responsibility than a resource. Spending most of their day in school, they added to the burden of work and were a significant drain on the family's food resources. Though Kuma had enough to eat, she did not receive the quantity or variety of food that she would have under more ideal circumstances.

Spacing of Children

Kuma and Aupai's children were not well-spaced. The couple did not observe the traditional postpartum sex taboo. This is a matter of great variation, discussion and criticism. Jealousy over other women is common and works against the taboo. On the other hand, pressure from other women to be a good mother and have children far apart is great. The Seventh Day Adventist mission views pollution beliefs as backward and Satanic. The influence of this view on the spacing of children is difficult to gauge. Kuma and Aupai were members of the mission. They gave their religious affiliation as the reason he violated pollution restrictions by coming to visit her at the birth house, sitting on the steps down below to talk to her. Given the lack of help from other kin, it was also partly pragmatic for them to violate traditional prohibitions

Having many young children increased the burden on the parents. Kuma's role, even during late pregnancy and the postpartum period, was mainly that of food supplier and caretaker, rather than that of recipient of these services.

Labor Shortage

The family suffered from a severe labor shortage. Their situation was worse than most because there were so many dependent children in the household. They had a very limited group of kin in residence who were willing to help.

The effects were multiple. Kuma was never relieved of much work at any time. Few preparations were made for the birth. According to the ideal, household necessities and personnel to care for mother, infant and other children should be arranged in advance. But for various reasons this is not usually done. Women do not want to draw attention to themselves at this point. They dread a flock of gossiping women joining them in the birth house and are also extremely vulnerable to magic, sorcery and mischievous spirits. Other people are not eager to risk the effect of pollution surrounding birth, which is one reason the birth house remains delapidated. Kuma's situation was even more haphazard than usual because there were no adult female kin to organize things for her immediately after the birth.

Her own children were frequently in the birth house, and she looked after them while attending to the new baby. Ordinarily they would have been allowed to visit one or two at a time, but not until the day after the birth and not for the entire day. One of the younger sister's jobs as attendant is to monitor the coming and going of visitors so the mother can rest. Most of the time Kuma had no such gatekeepers and no grandmother at home to mind her other youngsters. Under these circumstances there was no practical reason to remain in the tiny, unstable birth house. She returned to her own house in the village too soon after the birth and was severely criticized for it.

School, Mission and Work

The overall effects of modern changes increased the difficulties for Kuma and Aupai during this period. They had to support a child living out of the village at school. He was an important investment in their future and would eventually help to support them. At present his school fees were a major expense which required them to accumulate surplus fish and crab to sell at the market. They could certainly have used his help if he had been there. School occupied Aupai's teenage siblings, making them less available to work even to support themselves.

The school teacher's wife was an important person for Kuma at the time of the birth. Without the appropriate kin group, it was very fortunate for her that the teacher's wife had been trained both by her own mother, a midwife in her home village on one of the coastal islands, and at a government hospital. She delivered the baby skillfully and would have known what to do if there had been any complications. The couple's mission affiliation distanced them somewhat from Aupai's natal parents who were quite traditional. Their willingness to disregard the pollution restrictions which others took quite seriously made them the objects of stern criticism.

The main effect of cash economy participation was to draw people from the village to town. This meant the absence and dispersal of a large number of potential caretakers for Kuma and the new baby.

Feeding

Though work, rest, housing and help fell far short of the ideal, there was an effort to provide adequate food. The baby was small, indicating that Kuma's food supply during pregnancy was adequate but not plentiful. She was able to get enough to eat mainly through her own and her husband's efforts.

The adoption claim hinged entirely on feeding. Even though the adopting parents did none of the other work they should have, the community and the family considered their efforts sufficient. Whatever other problems intervened to complicate the lives of the natural and adoptive parents, both mothers made feeding the baby a priority. Seven months after his birth, the baby was plump, alert and cheerful, though perhaps still a bit small for his age.

SUMMARY

Murik life presents problems in providing what they consider to be ideal care for mother and infant. Compared with the difficult reality, the ideal period of indulgence and abundance for mother and child seems more like a wish than a realistic goal. In actual situations, they are experts at making contingent arrangements to carry on under difficult circumstances. Recent changes such as school, mission and cash economy add to the difficulties. Young parents must often overcome the problems of absenteeism, numbers of dependents and a heavy work load.

The traditional emphasis on food and feeding is an important part of managing these problems. In every arena of social life, caring, interest, membership and moral worth are expressed through giving food and feeding. Ideal care for mothers and infants requires a variety and abundance of high quality foods which for many reasons are not easily obtainable. Whatever the composition of the family and the available labor supply, the social significance of feeding places a high priority on the feeding of mother and infant. Thus, even in times of scarcity, illness and family instability, the outcome of multiple social pressures is to provide adequate food to mother and child.

ACKNOWLEDGEMENTS

I wish to thank the other participants in the ASAO Session on Infant Care and Feeding in 1982 and 1983, especially Leslie Marshall, Achsah Carrier, Dorothy Counts, Mary Maxwell Katz and Anne Marie Tietjen for their very helpful comments on earlier drafts of this paper.

I would also like to thank my husband, David Lipset, for his encouragement and insights into "things Murik".

Any errors and misinterpretations are, of course, my own and not theirs.

REFERENCES

Barry, H., M.K. Bacon and I.L. Child (1959). Relation of child training to subsistence economy. *Amer. Anthropol.* **61**, 51-63.

LeVine, R.A. (1974). Parental goals: A cross-cultural view. In H.J. Leichter (Ed.), *The Family as Educator*. Columbia University Teachers College Press, New York, pp. 52-65.

Whiting, B.B. (Ed.) (1975). *Six Cultures: Studies in Child Rearing*. John Wiley and Sons, New York.

Whiting, J.M. and I.L. Child (1953). *Child Training and Personality*. Yale University Press, New Haven and London.

CHAPTER 9

INFANT CARE AND FEEDING IN KALIAI, WEST NEW BRITAIN, PAPUA NEW GUINEA

DOROTHY A. COUNTS

INTRODUCTION

The problem of providing infants in developing countries with adequate nutrition is receiving widespread attention. Government health services in Papua New Guinea are cognizant of the importance of proper infant nutrition and are actively discouraging the practice of bottlefeeding. In West New Britain Province, posters describing the advantages of breastfeeding adorn the walls of local health clinics, and the paraphernalia required to bottlefeed an infant is available only from the clinics or by prescription. Government promotion of breastfeeding is a response to the vulnerability of modernizing nations, such as Papua New Guinea, to a host of factors that establish a context in which women in other parts of the world have replaced breastmilk with other foods (Harrell, 1981). These factors seem to accompany increased affluence, educational opportunities, urbanization, commercialization, availability of medical care (Gonzales, 1963; Ghosh *et al.*, 1976; Marshall and Marshall, 1980), and especially the wage employment of women (Nardi, ch. 16; Jimenez and Newton, 1979; Nerlove, 1974; Van Esterik and Greiner, 1981; Bradley and Peberdy, 1981). They include the notion that the female breast is primarily an erotic appendage; a concern that human milk is inadequate in quality or quantity and should be supplemented with cow's milk or semi-solid foods; the idea that infant feeding should be spaced several hours apart and eliminated at night as soon as possible; and the belief that adults and children require separate space and time (Harrell, 1981).

In 1971, the town of Kimbe was created as a marketing center for West New Britain. Since Kimbe's establishment, its presence has brought many of the appurtenances of modernization to the villages of the province. The effects of modernization on the social-cultural context of infant feeding among the people of an isolated rural community in northwest New Britain will be discussed.

METHODOLOGY

Since 1966, the author has spent 27 months doing field research in West New Britain. Most of this research has been based in the Lusi-speaking village of Kandoka, the largest village in the Kaliai electorate and the second largest community on the northwest coast of the island of New Britain (Figure 1). Between June and September, 1981, data were collected specifically on infant feeding and nutrition. Most of the analysis in this paper is based on data collected in 1981, supplemented by data obtained during earlier field research which was not focused on the subject of infant nutrition.

The research method used was participant observation. During four visits to West New Britain, the author and her family lived in the village, observing behavior and participating in community life. In 1981, we shared a living compound including a cook-house and fenced enclosure with another family. The woman of that household was barren, but she and her husband adopted a number of children; she consistently babysat for the children, including the still nursing infants of her brothers. Consequently considerable time was spent in the company of nursing mothers and their infants. Ample opportunities were offered to observe the interaction between women and their children and to discuss, in a relaxed and informal context, the subjects of human fertility and child care. These discussions were held with approximately 50 village women, nursing mothers and non-lactating women alike.

In 1981, there were 360 people living in Kandoka and its associated hamlet of Maiai. Twenty-five of these people were nursing infants who ranged in age from those born between June and September, 1981, to toddlers, aged three. The interaction between all of these children and their mothers was observed during the months spent in the village. Discussions were also held with 12 nursing mothers and others on infant feeding, wetnursing, the introduction of solid foods and other related topics.

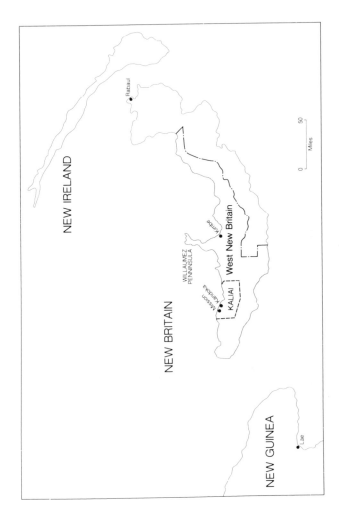

FIGURE 1   Map of West New Britain Province.

THE KALIAI AND MODERNIZATION

The people of Kaliai constituency of West New Britain Province are isolated. There are no roads or airfields in Kaliai. The only transportation is by foot travel along forest trails or by navigation of the rivers and coast. In 1966, at the initiation of this research in Kaliai, the only contact with the outside world, represented by the city of Rabaul some 300 miles to the east, was by commercial trading vessel, small mission launch or an occasional government work boat. These links were infrequent and undependable, and months would pass when no ship visited the area. Consequently, the people of Kaliai were minimally influenced by the economy, technology and values of modern industrialized society. This situation began to change rapidly in 1971, when Kimbe town was established at the eastern base of the Willaumez Peninsula. The town provides a readily accessible market for locally grown copra, is a source of supplies for local trade stores and is the site of the provincial capital, high school and hospital. In 1966, Kaliai had only one locally owned and run trade store, there were no outboard motors or cooperative youth clubs and few women had lived outside the village. The situation altered dramatically by 1981. In Kandoka there were six outboard motors, three trade stores and three young peoples' clubs, one of them a women's club. There were a number of village women who had attended or were attending high school, and one — the first woman from Kaliai — was a biology student at the University of Papua New Guinea. Although Kaliai is still far from being an industrialized or urbanized area, the influx of goods and ideas has increased from a trickle to a torrent, and villagers consider themselves to be citizens of a modernizing country.

To what extent has this modernization affected the lives of women and children, and particularly what effect has it had on child nutrition in Kandoka? New foods have been introduced into the diet of adults and older children, but few changes have been made in traditional infant feeding practices. For reasons discussed below, in 1981, there were no new social or economic activities competing with traditional domestic responsibilities for women's time, nor were there any social or economic pressures or incentives for earlier weaning or introduction of food supplements to breastmilk. (Also see Conton, ch. 6; Nardi, ch. 16).

BREASTS, MILK AND KINSHIP

It is difficult to know whether Kaliai men traditionally considered the female breast to be erotic. By 1981, thanks to the influence of the mission, movies seen in Lae and Rabaul, foreign magazines and other publications (one man had obtained a copy of the *Kamasutra*), younger men thought of women's breasts in terms of sexual play. The new eroticism is reflected in women's dress. In 1967, almost all village women wore *meri* blouses only on Sunday; in 1981, all women under the age of 30 consistently wore blouses, and more educated women wore brassieres even when they danced in ceremonies where they were supposed to dress in traditional costume. In spite of the introduction of foreign attitudes towards the female breast, Kaliai continue to think of it as *the* source of infant food. The importance of this function overrides any erotic symbolism associated with the breast. As evidence, lactating women continue to nurse children publicly, without self-consciousness, and men pay no attention. (For a similar case see Conton, ch. 6.)

The continuing emphasis on the female breast as the source of nourishment for infants is related to the fact that breastmilk is the substance of maternal kinship. The Kaliai are normatively patrilineal and emphasize agnatic links. Older Lusi-speaking Kaliai maintain that a child is composed entirely of its father's paternal substance *tanta aisuru*, "male substance" or "essence." (For a detailed discussion of this point and of Lusi notions of conception, see Counts and Counts, 1983.) "Aisuru" is a Lusi term used for those liquid substances which have the capacity to create social ties: semen or male substance, breastmilk (*turuturu aisuru*), and the fluid of the green coconut (*niu aisuru*). According to older Kaliai, a woman is kin to her child because she nurtures it. This critical nurturing is a sharing of substance that occurs, not when the fetus grows in its mother's body, but when she gives it her breastmilk. *Tutu*, the baby-talk name for mother in Lusi, is an abbreviated form of the word for breast.

This theory of procreation is changing, for younger educated Kaliai who have learned the modern European model of reproduction argue that children are related equally to both parents. Nevertheless, the relationship between nurturing, shared substance and parenthood is deeply ingrained in Kaliai kinship ideology. This ideology extends to kin other than immediate parents, for kin relations may be created by giving food as well as semen and breastmilk, an important aspect of adoption discussed below.

The nurturing role is a critical one for a Lusi woman because the primary responsibility of a woman is to feed her children and others who are dependent on her. The ability to create and express ties of kinship by giving food is not limited to women. Mother's brothers and adopting parents also establish their relationship to children in this way. Mother's brothers share substance with sister's children by giving them coconuts from palms which those children eventually inherit and which are the source of *niu aisuru*, coconut liquid.

Adoption, which is widespread throughout the Pacific (Carroll, 1970; Brady, 1976), occurs frequently in Kaliai. It is common for an individual or a couple to raise the child of a relative who has several dependent children. Adopting parents who ask for a baby while it is a tiny infant begin shortly after its birth to fulfill the two necessary duties of parents: to name the child and to feed it. They do this by giving food, especially fish, meat and drinking coconuts to its birth mother, for anything eaten by her is also consumed by her baby and these foods are considered to be especially strengthening. By providing for the nursing mother, the prospective adopting couple demonstrate their ability and intent to nurture the child and to meet their responsibilities as parents. Failure to meet either of these obligations — naming or feeding — causes the natural parents to terminate the arrangement. (See Barlow, ch. 8, and Tietjen, ch. 7, on the relevance of food for kinship in other Papua New Guinea societies.)

THE CONTAMINATION OF BREAST MILK

Mother's milk is considered to be the ideal food for infants. All village mothers breastfeed their infants. The child is put to the breast soon after birth, as suckling is thought necessary to produce milk. Women do not worry about the production of adequate milk, for milk flow can be readily increased by drinking lots of fluids, especially fish soup and green coconut liquid. Breastmilk is the only exclusively female effluvium that is not polluting. It can, however, be contaminated by male substance. The belief that sexual fluids, especially menstrual blood and the effluvia of afterbirth, may cause illness or death is reported widely throughout Papua New Guinea (Meggitt, 1964; Meigs, 1976 and 1983; Chowning, 1980; Goodale, 1980 and 1981; Kelly, 1976; Lindenbaum, 1972). Anthropologists have, however, seldom reported that semen may be considered to be potentially as dangerous as menstrual blood (Faithorn, 1975). This belief has significance for Lusi infant feeding practices.

As noted above, older Lusi maintain that a child is composed entirely of its father's semen. This point is debated by younger Lusi, but most agree that a successful pregnancy requires numerous acts of sexual intercourse, a notion that is widespread in Melanesia (Chowning, 1985; Jorgensen, 1983). Sexual activity between parents should continue until the fetus is "strong"; until it quickens. After that it is a matter of individual preference whether the parents continue having sexual relations. Semen does not seem to harm a child while it is in its mother's womb. After it is born, however, its parents should refrain from sexual activity lest the semen enter and contaminate the breastmilk through cords that are thought to link the uterus and breast. This contaminated milk is spoiled and may make a child ill, especially if the semen is from a man other than the baby's father.

This notion, that semen is polluting to young children when they ingest it in breastmilk, influences Lusi thought regarding postpartum sexual abstinence and the practice of wetnursing. All the Lusi with whom I talked agreed that the parents of a young child who is totally dependent on its mother's milk should refrain from sexual intercourse. This restraint should continue until the child has achieved a certain independence, variously expressed as when it is old enough to tell its parents of its dreams or when it is old enough to gather shellfish or try to spear small fish in the shallows: about three years of age. When a child has reached this stage, people anticipate no problems should a mother have another baby. In fact many women opined that this was the ideal spacing of children.

Wetnursing is a long established practice in Kaliai. Older informants said that, traditionally, a child whose mother had died or who was otherwise unable to suckle it was either passed around among nursing women or adopted by a lactating mother whose child had recently died. Today such children can be fed by bottle with formula obtained from the medical centers, and they may be kept by their parents or adopted by close kin. Wetnursing services are usually exchanged between women who are closely related: sisters, sisters-in-law, mothers and daughters. These women (and their husbands) already have a parental relationship with the child who will call them "mother" (or "grandmother") and go to them for food. Other women, especially those who do not have close relatives living in the village, would not consider wetnursing because, they say, the milk of another woman might be "dirty" (contaminated with semen) and make their babies sick.

Semen is not the only substance that can contaminate mother's milk. Older Lusi say that when they were young there were many foods — particularly red- or yellow-fleshed fish, giant fruit bats and wallaby — that pregnant or lactating women and their husbands might not eat because the fetus or nursing infant might assume specific undesirable characteristics of these animals. For example, if either parent ate the flesh of the flying fox, the child might be mentally defective or it might shake and tremble as the animal does, or it might be unable to sit at the normal time because the animal does not sit erect. A pregnant woman did not eat wallaby because the child might develop epilepsy and have seizures during the full moon and the mother might have false labor. Neither parent ate the flesh of this animal lest the legs of the child be weak and underdeveloped in imitation of the front legs of the creature. I know of no one who follows these taboos today, but the fact that both parents were prohibited from eating certain kinds of food suggests that the Lusi tacitly recognize several things:

A pregnant woman shares substance (including some part of the food she eats) with her developing child.

Food eaten by a man enters his semen. This, in turn, contaminates a lactating mother's milk when the couple has sexual intercourse.

Parents of infants were not expected to obey totally the post-partum taboo on sexual intercourse; otherwise there would have been no reason to place food restrictions on the father.

CHILDREN, FEEDING AND DAILY ACTIVITIES

Infants are fed on demand and children of all ages snack frequently during the day instead of being limited to two or three large meals. Children nurse as often and for as long as they like and are offered the breast at the first sign of discomfort. The practice of using food to comfort continues throughout life. In later childhood a weeping child who is crying for no obvious reason is asked, "Are you hungry?" When a person is ill or dying, concerned relatives attempt to provide him with his favorite foods. There is no notion that crying is good or healthy for infants or that picking up, feeding or otherwise comforting a crying baby will somehow spoil it. Both women and men attend immediately to an infant's needs but, as children grow older and less dependent and their needs become more complex, parents and older siblings become hardened to their crying. Many children of three or four years of age throw temper tantrums which are largely ignored. Indeed, to paraphrase Harrell's observation about a Taiwanese community, the howling of toddlers is so common in the village that one learns to tune them out as background noise, in a class with buzzing outboard motors and agitated pigs (Harrell, 1981).

Kaliai villagers seem to have little compelling need for personal privacy, and there is no feeling that adults require their own space and time apart from children. Infants share their mother's sleeping mat, and young children sleep beside their parents. Residence in Lusi-speaking Kaliai villages is ideally virilocal and married couples nowadays live in individual family households, but women are not isolated by this residence pattern. They spend much of their time outdoors gardening, cooking in open cook houses, sitting on their verandas or on the ground beneath or beside their homes. Women, especially sisters and sisters-in-law, share tasks, often under the direction of a mother or mother-in-law. Children in this environment move freely among women whom they call mother or grandmother and may go to any of them for food. Brown's characterization of women's labor accurately describes the Kaliai: Women's tasks are by-and-large repetitive, interruptable and nondangerous (Brown, 1970). The pace of village life is relaxed, so that the demands of a hungry infant do not disrupt a woman's routine; and the value placed on the defining act of motherhood — giving food to a hungry child — is so important that there are few things that could compete successfully with it for mother's attention. A nursing mother is not idle, for there are many handiwork tasks, such as mat making, basket weaving and making shell money, that a woman can do while she suckles her baby. Other work will be taken over by sisters and friends if it is urgent. Otherwise it waits while a child is being fed.

Village women do not engage in non-domestic labor that separates them from their children. There are no organized adult activities that exclude children other than those reserved for initiated males; these occur in a context of ritual ceremony and are not an everyday affair. Children ordinarily go almost everywhere with their parents: to church services, on trading voyages or on trips to the town of Kimbe, and to all-night festivals of singing and dancing where they may participate in the activities until they grow tired and fall asleep on the ground. Mothers do occasionally leave young children at home when they go to distant gardens, but not usually before a baby is at least six months old and its diet has been expanded to include solid foods. The decision of whether to leave a child or take it along, and the choice of caretakers, is idiosyncratic. Some women often leave their children while other women almost never do. Some men regularly babysit, but children are usually left in the care of a female relative: grandmother, aunt or older sibling.

PREGNANCY, WEANING, ALTERNATIVE FOODS

As noted above, breastfeeding usually continues for a minimum of two years, for it is thought that a child who is weaned earlier will be slow to develop. Sammy was pointed out to me as an example of the deleterious effects on a child of early weaning. His mother became pregnant and ceased nursing him when he was about a year old. By the time he was 18 months old all of his age mates were walking, and many of them were sufficiently adept to run after their older siblings. Sammy could still not stand alone, and his slow development was attributed specifically to his early weaning.

A nursling should be weaned after his mother misses her second menses and she is certain she is pregnant. If a child continues nursing longer, women agree, he will likely become ill. Instead of the breast, the child is given the liquid of green coconuts, water and soup as the liquid component of his diet. Most children are weaned because their mothers become pregnant again, but if pregnancy does not occur, the length of time a woman permits her child to nurse is a matter of individual preference. For example, one young woman tired of nursing her third child when he was two years old and weaned him by leaving him in the care of a relative while she went to the gardens for several days. The child screamed most of the first three nights, but by the end of a week the mother's milk had dried up and the child had adjusted to drinking coconuts and water. Another woman, middle-aged, permitted her eighth child to nurse until the child was about six years old and so preoccupied with her play and with other foods that she forgot about her mother's breast for a day or two at a time.

There is, except for bottles and formula provided by the medical center, no alternative to breastmilk for very young infants. Powdered and canned milk are sold in the nearby plantation store, managed by whites, but not at village trade stores, and no village woman buys milk for her children. The plantation trade store manager reported that she sold milk to the plantation foremen who added it to their tea and coffee and, occasionally, to teachers from nearby local schools who gave it to their children. Village trade stores do carry Milo (a chocolate tonic drink powder) which women mix with water for their children, especially if they are ill. Sick children are also given hot sugar syrup to drink.

Infants are given their first solid food at about six months. The first foods are usually very ripe bananas or cooked sweet potato that mother offers her infant from a portion that she herself is eating. Other foods are then gradually added to the child's diet. The staple diet includes cooked foods such as sweet potatoes, taro, tapioca and plantains which are either roasted over coals or grated and baked with coconut cream; wing beans; a variety of greens; yams, and recently introduced foods such as pumpkins, tomatoes, small green onions, and green beans. Sea foods such as fish of many varieties and shellfish are readily available and are eaten two or three times a week. Game meats, chicken and pork are available less often. Meals are ordinarily prepared once or twice daily, the main meal in the mid-afternoon or evening, and a smaller breakfast if there are no cold foods left over. If there is no prepared food available, hungry children eat coconut, bananas, sugar cane, other available fruits such as pine-apple, Malay apple, mango and papaya, or biscuits purchased from the trade store.

Children are not urged to "clean their plate." Instead, several children may eat from one dish, each child eating as much and as fast as it can in order to get its share. I have never seen a child of any age refuse food, even if it was not complaining of hunger. By the time they are four or five years old, children are allowed to consume almost everything that adults do, including areca-betel-lime mixture and tobacco. The exception is alcoholic beverages, which are restricted to adult males.

The most important change in child nutrition that has occurred in Kaliai since 1966 is due to the availability of European foods. In 1966 there were no successful village trade stores and, therefore, little access to European foods in Kaliai. Several times between 1966 and 1971, Kandokans attempted to organize a village store, but they were defeated by transport problems and closed in a few months. At that time the nearest supplier of wholesale goods for stock, and the closest copra marketing board where locally produced copra could be sold, was Rabaul, about 300 miles to the northeast. Copra often could not be sold before it rotted, and depleted trade good stocks could not be replenished due to infrequent and unreliable transportation. As a result, the stores soon failed. The establishment of Kimbe changed this, for there are numerous wholesale suppliers and a copra marketing board located there, and Kimbe is only a few hours travel from Kaliai by motorized canoe. Village entrepreneurs who own outboard motors are, therefore, no longer dependent on commercial transportation. The presence of Kimbe has made it possible for village trade stores to stock a steady supply of sugar, white rice, soda pop, tinned meat and fish, Milo, candy and biscuits. Ready access to Kimbe has also made it possible for villagers to sell their copra more easily, thereby increasing their cash income. As a result, in 1981, the diet of most children included at least one of these items three or four times a week. This was not true in 1966.

It is too early to assess the effect of introduced foods on the health of the Kaliai people, for there are costs as well as benefits. In 1981, purchased foods supplemented rather than supplanted garden foods in the local diet. In 1982, there was a serious drought in northwest New Britain. My correspondents in the village reported that, during this difficult time, many people relied almost entirely on purchased rice because their gardens failed. If this food had not been readily available, villagers might have gone hungry. On the other hand, the dental caries attributed to sugar-rich processed foods may more than balance their convenience during famine. In 1981, the people of rural Kaliai did not have ready access to dental care.[†] Extensive dental care was available, however, in the Kimbe area, and it is notable that dental health services statistics for 1981 indicate that there were significantly fewer fillings and extractions in West New Britain than in the more affluent provinces (Townsend, personal communication, 1983). These data suggest that, for people with the money to buy white rice, pop and sweets with empty calories, the presence of these foods is a mixed blessing.

DISCUSSION

Northwest New Britain is in the process of rapid social and cultural change, due primarily to the establishment of Kimbe. Introduced concepts disseminating into rural communities have changed peoples' ideas about sexuality and procreation. The female breast, for

---

[†]There were, however, 13,287 dental examinations conducted in West New Britain in 1981 when the province ranked fourth in the number of dental examinations conducted in Papua New Guinea (Townsend, personal communication, 1983).

example, is now a focus of erotic attention, and educated young people challenge the traditional notion that children are composed primarily of paternal substance. In spite of these changes, there has been no significant change in infant feeding practices. The ideology of nurture and the value of breastmilk as the essence of maternal kinship continues to be a basic principle of Kaliai social thought and organization that overrides other considerations, such as the eroticism of the female breast. Consequently, all infants are normally breastfed, preferably until they are at least two years old.

The only worry that village women express about the quality of breastmilk is the possibility that it might be contaminated by semen. The idea that male sexual substances may be dangerous to vulnerable people, including young children, has been reported elsewhere in Papua New Guinea (Hogbin, 1943; Faithorn, 1975). It appears that there is a constellation of assumptions found in Kaliai which is widespread in Oceania (Tietjen, ch. 7; Conton, ch. 6; Jenkins, Orr-Ewing and Heywood, ch. 3; Obrist, unpublished observations, 1983[†]; Gegeo and Watson-Gegeo, ch. 13; Morse, ch. 14). These assumptions include the following:

There is a physiological connection between uterus and breast that allows semen to enter a lactating woman's breast from her uterus.

Mother's milk is contaminated if it is mixed with semen. Such milk will make an infant sick.

This cluster of ideas seems to form the conceptual foundation for a number of customs related to infant feeding, including expelling the colostrum because it is dirty or poison and will make the infant ill; restrictions on wetnursing which may be practiced either only among close kin or not at all; and the postpartum sex taboo. Clearly the factors influencing a mother's decisions about the way in which she feeds her baby are complex and may include notions about kinship, conception and sexual contamination as well as concerns about nutrition and the appropriate allocation of her time and energy. Because these concepts may have practical consequences, especially for wetnursing and the wastage of colostrum, research on child feeding practice must include the broader cultural context. If it is grounded in cultural understanding, education designed to allay women's fears about the contamination of colostrum and breastmilk

---

[†]Obrist, B. (1983). The study of infant feeding: Suggestions for further research. Paper presented at the twelfth annual meeting of the Association for Social Anthropology in Oceania, New Harmony, IN.

by semen could result in the modification or elimination of behavior
that denies valuable nutrients to infants.

ACKNOWLEDGEMENTS

The research on which this paper is based was funded by grants from the U.S. National
Science Foundation in 1966; the Wenner-Gren Foundation and the University of
Waterloo in 1971; the Canada Council and sabbatical leave from the University of
Waterloo in 1975-1976; and by the Social Sciences and Humanities Research Council
of Canada (SSHRC) and sabbatical leave from the University of Waterloo in 1981.
This is a revision of a paper that was presented at a symposium on Infant Care and
Feeding in Oceania held at the 1983 meetings of the Association for Social Anthro-
pology in Oceania organized by Dr. Leslie Marshall. My thanks to her and to other
members of the symposium, and to Patricia Townsend of the Papua New Guinea Institute of Applied Social and Economic Research for their helpful criticisms of earlier
drafts of this paper.
All personal names used in this paper are fictitious.

REFERENCES

Bradley, C. and A. Peberdy (1981). *Reproductive Decision Making and the Value of
    Children: The Tolai of East New Britain.* Papua New Guinea Institute of Applied
    Social and Economic Research, Boroko.
Brady, I. (Ed.) (1976). *Transactions in Kinship: Adoption and Fosterage in Oceania.*
    ASAO Monograph No. 4, University Press of Hawaii, Honolulu.
Brown, J. (1970). A note on the division of labor by sex. *Amer. Anthropol.* **72**, 1073-
    1078.
Carroll, V. (Ed.) (1970). *Adoption in Eastern Oceania.* ASAO Monograph No. 1. Uni-
    versity Press of Hawaii, Honolulu.
Chowning, A. (1980). Culture and biology among the Sengseng of New Britain. *J.
    Polynes. Soc.* **89**. 7-31.
Chowning, A. (1985). Family fertility decisions among the Kove. In N. McDowell
    (Ed.), *Reproductive Decision Making and the Value of Children in Rural Papua
    New Guinea.* Papua New Guinea Institute of Applied Social and Economic
    Research, Boroko. In Press.
Counts, D.A. and D.R. Counts (1983). Father's water equals mother's milk: The con-
    ception of parentage in Kaliai, West New Britain. In D. Jorgensen (Ed.), *Ideo-
    logies of Conception in Papua New Guinea.* Special Issue of *Mankind* **14**(1), 46-
    56.
Faithorn, E. (1975). The concept of pollution among the Kafe of the Papua New
    Guinea Highlands. In R.R. Reiter (Ed.), *Toward an Anthropology of Women.*
    Monthly Review Press, New York and London, pp. 127-140.
Ghos, S., S. Gidwani, S.K. Mittal and R.K. Verna (1976). Sociocultural factors affect-
    ing breast feeding and other infant feeding practices in an urban community.
    Indian Pediat. **13**, 827-832.

Gonzales, N.L.S. (1963). Breast feeding, weaning, and acculturation. *J. Pediat.* **62**. 577-581.

Goodale, J. (1980). Gender, sexuality and marriage: A Kaulong model of nature and culture. In C. MacCormack and M. Strathern (Eds.), *Nature, Culture and Gender*. Cambridge University Press, Cambridge, pp. 119-142.

Goodale, J. (1981). Siblings as spouses: The reproduction and replacement of Kaulong society. In M. Marshall (Ed.), *Siblingship in Oceania*. ASAO Monograph No. 8. University of Michigan Press, Ann Arbor. MI, pp. 275-305.

Harrell, B. (1981). Lactation and menstruation in cultural perspective. *Amer. Anthropol.* **83**. 796-823.

Hogbin, H.I. (1943). A New Guinea infancy: From conception to weaning in Wogeo. *Oceania* **13**, 285-309.

Jimenez, M.H. and N. Newton (1979). Activity and work during pregnancy and the postpartum period: A cross-cultural study of 202 societies. *Amer. J. Obstet. Gynecol.* **135**, 171-176.

Jorgensen, D. (1983). The facts of life, Papua New Guinea style. In D. Jorgensen (Ed.), *Ideologies of Conception in Papua New Guinea*. Special Issue of *Mankind* **14**(1), 1-12.

Kelly, R.C. (1976). Witchcraft and sexual relations: An exploration in the social and semantic implications of the structure of belief. In P. Brown and G. Buchbinder (Eds.), *Man and Woman in the New Guinea Highlands*. Special publication of the American Anthropological Association. No. 8. pp. 35-53.

Lindenbaum, S. (1972). Sorcerers, ghosts, and polluting women: An analysis of religious belief and population control. *Ethnol.* **11**, 241-253.

Marshall, L. and M. Marshall (1980). Infant feeding and infant illness in a Micronesian village. *Soc. Sci. Med.* **14**, 33-38.

Meggitt, M.J. (1964). Male-female relations in the Highlands of Papua New Guinea. In J.B. Watson (Ed.), *New Guinea: The Central Highlands*. Special issue of *Amer. Anthropol.* **66**, 204-224.

Meigs, A.A. (1976). Male pregnancy and the reduction of sexual opposition in a New Guinea Highlands society. *Ethnol.* **15**, 393-407.

Meigs, A.A. (1983). *Food, Sex and Pollution: A New Guinea Religion*. Rutgers University Press, Baltimore, MD.

Nerlove, S.B. (1974). Women's workload and infant feeding practices: A relationship with demographic implications, *Ethnol.* **13**, 207-214.

Van Esterik, P. and T. Greiner (1981). Breastfeeding and women's work: Constraints and opportunities. In E.Baer and B. Winikoff (Eds.), *Breastfeeding: Program, Policy and Research Issues*. Special issue of *Stud. Fam. Plann.* **12**, 184-197.

CHAPTER 10

PATTERNS OF INFANT FEEDING IN KOVE (WEST NEW BRITAIN, PAPUA NEW GUINEA), 1966-83

ANN CHOWNING

INTRODUCTION

The government-sponsored campaign to persuade Papua New Guinea women to continue breastfeeding sometimes ignores the difficulties that can arise even in a village setting. Westerners may also assume that where prolonged breastfeeding is the norm, it is regarded as healthful, pleasant, and stress-free for both mother and child, even though reports have long existed of nonliterate societies in which this was not the case (for example, the Mundugomor as described in Mead, 1935). It is well-known that mothers may have problems because of ill-health, residential patterns, and the demands of work, but it is less often appreciated that they may choose or be forced to restrict access to the breast because of theories about nutrition and the reasons for illness and slow development in infants, as well as by their own and their husbands' desires to resume sexual intercourse. In individual cases, they may also simply regard breastfeeding as a burden. Among the Kove a complex of factors, including exposure to Westerners, often produces a situation which departs considerably from the ideal presented on posters at Papua New Guinea health clinics.

METHODOLOGY

Fieldwork consisted of about 18 months total, divided between three different Kove villages, Kapo (1966), Somalani (1968-69), and Nukakau (four stays between 1971 and 1983). The research was not oriented towards the topic of infant nutrition; data on the subject were

G

collected in the course of an investigation of social change and social organization. Being almost constantly in the villages, I was well situated to observe the treatment of babies and children, as well as to inquire about diet, food taboos, and related matters. At the same time, women frequently asked questions in turn about what they thought to be Western practices, or commented on them. From the start of the fieldwork, I tried to convince them that they were wrong in thinking that women of European descent do not breastfeed. Whether they believed this or not, the efforts of medical personnel and observations by Kove working in towns did change their ideas on this matter.

In most cases, the ages of infants could only be estimated, and the developmental stages which the Kove recognize are used in this report. Where specific ages are given, I had either been present when the child was born or had seen its registration card for the maternal-child health clinic operating from the Catholic mission in 1983. Birth dates given by parents were rarely trustworthy.

RESIDENCE, FOOD, AND WORK

The Kove (or Kombe) live on the north coast of West New Britain, immediately to the east of the Kaliai or Lusi described by Counts (see ch. 9, Figure 1), except for two breakaway villages located west of Kaliai. The Kove language is so closely related to Kaliai as to be little more than a separate dialect, and superficially the two cultures are almost identical. Below the surface, however, a number of significant differences exist, and these have been strengthened by divergent histories of contact with the outside world, including mission influence and ease of access to government stations and urban centers.

Early in the century most Kove villages were located on the mainland of New Britain, but pacification removed the threat of enemy raids, and almost everyone moved to tiny islands, well away from the mosquitoes of the mainland. An already strong maritime orientation increased, so that the Kove became renowned as traders and travelers uninterested in gardening (and believed by many outsiders to subsist wholly on fish and coconuts). In fact, the gardens were usually well inland, masked from view by mangrove swamps and coconut groves, and reached only by long canoe trips. Only in the past few years have motors and gasoline for them sometimes been available for those who can afford to pay. A few villages remained on the mainland, and new settlements have been started by refugees from the overcrowded islands, but for most Kove home is a tiny sandy islet, with no fresh water or vegetation apart from coconut palms and a few other trees. Most islands are surrounded by reefs on which women and girls collect shellfish and other seafoods at low tide, and on which boys spear small fish. Grown men exploit deeper water to obtain larger fish, turtles, and an occasional dugong. Bad weather, however, often makes deep sea fishing impossible, so that the steady supply of animal protein comes from females' collecting at the reefs, the shore of the mainland, and the mangrove swamps. Wild pig caught in traps set up along the garden fences is also eaten occasionally. Domestic pork is reserved for special occasions. Even by coastal Melanesian standards, the Kove diet is rich in animal protein, which they expect to eat every day. Seventh Day Adventists (a substantial portion of the Kove population) are, however, forbidden pork, all shellfish, and many other seafoods such as shark, stingray, and turtle.

The traditional starch staple was taro, heavily supplemented by sago, which is manufactured by men. Clearing and fencing the gardens are also male tasks, as is most of the work of housebuilding. Women collect firewood and water, do most of the garden work, cook, and manufacture mats, baskets, and shell money. Garden crops have expanded within the century with the introduction of sweet potatoes and manioc (cassava), which have largely replaced taro since the taro blight of the early 1960s and several subsequent bad droughts. When in season breadfruit, cultivated or wild, substitutes for tubers. These foods comprise the starch staples, *haninga muru*, which supply the bulk of the calories and are normally eaten two or three times daily. No other indigenous foods substitute for them. Bananas are greatly liked and eaten when available, but are not considered filling. Coconuts, which have been abundant since the forced plantings of the 1920s, are eaten as snacks, and grated coconut is added to all cooked dishes apart from those baked directly on the fire. Coconut alone, however, is not satisfying. Other vegetables, fruits, and nuts are occasional additions to the diet. By comparison with other Melanesians, the Kove grow and eat few greens. Papaya (pawpaw), mango, and pineapple are given to children when available, the first being particularly suitable for those unable to chew well. Coconut milk is drunk daily if men or older children are available to climb the palms and harvest the nuts.

To obtain necessities other than coconuts and shellfish, including fresh water, women must travel to the mainland or occasionally to other uninhabited islands. Usually many women and a few men go together for a day-long trip on one canoe. The likelihood of exposure to heat, rain, wind, biting insects, and the irritating fuzz of vegetation is high, and makes mothers reluctant to take babies, who may get sick from any of these causes. Furthermore, because the Kove lack both the large net bags of the mainland of New Guinea and the barkcloth slings of their neighbors to the east, very young babies are carried in the arms, and slightly older children cling to the back, arms around an adult's neck and one of the adult's hands under the child's buttocks. Thus, not only is it dangerous for the baby to take it to the gardens, but the mother is also prevented from carrying out her tasks unless she alternates babysitting with her husband or takes along a babysitter in the form of an older child. If it is at all possible, she prefers to leave the baby in the village. Many women report that they do not go to the gardens until the baby is old enough for solid food. Such a decision means that some other woman (most likely a close kinswoman) must be willing to take over the new mother's duties in the garden. By contrast, it is a minor burden to fill up water for several households at once, and most women going to the water source on the mainland carry containers for other women. Firewood gathered from the swamps can be supplemented with readily available driftwood and coconut husks. What women kept at home by babies always mention is neglect of their gardens.

EMULATION OF "EUROPEANS"

In the years between the two World Wars, most Kove men went away to work on plantations and developed a taste for rice, which they even grew for a brief period. The Kove think of rice as the principal starch staple of so-called Europeans, whites of European descent, and value it as much for this reason as for its taste and convenience. The Kove generally believe that Europeans grow faster as well as larger, are healthier, and live longer than themselves, and think that a major reason is the superiority of their food. Not only is it "cleaner", softer, and tastier, but it is more nutritious. All foods provided to laborers and for sale in stores is automatically superior to local fare in at least some respects. Ideas about what Europeans actually eat are quite erroneous. Fruits and vegetables are thought to play no part in their diet. The main, if not the only, meats are assumed to be canned mackerel and corned beef, although it is known that Europeans relish crayfish, oysters, and deep sea fish such as bonito when they are living in Papua New Guinea. When possible, the Kove will sell these foods to Europeans so that they can obtain cash with which to buy rice, crackers, flour (to be fried in beef dripping), coffee, tea, and sugar — the basics, along with the canned meats and fish, of a European-style diet. Flour, crackers, and fried flour are all considered *haninga muru*, as is rice. Papua New Guinea white rice (the only kind the Kove eat) is enriched, certainly not a bad substitute for sago and manioc for those who can afford it. Its popularity has grown steadily in recent years, and children are often reported to prefer it to any indigenous starches. Other foods are considered particularly appropriate for children, notably hard candy, cookies, and snacks such as cheese-flavored Twisties, but except in towns they are not likely to constitute a daily part of the diet.

The characteristic European-style beverage is tea, drunk very weak without milk but with a great deal of sugar. Weak coffee is liked but is drunk much more rarely, probably because of its expense. Both beverages are forbidden to Seventh Day Adventists, but others consider that the perfect breakfast, for small children as well as adults, consists of plain crackers dipped into sweet tea. Beer, which is drunk only on special occasions, is also considered nutritious, and is given to babies "to make them fat". (The Kove generally correlate fat with good health.)

The most important consequence of ideas about European food and feeding habits has been the widespread belief that bottlefeeding is better for babies than breastmilk. The supplies for bottlefeeding were difficult and expensive to obtain, even before the 1977 prohibition on the sale of bottles, but in the period from the mid-1960s to the mid-1970s some women were bottlefeeding because they or their husbands thought it was better for the baby. A man elated at having twin boys after three girls forbade his wife to breastfeed them (and then said she could do it "just a little") because he did not know what sickness was in her breasts. Although the supply of milk they had been given at the mission hospital was said not to be plentiful, he hoped to keep them on the bottle.

TRADITIONAL INFANT FEEDING

Birth

Traditionally a newborn baby was not nursed before it cried, and then not by its own mother, for two reasons. Colostrum is considered

worthless and even repulsive — "like water and then pus," as one
woman said. The mother should wait until her milk appears. Second,
the blood shed in and after childbirth is considered highly polluting,
more so than menstrual blood, and it is better for the child to avoid
too much contact with the mother until the blood flow ceases. During
this period the mother stays in a shelter on the ground, so that she
does not contaminate the house floor. At present, many women give
birth at the Catholic mission hospital, where the baby is put to its
mother's breast much earlier than in the village, but the same women
will follow traditional practice if their next baby is born in the village.
They point out that in the hospital, the Sisters (nurses) give an injection to end the bleeding. In the village, there are always a number of
lactating women, and anyone who is willing to do so may wetnurse
the neonate. Preference is given to a close relative of the baby, who
will not feel imposed upon, and to a woman with plenty of milk. In
two cases witnessed by the author, the women began to nurse their
own babies within four days after the birth.

   In practice, women do not always wait for the baby to cry (in
apparent hunger, as opposed to any initial wail). One woman said
that she felt "sorry" for her baby and fed it before it cried. When
there is reason to be concerned about a baby's survival, efforts are
made to get it to nurse soon. In two cases of small, seemingly premature infants observed by the author, concern was expressed over
the failure of the babies to nurse or to cry 12 to 15 hours after the
births.

Dependency on Milk

For the first months of its life the baby is given nothing but milk. Precisely how long this period lasts varies, with different women giving
different developmental stages for the introduction of solids. A
grandmother who is reputed to have borne 13 children and who is an
acknowledged expert on matters having to do with fertility and pregnancy said that a child received its first solid food when its skin darkened and when it smiled or laughed (the same word is used in Kove).
The foster-mother of a child whose mother had died in childbirth also
said that it could be weaned at three months, although her time
assessment might not have been accurate. The only time I saw a child
as young as three months being given solids, the father said that the
reason was that it cried and seemed dissatisfied on milk alone.

Most women correlate the introduction of solid food with sitting, and all children who could sit alone seemed to be receiving some solids, though an occasional mother suggested that it would have been preferable to wait.

The period during which the child is wholly dependent on milk is hard on the mother. Either she has to stay in the village the whole time (something the Kove find imposes severe psychological strain), take the baby with her and risk its health, or find a trustworthy wetnurse. Furthermore, if she stays home she will constantly be called upon to act as a wetnurse herself because of her availability and because the mother of a baby so young can be trusted not to have resumed sexual relations with her husband. When a baby is older, the mother cannot be trusted to have uncontaminated milk, and many parents are wary of using her services.

Wetnurses

A few mothers insist on feeding their own babies alone, and take the consequences of restricted activity for themselves and, often, periods of hunger for a baby left in the village. Other women, or their husbands, impose other restrictions; for example, one man would only let his baby nurse from a woman whose own child was growing well. Most women, however, make considerable use of wetnurses. These are not usually the actual caretakers, who are most likely to be grandparents, but may be any of a variety of kin or neighbors. Unless she is closely related to the mother, the wetnurse is not likely to want to undertake the care of a second baby in addition to her own. What often happens is that in fact the wetnurse is not asked in advance if she is willing to serve. If women know that a particular lactating mother is staying in the village, they may simply instruct their caretakers to take the child to her when necessary. Often it is regarded as no particular burden, and the wetnurse may not give the child more than enough to quiet it. However, the failure of mothers to coordinate their plans or to consult the potential wetnurses in advance can place intolerable burdens on them. One new mother, for example, was brought three babies in addition to her own to feed. One mother who was frequently called upon to feed two other babies complained that she was going to stop staying home because her own baby was being deprived, especially because the other two babies were bigger and older and "finished" her milk. Later the same mother told of weeping at the appearance of her baby after it had been left behind deprived of milk. Sometimes a potential wetnurse will simply refuse, and one or more babies wholly dependent on breastmilk can be left for many hours with nothing but water or coconut milk (see below). In such a situation, a mother was noted to be mourning over the state of her baby at the end of a day, finding its belly shrunken and its spirits depressed. It was partly to avoid such situations that bottles were sometimes used, and in 1966, a woman was observed feeding two babies alternately with what looked like very dilute milk.

Solid Foods

Traditionally the first solids were taro, sweet potato, and ripe banana. The first is always premasticated or mashed between the fingers. Nowadays rice is likely to be the first food. Other foods are offered at varying times. Observation does not confirm statements that mothers wait until children are walking before giving "hard" foods such as baked taro and manioc. Children who have only a few teeth are often given such foods to chew on. Fish, shellfish and trepang, papaya, scraped coconut, sugarcane, greens, and crackers are given well before the child is a year old. By the time he is walking he usually receives the full range of adult foods, apart from those taboo to very young children and their parents. The principal ones of these are fish that live in holes in the reef; eating them, on the part of any of the three, results in soul loss and serious illness on the part of the child. There are several other similar food taboos, but none of them greatly affects the range of foods available. A child is allowed to express personal preferences. Adults continue to prepare certain foods for children under three, removing the bones from fish and peeling sugarcane, breadfruit seeds, and coconut "apples" (the oily sprout of the coconut, a favorite food of young children). Children are allowed all adult food except for those few traditionally reserved for older men. Children normally eat a full range of food before they stop breastfeeding.

Liquids

Babies are given water only when there is no alternative, or occasionally to quiet hunger. The preferred substitute for breastmilk is coconut milk. Traditionally a very young and soft-skinned nut would have a hole pierced in it so the baby could suck from it as from a breast. Juice cups and funnels are now used. As soon as they can drink from a cup, babies are offered sweet tea when available. Milo, containing powdered milk and vitamins, is liked for its taste alone but is too expensive to be used often.

ATTITUDES TOWARD BREASTFEEDING

Regardless of how much food it is eating, women generally agree that unless special problems arise, a baby should be breastfed until it is walking well. The expressed ideal is that a woman should abstain from getting pregnant and weaning her child until it is collecting shellfish on the reef, if a girl, or spearing fish, if a boy. One grand-mother, a rare exception, boasted of having observed this three or four year interval with her children. Much more commonly, women expect to breastfeed for about two years, wholly for the sake of the child, but they very often complain about the process and make it somewhat difficult for the child.

The Kove do not regard crying on the part of a baby as necessarily a sign of distresss. As one woman said, "Crying is baby's speech," and it may be conveying a number of emotions. Crying is often deliberately provoked by parents and others, who slap, frighten, or otherwise annoy a baby until it starts to cry, and then cuddle it. Spon-taneous crying, not induced by another person, is often attributed simply to a bad mood, especially if the child has just awakened, or to heat, so that it is commonly suggested that a crying child should be bathed. Continuous crying is considered to be harmful and to cause sickness. A child who wails steadily is eventually distracted in a number of ways — for example, by being offered food or the breast; or by being carried, patted on the rump, or jiggled; or by being handed a plaything. Offering the breast is by no means the first reaction. Even if the mother is present, she may allow the child to wail and fumble for her breast or, if she is standing, pull at her skirt, for a considerable period before feeding it (10 or 20 minutes, in some cases). The increasing fashion of covering the breasts makes it easier for a seated woman to deny a baby the breast than was once the case. Even if she is bare-breasted she may cover her nipples with her hands or use one to tease the child, striking its face with it or withdrawing it shortly after the baby starts sucking. The mother usually seems to satisfy the child once she finally decides to feed it, but it may become frantic while waiting. This semi-rejecting behavior is more common with older children, but it is often observed with those far too young to be weaned.

The mother of a child of 1 ½ often remarked how tired she was of breastfeeding, denied her son the breast for long periods, and reproached him for eating a lot of food and still wanting to nurse. Her attitude illustrated the lyrics of a song popular in Kove: "When I was a single girl, I wanted to marry. Now I'm married, it's very hard work — cooking food, the baby cries, I get no sleep." All Kove women want to be married and to have children, but childcare, especially prolonged breastfeeding and all it entails in social restrictions, is regarded as a duty at best and a burden at worst.

Normal Weaning

There is little criticism of a woman who waits until her baby is walking well before getting pregnant again. In the absence of pregnancy, however, the mother usually waits until the child has all its milk teeth and is, in what is said to be a correlated process, talking. Then she can wean the child and resume her normal activities, including sexual intercourse, secure in the knowledge that she has given the child a good start in life. The usual way to accomplish weaning is separation, as by sending the child to stay with a grandparent in another village for a few days. Other methods include putting red peppers or ginger on the nipple, or painting it red so that it looks frightening. Once a child is talking well, and has all its baby teeth, it no longer needs milk. Sometimes a last-born is suckled for a total of three or four years, but nowadays a child who breastfeeds for so long is likely to be subject to much teasing from adults and children alike. Such prolonged suckling is most often seen on the part of mothers assumed not to be engaging in sexual activity, such as widows. However, there seems to be little worry about the effects on the child of ingesting "polluted" milk if this milk forms an inconsiderable part of its diet.

The mother of an infant only a few months old may well complain that she is tired of breastfeeding. If she weans the child too soon, however, she is thought to condemn it to death, and can only do so without blame in exceptional circumstances. The most common reason for early weaning is that the mother gets pregnant again, and the only thing worse than getting pregnant too soon is continuing to breastfeed after doing so. The other reasons for early weaning, or failure to breastfeed at all — failure of the milk supply, death of the mother, and adoption of the baby — entail no social condemnation, and are handled in different ways.

Contamination of Milk

In a discussion of conception beliefs among the Kove's neighbors, Counts and Counts (1983, p. 58) point out that Kaliai "ideas about conception and reproduction ... do not constitute part of a tightly structured, internally consistent ideology," even if changes brought by mission and school are discounted. The same can be said of Kove ideas. An example of such inconsistency is the following set of Kove beliefs: coconut and coconut milk help produce semen, semen contaminates breastmilk, coconut milk is the best substitute for breastmilk. In contrast to the Kaliai, the Kove do not equate semen and milk, but the partial equation of coconut and semen has some behavioral consequences. It is considered best for a nursing mother not to eat coconut, and the same woman who laughed at a friend for suggesting that it might be possible to get pregnant just from eating coconut and white-fleshed fish (also a semen producer) explained that her baby was exceptionally healthy because she avoided both of these foods while the baby was still dependent on milk.

The usual belief is that, where pregnancy-causing magic is not involved, pregnancy results from the coagulation of the mother's blood by the father's semen. Beliefs, similar to those of the Kaliai, that cords through which semen can pass link uterus and breasts (Counts, ch. 9) are implicit in Kove. Pregnancy is always said to begin in the breasts, as shown by their shininess and darkening of the nipples. While the fetus is "in the breasts," it is still being formed by additional injections of semen, and it is only during this period that it can be aborted without danger to the mother's life. Then it moves down to the womb. While there, it constitutes a greater danger to the suckling baby than do occasional acts of intercourse that do not result in pregnancy.

Counts and Counts (1983, p. 54) suggest that for the Kaliai, "the ideal of postpartum abstinence is not, and never has been, reality." Such is not the case in Kove, at least for the mother. Men, however, are acknowledged often to have less concern for their babies than for their own sexual desires. One Kove man killed his wife and another pushed his wife into a fire and burned her badly when the women refused intercourse because of the youth of their babies. Persons who do yield to their own or their spouses' desires are condemned by both sexes, because semen contaminates the breastmilk and makes the baby sick.

Many Papua New Guinea societies have beliefs that semen contaminates breastmilk. Some restrict the sexual activity of the father as well as the mother while the baby is young. What seems to be more distinctively Kove is a belief that pregnancy is even more polluting to breastmilk than intercourse (though other people, such as the Sengseng of West New Britain, think that a baby sickens if it is nursed by a pregnant mother). To the Kove, pregnancy is a dangerous and polluting condition in other contexts as well, so that a pregnant woman may not approach anyone who has a healing wound, such as a newly slit ear or superincised penis, or anyone with elephantiasis, lest they develop bad sores.

It is almost certainly a combination of shame at misbehavior and the hope that the consequences will not be fatal which leads many women to deny that they are pregnant and to continue to breastfeed during this period. The lack of a substitute for mother's milk, of the sort available to women living in towns, may be a factor as well. The use of blouses enables some women to conceal pregnancy longer than would have been the case in the past. If pregnancy is detected, other kin of the mother are likely to intervene forcibly and compel weaning or even take over the child. One childless woman, whose own baby had died, reported that she had "adopted" two other children for this reason. Her brother's wife had become pregnant when the baby was not yet sitting unaided. She took it over the mother's denials that she was pregnant, and reared it on a combination of sweet potato cooked in coconut cream and coconut milk to drink. The woman attributed the death of her sister-in-law's subsequent baby to the anger of her dead mother's ghost over the too early pregnancy. This same woman also claimed to have taken over the child of her husband's brother because the mother got pregnant when the previous child was not walking well, but she made no mention of the second child's needing special feeding.

Although it is frequently said that a child weaned too soon will die, no actual cases of this were recorded. Retarded development and sickly looking children are said to result from the mother's getting pregnant too soon. The commentators do not necessarily distinguish cases in which the baby was weaned early from those in which the cause was contamination by semen in the milk or the mother's pregnancy. Fever and jaundice are specifically the result of violation of taboos by either parent, but thin ribs and a protruding belly can have either cause.

As has been noted, if the parents are living in town with access to "good" food, there is thought to be no harm in early weaning. The Kove constantly cite the example of the town Chinese, the paradigm of those whose babies are close together yet healthy. Furthermore, one grandfather said that even in the village, it is no longer necessary to observe either sex or food taboos because European medicine is now available to counteract any harm caused by breach of them. He then amended his statement to say that mothers still needed to observe them, but not fathers.

Failure of Milk

Only a few mothers reported difficulties resulting from inadequate milk. In one case, the author observed a woman wetnursing a baby in the presence of the real mother, who explained that her milk "lacked food." Another woman complained that her breasts were "too soft" to be good. Most women who suffer from inadequate milk are given the materials for bottlefeeding at the mission hospital.

Variant Theories about the Need for Milk

Most Kove are well aware that an infant needs a good supply of milk, and one case was recorded of a baby who died following the mother's death "because there was no one to nurse it." Nevertheless, other ideas exist; one mother suggested that milk is simply desired by babies because they develop the habit, rather than actually being needed. Sickliness and slow growth may also be attributed to other causes than malnutrition, such as attack by spirits. One man said that when his wife had swollen breasts after the birth of their seventh child, the infant eventually had "arms and legs like my finger, and no hair," but he did not seem to have realized that the problem was malnutrition. (The baby was receiving some wetnursing.) It was rescued by a passing government patrol, but he expressed no guilt over his failure to take it to the mission hospital.

Most Kove have no idea of the origin or composition of powdered and canned milk, and the older ones would not automatically think of seeking it as a substitute for human milk. If actual or potential difficulties (as with the birth of twins) were not noticed by mission personnel, a substitute might not be requested by parents.

By contrast, younger women and men assumed that formula was more nutritious than breastmilk because it is of "European origin" and because until recently they thought that expatriate babies received only formula. The campaign promoting breastfeeding, coupled with the realization that all expatriate babies are not bottlefed, has changed attitudes, and bottlefeeding not initiated by hospital personnel is increasingly confined to cases of adoption.

Problems with Bottlefeeding

Judging from its appearance, formula may sometimes be too dilute. The reasons are presumably the usual ones of ignorance of the necessary strength coupled with a desire to conserve supplies. A more consistent problem is lack of hygiene in preparation of the formula and handling of the bottle. One woman told of having been threatened with jail when found by a visiting patrol using visibly dirty water to mix the formula, but almost no one bothers to boil water or really clean the materials. The Kove rarely use soap to wash any cooking or eating utensils, simply scrubbing them with sand. Although pigs, dogs, and small children defecate within the village area, if a bottle, feeding cup, or spoon is dropped on the ground, it is usually simply brushed off before being handed back to the child. The most careful mother observed did rinse the nipple before returning the bottle, but when others bottlefed her adoptive baby in her absence, they used no such care.

Adoption

Adoption is common in Kove, probably more so now than in the past. The usual reason is that a marriage is childless or has not produced a reasonable balance of the sexes (Chowning, 1985). Planned adoptions usually involve an agreement to take over the offspring of a close relative, ideally the sibling of husband or wife, shortly after birth. Adoption preferably takes place when the child is too young to remember his true parents, and it is usually specified that the child is still nursing when taken. Often no mention is made of wetnursing or a substitute for breastmilk, and it seems that the adopters waited until the child was old enough to survive without milk. In such cases, however, they are heavily indebted to the mother who suckled the child, and the adoption is not complete until she has been "paid for the milk." Just for this reason, many putative adoptions never become finalized. The Kove feel that there is a great difference between occasional wetnursing, which most women are willing to undertake as a favor with the expectation that it will be returned or for food, and the "growing" of a child by a woman who is solely responsible for the bulk of its feeding over a long period. Adoptions may take place very early to avoid the claims of the true mother, and in recent years they are likely to involve bottlefeeding if it can be arranged. A schoolteacher and his wife managed it by getting the supplies from the Seventh Day Adventist mission, where they were working. Another woman reported that she and her husband prepared for the adoption by buying a kerosene stove, bottles, powdered milk, and sugar in Rabaul (in 1971) and bringing them home, planning to rear the child entirely on the bottle. She insisted that she had relied almost completely on bottlefeeding, with only an occasional nursing by the real mother, to keep breastfeeding to a minimum and to establish her claim to the child.

Adoptions of children who have been breastfeeding sometimes take place too soon. In a recent case, the true mother left on a trip after handing over her child, during which time her milk dried up. She then was sent a message saying that her baby was very thin and maltreated. She returned home, took back her baby, managed to re-establish her milk, and looked after him until he was sturdy and had to be returned to the adoptive parents.

Many adoptions are not planned, but result from the death of a mother in childbirth, or from intervention in infanticide. A mother may also die so young that the child has not yet reached weaning age. Anyone who intervenes is usually prepared to obtain milk from the mission or to breastfeed the child.

Only if a woman has had a series of children die may it be decided that it would be best to take the next one from her at birth and give it to someone else to try to rear. In several such cases, it was not suggested that the mother's milk was at fault, but that close association with her might be dangerous to the baby, especially a sickly or undersized one.

In none of these cases was it said that a non-lactating woman who adopted a baby might induce lactation in herself, though it is common practice for babysitting grandmothers, and even grandfathers, to placate a hungry baby by allowing it to suck on a nipple. In one recent case, however, the grandmother did breastfeed. The grandmother initially gave the baby to other village women to suckle, but then dreamed that the ghost of her dead daughter told her to do the job herself. Though she was about 50, with her youngest child a teenager, she induced lactation, and was still breastfeeding the child after 15 months.

PROSPECTS FOR THE FUTURE

With increased access to cash and a proliferation of trade stores, the Kove are moving towards more dependence on commercial foods coupled with neglect of gardens, especially in the villages located near the township associated with a timber project operated by South Koreans. Men in the more traditional villages report that women living near the township are now doing all the garden work without help from the men, who prefer to depend on cash. This factor alone might push the women towards earlier weaning. Most Kove women no longer think that it is "modern" to avoid breastfeeding altogether, but they are well aware that town dwellers of all sorts do not breastfeed nearly as long as Kove women are expected to. As long as they still believe that foreign foods are more nutritious than their own, they are likely to see only advantages in early weaning coupled with the use of trade store foods to supplement and then replace milk.

The health campaigns that actually reach the villagers, usually from the Christian missions, say little about nutrition apart from advocating breastfeeding and, in the case of Seventh Day Adventists, attacking the consumption of pork, shellfish, tobacco, tea, coffee, and betelnut. Occasional suggestions from indigenous health specialists such as aid post orderlies are largely ignored. They do not carry enough prestige to be heeded when they advocate the eating of more vegetables. No one seems to realize that villagers are disadvantaging themselves and their children when they decide to sell their fruits and nuts to the Koreans and to foreigners who come to the market at Kimbe, the urban center for West New Britain. A few individuals have decided on their own that a varied diet is preferable to one in which the same starch staples, whether manioc or rice, are eaten every day. Some of the same people deplore the practice of encouraging small children to prefer "European" foods such as rice and crackers which are expensive and not always available. At village meetings, older women scold the younger ones whose husbands have not paid their Local Government Council taxes because the women are spending all the cash on store foods and ignoring the vegetables that they should be growing and eating. Apart from fish, women continue to be responsible for what is offered at a meal.

The consequences of changing diet can be seen in Kove children living in urban centers, some of whom are much thinner than their age mates in the villages, as might be expected when they lunch on soft drinks and Twisties and refuse vegetables from the market. At present, in the more remote villages the feeding of young infants continues much as it did in the past, except for a decrease in the amount of bottlefeeding that was practiced prior to the 1977 ban. The old problems remain: often unsatisfactory arrangements for the feeding of babies whose mothers have to leave the village to work, excessive demands for wetnursing on those women who stay at home, and pressure on parents to avoid sex for up to two years for the sake of their babies. At least nowadays, the result among married men is much adultery, leading to bitter quarrels. The new problem that seems likely to increase is the substitution of trade store foods for those traditionally eaten. The age of weaning will probably drop as fears about pollution continue to decrease, and the desire for cash will certainly lead to the neglect of gardens and the continued sale of tree crops. It is probable that the children at risk will not be the infants completely dependent on milk, so long as their mothers are well nourished, but those who are being weaned onto a limited range of foods of dubious value. While Kove continue to think that a baby is well served by being fed fried flour and sweet tea, it seems unlikely that all the newly available medical resources will improve infant health and survival. Only a strong campaign waged within the village by those who have prestige in Kove eyes is likely to combat the present trends. As it is, bottlefeeding may be a lesser danger to infant health than the feeding practices now developing.

ACKNOWLEDGEMENTS

Fieldwork in 1966, 1968, and 1969 was financed by the Australian National University, and in 1971-2, 1972-3, and 1975-6 by the University of Papua New Guinea. In 1983, I was on research and study leave from Victoria University of Wellington, with additional support from the Internal Research Committee of that university. In 1978, I spent three weeks in Kove as a representative of the Department of Environment and Conservation of the government of Papua New Guinea.

REFERENCES

Chowning, A. (1985). Family fertility decisions among the Kove. In N. McDowell (Ed.), *Reproductive Decision Making and the Value of Children in Rural Papua New Guinea*. Papua New Guinea Institute of Applied Social and Economic Research, Boroko. In press.
Counts, D.A. and D.R. Counts (1983). Father's water equals mother's milk: The conception of parentage in Kaliai, West New Britain. *Mankind* 14(1), 46-56.
Mead, M. (1935). *Sex and Temperament in Three Primitive Societies*. William Morrow, New York.

CHAPTER 11

INFANT CARE AND FAMILY RELATIONS ON PONAM ISLAND, MANUS PROVINCE, PAPUA NEW GUINEA

ACHSAH H. CARRIER

INTRODUCTION

The purpose of this chapter is to describe the basic techniques of childcare practiced on Ponam, a small island off the north coast of Manus Island, Manus Province, Papua New Guinea. In order to place these practices in their social context, the description of infant care and feeding is preceded by a discussion of marriage and husband-wife relationships and is followed by a discussion of both husbands' and wives' roles in assuring that infants are well cared for while the household is maintained.

The great importance of proper infant care to the Ponam people can best be illustrated by a brief summary of certain of my experiences during the research period:

My husband and I first went to Ponam in October, 1978, and stayed there until February, 1980. We returned in December, 1980, for two months when I was pregnant with my first child, and returned again when my son was three months old, one and a half years old, and two and a half years old. I learned some things about child care during my first period of fieldwork simply by observation; I learned more when I went back the second time, as women tried to teach me the things they thought I would have to know; but none of this prepared me for what happened when I went back for my first visit with my son.

When we lived on the island my husband and I tried, like most anthropologists, to do things more or less as the people did — to eat their food, follow their customs and learn their manners — but, of course, we were never completely successful at this. I never became a good cook; my husband never caught many fish; we were rude occasionally without knowing why. However, people were, for the most part, extremely forgiving. They explained how we should do things, then said that different places and different people had different customs and that our errors were understandable, forgivable, and not usually very important. But they were not such cultural relativists about child care.

189

When we arrived in the provincial capital in September, 1981, people were waiting to take us to the island. The women began right away to correct my behavior and insist that I do things properly, in a way that they had never done before. The very first thing they said to me was: "The sun is hot. Where is the baby's umbrella?" When they learned that I did not have one, they said I must go to the store right away and buy one. When I came back with a colored one they said that I should have bought a black one instead, but they did not insist that I go back and exchange it.

This intense and demanding concern about the way I cared for my child continued unabated for the length of my stay. Women insisted that I do things properly: That I feed my son, wash him, dress him and carry him properly at all times. They were angry if I did not and would tolerate no excuses. Several times as I walked down the street women came running out of their houses to yell at me for carrying him improperly or not holding my umbrella at the right angle to provide him with shade. They later reminded me about it saying, "I don't want to see you do that again."

However, some of the techniques of child care that Ponams followed were seen as purely part of Ponam custom, not obligatory for someone of different ancestry, and women made it clear which these were. For example, sometimes my son would nurse too fast and begin to choke or cough. Instantly someone would shout, "Stand him up." Later they would explain that if a baby chokes, its mother should make a choking noise herself which would move the choking from the baby to the mother and thus protect it. Although people explained this to me numerous times, they always told me to stand the baby up first, and they never absolutely insisted that I do the sympathetic choking. Further, they began their description of the choking technique with, "We Ponams do this ...", indicating that the action was customary rather than obligatory.

Apart from these self-consciously customary techniques, Ponams were intolerant of deviations from their standards of child care. My attempts to justify my actions by saying that Americans do things differently were unacceptable. If Americans did things differently then they were wrong, and I should learn to do them the right way. Whether or not I learned and followed the uniquely Ponam customs which I as an anthropologist had come to study, was my own business. People were willing to teach, but were not particularly concerned about failure. That would be, after all, my own problem. However, a failure to learn and follow the universal necessities of child care would be not only my problem, but my child's problem as well, and Ponams would not tolerate that.

Although it was difficult to endure their constant criticism and advice, the Ponam women invariably were more skilled at mothercraft than I. They quieted my son when he cried, fed him, washed him, and amused him more quickly and easily than I. Further, there was almost nothing, apart from very

---

†During the years 1979-1982, there were two neonatal deaths out of a total of 30 live births and 1 stillbirth. Both of the children who died were twins. The infant mortality rate was thus 66 per 1000 live births, which was approximately the rate for Manus as a whole and considerably below the rate for most other provinces in the country (King and Ranck, 1982).

early feeding, that they did which would contradict Western medical opinion (though much, like the mother's sympathetic cough, would be thought harmlessly irrelevant). Certaintly their children were very healthy, given the conditions in which they lived.[†]

METHODOLOGY

The data upon which this chapter is based were collected during approximately 18 months of participant observation on Ponam Island from October, 1978, through the present. During this time period, the author has also continued to have regular contact with Ponams residing in Port Moresby, the capital of Papua New Guinea. Infant care and feeding practices were a focus of the fieldwork in the period following December, 1980.

SOCIAL BACKGROUND

Ponam is a tiny raised coral island about $2\frac{1}{2}$ miles long and seldom more than 200 yards wide, located about 4 miles off the north-central coast of Manus Island. It is relatively infertile and Ponams are not sufficiently interested in gardening to undertake the intensive cultivation techniques necessary to raise crops there. Instead they fish, taking seafood from their extensive reef for their own consumption and for trade at the weekly markets shared with mainland Manus gardeners. Inter-village trade in produce, seafood and manufactures was once crucial to the Manus economy (Schwartz, 1963), but has become less important since the Second World War, as cash, imported commodities and wage labor have penetrated and undercut the indigenous economy (Carrier and Carrier, 1985). Ponams still trade at local markets, but now they also buy many imported and local goods for cash. Because they can earn almost no money at home, however, they depend on remittances sent by migrants working elsewhere. In 1979, about 200 of the slightly more than 500 Ponams were migrants (Carrier, 1981).

There are 14 patrilineal clans on Ponam which are important for island politics and property but not for child care. The island's nine totemic (*kowun*) matriclans are relevant, however, because membership in them determines the set of food taboos (*hambrun*) to which Ponams, and particularly Ponam children, are subject.

Islanders live in a single village divided into hamlets, most of them centered on a clan men's house. Ideally residence is patrivirilocal, with a young man bringing his wife to live in a new house near his father's house in the hamlet surrounding his clan men's house. Although a young couple lives near the husband's father, the two families do not form joint households. On the contrary, the young couple should become economically and socially independent as soon as possible. The young housewife should take, and be allowed to take, full responsibility for her own household without depending on her in-laws for advice or labor. The advice, at least, usually continues far longer than she wants. Thus, from the time of their establishment at marriage, Ponam households are largely independent and, as a consequence, children are cared for primarily by their own parents.

Because each household is largely independent of others, couples depend heavily on each other. Mutual compatibility is of considerable importance. Such compatibility is promoted by the fact that both men and women marry late to partners of their own choosing. In 1979, only 43% of women and 32% of men age 20 to 29 and 76% of women and 82% of men age 30 to 39 had ever married. There is also a high rate of singleness: 16 % of women and 13 % of men age 40 to 59 had never married. The vast majority of residents who marry, marry endogamously. In 1979, 96% of male and female residents over age 20 had Ponam fathers.

In the past marriages were arranged and even today it is proper for the groom's family to make a formal request to the bride's family for her. All the family on both sides should agree to the marriage before the engagement is announced, but usually they do this only after the couple themselves have made clear their wish to marry. Couples cannot be forced to marry against their will and it is difficult to prevent them from marrying if they are determined. When families are reluctant to arrange things properly, the couple may simply move in together.

As the high rate of singleness suggests, Ponam think that a single life is preferable to a poor marriage. Single people are not excluded from social, political or economic affairs, though it may be more difficult for them to play a full part. Some have difficulties in old age without children to care for them. Parents do not place great pressure on their children to marry, but often attempt to dissuade them from making unsuitable matches. This attitude is possible in part because marriage is not brokered by bigmen, and marriage manipulations are not the cornerstone of political or economic success. It is important for people to choose their marriage partners wisely because, when Ponams converted to Catholicism in the 1920s, they accepted that individuals may have only one spouse. They interpret this doctrine as forbidding widow and widower remarriage, divorce and polygamy. They are lax about applying the rule about divorce and remarriage to those married to non-Ponams. They hold strictly to the rule in the case of island marriages, however, and do not even allow couples to separate for more than a few weeks at a time.

Each married couple's household is an independent economic unit which may cooperate with others in a number of enterprises, such as fishing, housebuilding, and exchanges, but does not share responsibilities with them. Although husband and wife are jointly responsible for the success of their household, their labor is divided. Women are responsible for work inside the house: for food, for children and for maintaining household goods. Men are responsible for what is outside: for property and for fishing. Where these spheres overlap, husband and wife share responsibilities. Thus, both men and women tend the land and coconuts, visit the market and fish in the lagoon. They also share responsibility for the household's success in ceremonial exchange.

Ponam has no gardens and thus Ponam women are not burdened with this task, which takes so many Papua New Guinea women away from home and their children. Ponam women's major work is housekeeping, which means that they are available at home most of the time. Because fishing and marketing can be done by both men and women, most mothers of young children are able to rely on their husbands to supply food during their child's first months. The only exclusively female task to take women away from home is cutting firewood, which takes only one or two days of work per month.

Women depend very heavily on the help they receive from their husbands, at least during the early years of their marriage. (Children do not contribute significantly to the household until they leave primary school, at which attendance is universal through grade six.) Although a man's primary obligation is work outside the house, he may do housework as well. While caring for the men's house, men regularly sweep, clean and cook. If a man's wife is ill or away from home, he may do these or any other pressing jobs without resentment and without risking censure or loss of pride that normally provoke men and women to do their proper jobs. The maintenance of the independent household for which he is jointly responsible is more important than the sexual division of labor.

Consonant with this, women tend to work alone when they do jobs which can be done by one. While they work collectively to prepare for exchanges and feasts, they do not routinely help each other with basic housework. A few women may get together to cut firewood or scrape coconuts (each one taking her own share), and neighbors often sit together to do their individual handwork, but each woman is responsible for her own housework and spends most of her time at home alone or with her husband and children. This does not mean that women live in a kind of *purdah*, or that they are unfriendly and uncooperative with one another. There are numerous occasions and duties, usually associated with ceremonial exchange, which bring women and children out of their house and into the sociable company of others. Kin are usually willing to help a household in trouble. However, in the normal course of events, a woman is expected to stay home and work for her family, rather than to ask others to do her work for her.

INFANT CARE

Ponams believe that infants who do not yet sit or crawl only fret or cry when something is definitely wrong and, therefore, parents should always do whatever is necessary to calm and comfort them. Unlike American parents, for example, Ponams do not believe that infants learn patience and self-reliance by being left alone or left to cry. They believe instead that infants who cannot speak or show any comprehension of language, cannot receive instructions. Therefore, they cannot learn social or moral responsibility and it is wrong to try to force them to do so. Parents never allow infants to cry it out, and ideally never allow them to cry at all. The best parent does not simply respond to a child's cries but anticipates them, providing all that the child wants before it can cry for it. Parents admit that this is difficult, in part because babies are so variable; not all babies are like each other, and no baby's wants remain constant throughout its infancy. Therefore, a parent must remain vigilant, observe the child's pattern of behavior and learn to anticipate it. When a mother fails to anticipate her child's needs, and it cries, she should act instantly to right the wrong. It is striking that when an infant cries, its mother runs to pick it up. She does not walk.

Although infant care is difficult, there is no excuse for failing to do it well. Just as infants lack the capacity for moral responsibility, so they lack personality. There is no such thing as a difficult infant, only careless, lazy or ignorant parents. This is not true of older children who can walk and run and, most importantly, can talk. Having learned language, they are capable of receiving instructions and thus are capable of disobedience. They may be bad out of their own intention or because their parents did not teach them to be good. However, infants cannot understand speech, cannot be told how to behave, and therefore cannot be disobedient. They have needs but no obligations. Instead it is the parents who have the obligation to care for the infant.

Breasts and Bottles

All Ponam women begin feeding their children at the breast, nursing on demand without concern for timing the length of feeds or the interval between them. They say that a healthy infant will nurse as much as it needs to. They accommodate themselves to this by offering the breast frequently whenever the child is awake, but never trying to coax an unwilling infant to nurse. Women prefer brief and frequent feeds, usually on one breast at a time. Only when a woman feels engorged or tries to soothe a reluctant baby to sleep does she try to encourage long nursing at both breasts. Usually when a baby seems restless or cries, the mother puts it to the breast for a few minutes or until the child seems to lose interest. Except when particularly busy she is willing to nurse again in an hour or less.

Ideally a mother should always be free to nurse her infant at any time and therefore should not leave the child's side for the first few months unless it is absolutely necessary. Once the child can see and hear and smile it can go without nursing for two or three hours and may safely be left for this amount of time. Women do not leave often and do not require the child to go without milk for this amount of time if they are with it. After about six months, when solid food becomes a substantial part of the diet, a child can be left for a morning or afternoon, though not for a whole day. Still this is not done on a regular basis. They continue to be available to nurse more or less on demand until the child itself begins to break away.

No island women bottlefeed in order to gain freedom from their children, and generally they disapprove of those who do so. They say that breastmilk is best for babies because it is provided for them by God. They view formula as evidence of laziness, as a way of getting out of work which a woman should do herself. Furthermore, no island woman has enough money to bottlefeed, as all are well aware.

Despite the general air of disapproval, three children (with urban working parents) were bottlefed on the island during the research period. One illegitimate girl was fed by her grandmother while her mother worked in another province. One set of twins[†] continued the bottlefeeding they had begun in the city when they and their mother returned to Ponam for a six month holiday. This second mother began giving her twins supplementary formula by cup and spoon when they were about a month old because they were gaining weight slowly and she believed she did not have enough milk. She said that they preferred formula to breastmilk, ate well and gained well, and she soon substituted bottle for cup and spoon. When she did this, first one twin and then the other rejected the breast altogether.

Both of these mothers bought bottles on prescription as is required by Papua New Guinea law. None of the children suffered any of the difficulties often associated with bottlefeeding in poor communities, at least in part because Ponam has abundant fresh water and women are very concerned about cleanliness. Ponam housewives are expected to keep a thermos of boiled water on hand to provide tea for meals and visitors, and thus it was little trouble to provide some extra for the babies' formula. Furthermore, both of these mothers were sufficiently wealthy that there was no temptation to water down the formula.

Unlike the mother of twins, most of the Ponam women breastfeed in the capital city, Port Moresby. Generally migrant men do not marry until they are able to support a wife and children with their own wages, and working women are expected to resign at marriage or at least when they begin their first pregnancy. Thus, they do not need to bottlefeed in order to take employment. However, urban women are more attracted to the idea of bottlefeeding than rural women and often wean their children directly onto bottles of fruit juice or powdered milk rather than onto a cup. Urban women also often wean earlier than island women, usually at less than one year and before they have conceived another child. Because urban women are relatively wealthy, they usually have plenty of food easily available to

---

[†]Ponams have an unusually high rate of twin births. In 1979-1982, 3 out of 28 deliveries were twins. Although women complain about the burden of overwork which twins create, insufficiency is not usually thought to be a problem.

offer their children, and thus do not need to rely on the breast as a pacifier in the way that island women do.

Wetnursing, a common practice in much of Papua New Guinea, is not much practiced on Ponam. Women have no objection to it in principle, but cannot easily find another mother who adheres to the precise combination of maternal and paternal totemic avoidances which their child requires and, therefore, seek wetnurses only as a last resort. In any case, a nursing mother should never leave her child for longer than the nursing interval, and thus the problem should not occur regularly.

Solid Food

Ponams make a qualitative distinction between breastmilk (*sus*) and food (*kana*). Milk is not another kind of food for babies, but is in a category of its own. Infants can survive on milk alone but such a diet is not always satisfying. Mothers are alert for the pattern of fretful crying which indicates that a baby is hungry. A healthy infant who is satisfied with its diet sleeps most of the time. It should wake to nurse, be bathed and played with for a little while, and then go back to sleep again or at least lie quietly. A child who does not do this, but instead nurses, turns away from the breast, sleeps for a moment, then wakes crying only to repeat the pattern again, is considered to be hungry and in need of solid food regardless of its age. Parents expect this to occur during the first month of life and speak with pride of the child who demands food in the first few days of life. Perhaps Ponam women rarely complain of insufficient milk because they interpret as hunger the kind of fretful crying which is elsewhere seen as a symptom of insufficiency.

The most common first foods are winter squash and papaya. These are cooked, mashed, cooled until tepid and fed with a cup and spoon, often reserved for the baby's use. Ripe bananas and mashed green coconut are possible, though not preferred foods. Root crops are said to cause constipation, and the staple adult foods, fish and fried sago, are too difficult to chew.

The object of this early feeding is to keep the child happy and comfortable. Mothers believe that healthy infants will eat and drink as much as they need of their own accord. They regularly offer food but never attempt to force them to eat more than they want. However, mothers anticipate that a child in its first month will eat one or two teaspoons of food twice a day, and offer this when they can. The amount of food offered is gradually increased, and a child of four or five months will be offered at least a quarter to a half cup of food twice a day.

Nerlove (1974) argues that in societies where women make a substantial contribution to subsistence production, they are more likely to introduce supplementary feeding in the first month of an infant's life than in those societies where women do not. The Ponam case does not support this conclusion (nor were the data for the neighboring Manus people included in her sample). Ponam women work primarily at housekeeping and ceremonial tasks, not at subsistence production, and this seems to have been true in the past as well. Previously, however, women had the demanding job of making shell money, which has now been replaced by cash. It may be that their commitment to manufacturing rather than subsistence production was an inducement to supplementary feeding. However, this does not explain the early feeding today.

Protection from the Environment

An infant should never cry, and if parents are properly alert it will almost never need to. In addition to being well fed, infants must be protected from the elements, which might cause sickness or discomfort. They should not be exposed to direct sunlight but should always sleep in a shady place and be shaded with an umbrella when going out. They should also be sheltered from wind, rain, dust, smoke and so on.

Babies are considered more susceptible to heat than adults and washing is thought to be the most effective way to keep them cool. A baby who has been nursed and fed and still frets is probably hot and should be washed, or at least wiped with a damp cloth. This is particularly true of babies who wake in the night. Washing is felt to help prevent sores on the chest and neck caused by allowing the baby's saliva to lie on the skin, and it helps prevent diaper rash. Ponam babies wear diapers often during the first few months, when they are frequently carried in a sling or sleep on a pillow bed. In addition to regular washing, prompt diaper changes are considered necessary for the baby's comfort and well being.

Ponam women carry their babies in a cloth sling. A newborn is carried cross-wise so that it can sleep. Once its back is strong enough it is tied into a sitting position so that it can see. Once the child can cling to its mother's back, she carries it there instead of on her chest. Ponams point out that being carried this way is good exercise for the infant and encourages more rapid physical development than is possible for infants who spend all their time in bed. The reported drawback to the practice is that if babies are handled too often they become hopelessly dependent. Therefore, a baby should be put down whenever it is possible to do so, particularly when it sleeps. Although night feedings are proper, a mother should not allow her baby to become accustomed to sleeping next to her, especially with her nipple in its mouth like a pacifier.

Ponams tell a story about a man who had to clean and smoke his own fish when he returned from night fishing because his son was so used to sleeping next to his mother at night that he cried whenever she got out of bed. The story illustrates not only the importance of avoiding over-dependence, but also the priority put on infant care. Although cleaning and smoking fish are women's jobs, they should not be allowed to over-ride her primary obligation to her child.

THE OLDER CHILD

When a child reaches six months or so, the pattern of care changes. Parents are still extremely solicitous of children less than about two years old, but they no longer treat them quite so carefully. Babies who can sit up and begin to move about are strong enough to endure sun and wind and rain and rough handling. However, they still need constant supervision. Thus mothers remain tied to the village, and trips to the market or to gather firewood require her to find another caretaker. Fathers or mature children normally perform this task, and it is inappropriate for a woman to rely too often on others outside her family. When she does seek help she normally turns to another woman without children to care for, usually her mother, husband's mother, unmarried sister or husband's sister. The numerous mature unmarried women are a great help to their families in this way. Child-care difficulties may arise for those women with several young children, married sisters, infirm elderly parents, or husbands who are too busy with their own work. An infant or toddler will not be left in the sole care of another child. Adults and mature teenagers are preferred caretakers.

Once a child begins to reach, sit up and move about, it is allowed to reach for the food that others eat. Once it can do this, its mother will start to give it pieces of food when she or others are eating. The most common foods given at this age are bread, hard crackers, fried sago, banana and green coconut. Fish, soft meat, and stringy vegetables are introduced gradually as the child approaches its first birthday. During the early reaching stage the child's mother will continue to cook special meals and feed the child with a spoon, but she stops when the child can eat by hand, generally by one year. During this six months the child also learns to drink from a cup, being given water, tea, chocolate drink and fish soup.

When a child is able to feed itself, it progresses rapidly toward a normal childhood diet. It is rarely forced to eat anything, playing with what others eat, picking up what it wants and being fed from its mother's and father's plates. Once a mother is sure that her child can eat without spilling, she gives it a plate of its own, usually containing about a cup of sago or rice and a few ounces of fish, and requires the child to finish it.

Women usually serve hot meals of sago and fresh or smoked fish, or rice and tinned fish, in the morning and the evening. Those with young children also try to provide cold snacks (leftover sago or rice and fish) in the middle of the day as well. When plentiful, this food is left in a cupboard where any child can find it. Women try to ensure that they can feed anyone as much as they want to eat at any time, although this is not always possible. Hunger is thought to lead to serious illness, fainting and rapid starvation, and people gain no prestige being thin.

Each Ponam belongs to one of nine totemic matriclans and contact with any of his or her clan's totem plants or animals causes degenerative disease such as cataracts, deafness, tooth and gum decay, and skin diseases, unless the proper curative prayers are pronounced by the senior women of his or her matrilineage. The members of each clan are forbidden to eat several species of fish and one or two plant or animal foods. In addition, each matriclan identifies a number of plants and animals as forbidden only to young children on the grounds that they will cause speech impediments, delay toilet training or otherwise hinder the development of a fully social being. A young child is also required to avoid the totems of its father's as well as its mother's matriclan for the first few years of life. Once a child is about three years old, however, it begins to eat as an adult member of its matriclan.

Just as a mother must be careful of what her child eats, so a pregnant or nursing woman must watch her own diet and avoid anything which her child cannot eat, that is, all the foods forbidden to her own and her husband's matriclan.

BREASTFEEDING, CONCEPTION AND WEANING

Unlike Port Moresby women, island women usually breastfeed until they conceive again, usually when the nursing child is between 15 and 30 months old. Newly married women tend to space children more closely than do those with many children.

The frequency of nursing decreases after six months, when the child begins to eat a substantial amount of solid food, and drops again radically when the child begins to walk, at about one year. During the first six months or so the mother tends to nurse her child frequently throughout the day. Once the child is eating adult food she feeds it to distract it from the breast when food is available, usually at mealtimes. She nurses it when there is nothing else to eat. Once a child starts to walk, however, nursing between meals drops off as well, for the child soon begins to spend its time with other children instead of sitting around with its mother, and the mother refuses to nurse if she is busy. As a consequence the frequency of nursing drops to as low as six or seven times a day, which often seems to be as much for the mother's comfort as the child's. A toddler rarely comes to its mother and asks to nurse. More often the mother swoops down on the unsuspecting though not unwilling child to offer it the breast. With such a decrease in nursing frequency, fertility is likely to return (Harrell, 1981). If she is not avoiding intercourse in order to postpone the next child, the mother will conceive again, at which time she must wean her nursing child entirely.

Old women complain that the current birth interval, commonly two years, is too short. In their opinion a mother should not conceive again until her nursing child can run and talk and be away from her for most of the day. As children begin to talk at about two years, the ideal interval would be nearly three years. This long spacing is deemed necessary for two reasons. First, if a child is forced to wean when it is too young, it never grows strong. Second, if a woman has children too close together, it is impossible to care for them properly and she never has a break between pregnancy and intense nursing in which to get fat again.

The old women's complaint that poor birth spacing has led to a decrease in the weaning age suggests that Ponam is following the trend toward earlier weaning that has been noted in other parts of Papua New Guinea (Doan, unpublished observations, 1983)[†] and throughout the world (Jelliffe and Jelliffe, 1978). Nardi (ch. 16) argues that an increase in women's workload has been a chief reason for the decline in the weaning age in Western Samoa and suggests that this underlies earlier weaning in other areas as well. Another explanation must be sought for the apparent decline of the weaning age on Ponam, however, for there has been no significant increase in women's workload. The Ponam women's conclusion that early weaning is the consequence of early conception is likely to be correct, for there have been changes in the domestic relations of husband and wife which are likely to increase fertility.

Prior to the Second World War there was considerable segregation of the sexes. A husband ate and slept in a men's house, not with his wife and children, and there was considerable social support for couples who wanted to avoid conception. However, sexual relations changed dramatically in Manus around the time of the Second World War (see Mead, 1956) and men and women began to adopt a style of life modeled on the western nuclear family. Now only men who are trying to avoid intercourse with their wives sleep regularly in the men's house, and Ponam women say that this is the only effective way to prevent pregnancy. Because there is now little male pressure to remain in the men's house, men are likely to return to their own houses if the men's house is full of guests, or if it develops a leaky roof or for many other reasons. When this happens, according to Ponam women, couples will be unable to resist each other, no matter how good their intentions are. Whenever women wean following conception, as is common in Papua New Guinea (Obrist, unpublished observations, 1983),[‡] it is likely that any increase in contact between husband and wife will contribute to a decrease in the weaning age.

Once a woman is sure that she is pregnant (after missing her second period, or sooner for experienced mothers), she weans her

---

[†]Doan, H.M. (1983). Infant care and breastfeeding in selected villages in Papua New Guinea. Paper presented at the twelfth annual meeting of the Association for Social Anthropology in Oceania, New Harmony, IN.

[‡]Obrist, B. (1983). The study of infant feeding: Suggestions for further research. Paper presented at the twelfth annual meeting of the Association for Social Anthropology in Oceania, New Harmony, IN.

previous child. Nursing is considered dangerous because the accumu-
lating blood which forms the fetus poisons the milk, making the
nursling lose weight and shrivel up.

Most women describe weaning as untraumatic. A child may cry for
a while, but as long as it has enough water it does not suffer harm.
Weaning does not usually occur until the child is well established on
an adult diet and has gained a considerable degree of independence
from its mother. Several different weaning methods are used. Some
mothers refuse to allow the child to nurse. Those who find this too
difficult send it to stay with a relative who will give the child attention
and food. In extreme cases, women may rub their nipples with betel
pepper or other hot substances to frighten the child away.

CHILD CARE AND WOMEN'S WORK

Infant care is time-consuming and exacting for Ponam women as it is
for women everywhere. The mother of an infant is tied closely to
home. She should never take a child to any place where it cannot be
cared for properly, and for at least the first six months she cannot
leave an infant for more than an hour at a time. A mother must plan
her life very directly around her child's needs. She nurses her child
every hour or so, cooks special food for it two or three times a day,
bathes it in warm water three or four times a day, changes its diapers,
washes its laundry, and listens for its cries. All of her other interests
and obligations, and the demands of her husband and older children,
must take second place to these needs.

The older baby is less demanding, but still requires constant
attention until it is about two. At this age, and most dramatically after
the birth of another sibling, the child begins to move away from its
mother and away from the house. This is particularly true of boys
who, once they reach about four years old, rarely come home in the
daytime except for meals.

In some societies, women frequently assume responsibility for the
care of each other's children (Conton, ch. 6; Chowning, ch. 10;
Morse, ch. 14; Katz, ch. 15). In such circumstances, it is not difficult
for a mother to find a caretaker for her child, and the major
hindrance to the resumption of normal work is the fact that she may
not be separated for long from her nursing child. It would be wrong
to assume that such a communal approach to child rearing holds true
everywhere in Oceania, however. Ponam women normally marry

H

island men and thus remain with their own kin with whom they have extensive ties, but these ties do not extend to communal responsibility for domestic affairs such as childcare.

A Ponam husband and wife are jointly responsible for the independent nuclear household which they form on marriage, and very rarely share that responsibility with anyone else. Husband and wife cooperate with others in tasks which require many workers, such as fishing and housebuilding, but they do not share labor freely in other circumstances. Laziness and frivolity are frowned upon and to seek help from outside the household constantly is a sign of laziness and poor management.

Thus, when a Ponam woman needs help in caring for her child or in doing jobs incompatible with childcare, she turns first to her husband and other children (although pre-school children do little work, all school age children are busy in school, and more than half of all adolescents are away from the island to continue schooling). Unless she has a grown daughter who can help, a husband should take over entirely the work which he and his wife normally share while she is housebound with a child. He should also do all of the routine caretaking while his wife bathes, makes brief visits to neighbors and so on, and even occasionally care for the child for longer periods. This allows her the freedom to tend to her culturally-prescribed primary obligation, the care of small children, while helping to maintain the desired independence of the household.

ACKNOWLEDGEMENTS

I would like to thank the governments of Manus Province and Papua New Guinea for their permission to do this research, the University of London Central Research Fund and the Papua New Guinea Department of Education for helping to fund it, and the people of Ponam both for allowing me to do the research and for making it so enjoyable.

REFERENCES

Carrier, J. (1981). Labour migration and labour export on Ponam Island. *Oceania* **51**, 237-255.
Carrier, J. and A. Carrier (1985). Colonization and adaptation: Ponam Island, 1900-1980. In J. Clay (Ed.), *The Plight of Peripheral People in Papua New Guinea, Volume II: The Island Situation*, Report No. 17. Cultural Survival, Cambridge, MA. In press.

Harrell, B. (1981). Lactation and menstruation in cultural perspective. *Amer. Anthropol.* **83**, 797-823.

Jelliffe, D.B. and E.F.P. Jelliffe (1978). *Human Milk in the Modern World.* Oxford University Press, Oxford.

King, D. and S. Ranck (Eds.) (1982). *Papua New Guinea Atlas.* Robert Brown Associates, Bathurst.

Mead, M. (1956). *New Lives for Old.* Victor Gollancz, London.

Nerlove, S.B. (1974). Women's workload and infant feeding practices: A relationship with demographic implications. *Ethnol.* **13**, 207-214.

Schwartz, T. (1963). Systems of areal integration: Some considerations based on the Admiralty Islands of northern Melanesia. *Anthropol. Forum* **1**, 56-97.

CHAPTER 12

WOMEN'S WORK AND INFANT FEEDING: TRADITIONAL AND TRANSITIONAL PRACTICES ON MALAITA, SOLOMON ISLANDS

K. GILLOGLY AKIN

INTRODUCTION

Studies have indicated the nutritional suitability of human milk for the human infant (Gaull, 1979) and its great importance for immuno-logical protection of the infant (Pitt, 1979). In view of this, much concern has been expressed in both popular and scientific writing about the effects of bottlefeeding on the infant, the causes of the trend toward bottlefeeding in the Third World, and the construction of social policy to counteract this trend.

The choice of bottlefeeding or breastfeeding may be influenced by the demands on the mother for her economic contribution to the family (Morse, 1982). Popkin and Solon (1976) and Popkin (1980) examined time and income constraints on the family, the resulting intrahousehold reallocation of duties when the mother worked for wages, and the compatibility of the mother's employment with child care. This was in a situation of wage labor. Other studies (Thompson and Rahman, 1967; Huffman et al., 1980) have shown that infant care was affected by the mother's work in rural communities, especially during planting or harvesting, which require intensive labor input in a limited period of time.

A continuing assumption is that only urban mothers are affected by the incompatibility of breastfeeding and work (Van Esterik and Greiner, 1981) and that woman's major role is as mother (Jelliffe and Jelliffe, 1978). However, it has been demonstrated that in societies in which women contribute substantially to subsistence activities, child care is adjusted to the mother's other responsibilities — often by early supplementation of breastmilk (Nerlove, 1974). Unfortunately, the patterns of supplementation with traditional foods and by traditional means have only rarely been studied.

In general, traditional infant feeding practices in the context of culture and society have been inadequately described. This study was undertaken to document traditional practices in their social and cultural context in a conservative rural area of the Solomon Islands, and to describe changes in these practices which have occurred among more acculturated members of this ethnic group. The major focus of the study was on women's activities.

ECONOMIC, SOCIAL, AND CULTURAL CONTEXT

Approximately 10 ethnolinguistic groups live on Malaita, a volcanic island 100 miles long and 25 miles wide. (See Figure 1.) There are about 3,700 Kwaio speakers in the East Kwaio district, and two major communities: pagan and *sukuru*.[†]

In 1979 through 1983, when the author was resident in East Kwaio district, the pagan Kwaio were largely a conservative people. They practiced the traditional religion and shared an emotional commitment to the ways of their ancestors. The sukuru Kwaio practiced Christianity of several denominations. Some individuals were recent converts, others were members of families that converted one to three generations ago. This group was less homogeneous than the pagans, but the sukuru Kwaio generally shared an outward, future

---

[†]*Sukuru* has been chosen as the term for the community usually referred to as Christian to avoid the implication that the life-style of this community, in and of itself, is an expression of Christianity as practiced in the Western world. ("Sukuru" comes from the English "school", referring to mission schools.)

The term, pagan, is used because of its conciseness and accuracy over other available terms and because of its consistency with other ethnographic literature (Keesing, 1982). Typically, the pagans are referred to as "non-Christians", which has the disadvantage of defining the pagans in terms of what they are not, and implies that they are something other than what they ought to be. To refer to them as "traditional" Kwaio is inaccurate because "traditional" has political implications that cut across the boundary of the two religious communities. There are traditionalist Christians throughout the Solomon Islands (Keesing, 1982; Burt, 1982). What is left are Kwaio terms used to refer to the pagans, by both Christians and the pagans themselves. These are: *ta?a wikiti* "wicked people", *ta?a gula i buri* "backwards people", and *ta?a ?itini* "heathen people". The term pagan has less emotional overtones than does heathen in the Solomon Islands.

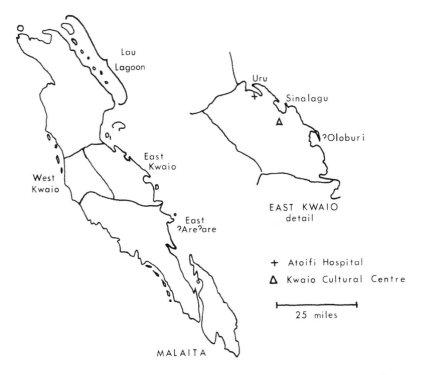

FIGURE 1   Map of Malaita with insert of the East Kwaio area showing Uru, Sinalagu and ʔOloburi.

orientation, and a desire for what they thought was a better way of life through acceptance of new ideas and Western technology. Thus, the area provided an interesting situation for the comparison of traditional and introduced infant feeding practices.

Mothers and caregivers in the two communities dealt with the problems of food provision and child care in different ways due to differing duties and economies. For some modernizing women, concerned about infant nutrition and performing one's duties well, and with the cash to buy the necessary supplies, supplementary bottle-feeding appeared as a viable addition to traditional infant feeding practices. This can be understood in the context of traditional practices and beliefs.

Settlement and Standard of Living

Inland traditional. The majority of the pagans lived in the "bush" (inland) in small isolated hamlets of one to fifteen inhabitants, with an average of nine people. Settlement was generally patrilocal.

The traditional subsistence system was one of swidden agriculture, supplemented by pig raising, foraging, hunting, and fishing. In the past, dry taro and yams were the staples, but following a taro blight in the 1950s, sweet potato replaced taro as the major crop. This was the major recent dietary change.

Emanuel and Biddulph (1969) estimated that the diet consisted of 85 % staples, 10 % protein foods and 5 % greens. A number of foods were forbidden or restricted to pagan females: hunted, stolen or ritually slaughtered pig, many species of fresh and salt water fish, opossum, fruit bat, eels, and about half of the species of birds eaten. Bananas were also forbidden to pagan women. Edible insects, common in the pagan woman's diet, were reported to be as good or better sources of protein than fish (Friedlander, Page, and Rhoads, personal communication).

The inland area was isolated due to cultural and geographical factors. Few inlanders (none of them women) knew the trade language, Pijin; virtually no one knew English. In 1979, the cash income was, on the average, about S.I.$10 (about U.S.$10) per adult male per year, earned by young men working on plantations on other islands (Keesing, personal communication). The main sources of foreign goods were young men returning home with cargo (rather than cash) and gifts from sukuru kin. The reciprocity system soon redistributed these goods, and so no noticeable differences in wealth existed among inland pagans. Shell money was far more important than cash in the local economy.

Social services were practically non-existent in the bush due to its great inaccessibility. Medical care was provided by a part-time nurse on the coast and occasional medical tours inland. The Adventist Mission Hospital at Uru Harbor was infrequently used by pagans due to intervening rough terrain and religious differences, although the Kwaio wanted Western medicine.

Coastal sukuru. Only patterns differing from those of the inland pagans will be mentioned here.

The majority of sukuru villages were on the coast. Settlement there was much denser than inland: 50% of the population of East Kwaio lived on less than 25% of the land. Sukuru villages were much larger than those of the pagans, ranging from 10 to 180 people nucleated around a church (Keesing, 1967).

Sukuru people had better access to cash and to modern goods and services. They had closer contacts with town and role models for change, partially by virtue of their religion, as early missions provided education and health care. They made more use of the mission hospital and the government clinic. Some of the coastal sukuru people depended on imported food to supplement their diets because they had the cash to buy it and because they were pressed for horticultural land. Cash was earned by copra-making, and more recently, cattle raising. A great deal of the community's cash came from relatives working out of the district. Figures for cash income on the coast were unavailable.

Nutrition and Health

A few nutrition surveys have been carried out in North and South Malaita. These populations are not strictly comparable to the inland Kwaio, or even to the coastal Kwaio, as the subjects of the surveys were coastal dwellers heavily involved in the cash economy and suffering from land shortage. In South Malaita, 3%-5.8% of the subjects were found to suffer from borderline malnutrition and 0.8%-1.5% from severe malnutrition, according to their Middle Upper Arm Circumference (MUAC)[†] (Gude, 1979). The Malaita Nutrition Survey (Department of Community Medicine, Papua New Guinea, 1979) reported an average of 11.6% mild or borderline malnutrition and 1.1% severe malnutrition, using MUAC. Using weight-for-height independent of age, moderate to severe malnutrition was found in 1%-5% of the study population, and mild malnutrition in 15%. Mild malnutrition ranged from 23% in 1-2 year olds to 7% in 3-4 year olds.[‡] Severe malnutrition was nearly always found to be due to specific circumstances such as severe illness or death of the mother.

A survey of the illnesses of Kwaio children in 1966 found skin infections, pediculosis and scabies, burns, respiratory illness,

---

[†]The standards were: above 13.5 cm., healthy; 13.5-12.6 cm., mild malnutrition; and below 12.5 cm., severe malnutrition.

[‡]The study population consisted of 634 children in North Malaita and 625 children in South Malaita, all under 5 years. Harvard Standards were used in the weight-for-height measurement.

hepatomegaly and lymphadenopathy to be common; enlarged spleens were common but acute malaria was rare; 4% had symptoms of yaws. Among the pagan bush Kwaio the low incidence of gastrointestinal disease was "striking", as this was the fourth leading cause of morbidity and the third of mortality in the Solomon Islands at that time (Emanuel and Biddulph, 1969). In comparison, at the time of the present study there were fewer burns, no yaws, and more acute malaria.

Gastrointestinal illnesses were observed more often among children in coastal sukuru villages than in the bush. Compared to pagan settlements, infectious diseases spread more quickly in the densely populated villages. Drinking water and the position of latrines were serious problems. This was observed by the author and by Malaita Province medical officers who toured the area.

Religion

Traditional.   The ancestors laid down laws for the behavior of their descendants, and punished the infraction of these laws with illness and death. Of special importance to this study are taboos regulating the actions of women.

Figure 2 shows the layout of a typical Kwaio village. The *tau* "men's house" and *ba?e* "shrine" were *abu* "sacred" areas which only males entered. These were on the uphill slope of the village.

The *bisi* "menstrual hut/area", *goo?itafu* "latrine" and *falegwari* "birth hut/area" were female abu areas. Males were not allowed to go to these areas once they had eaten food with men who had eaten sacrificial pork or had gone to the men's abu areas. Thus, a male child accompanied his mother to the bisi until he ate with men of his settlement, at which time he was weaned.

When a woman menstruated, she immediately left the dwelling house area and her possessions. She lived in the bisi. She was allowed to walk and forage in the forest and visit the women's area of other hamlets. She had a separate garden from which she harvested food. Women visited her in the menstrual area, and they shared food there. Most females moved freely between the secular and women's abu areas as long as objects from the two areas were not carried back and forth.

A woman planted a separate garden for birth while pregnant. She also built two connected shelters for herself and the unmarried girl

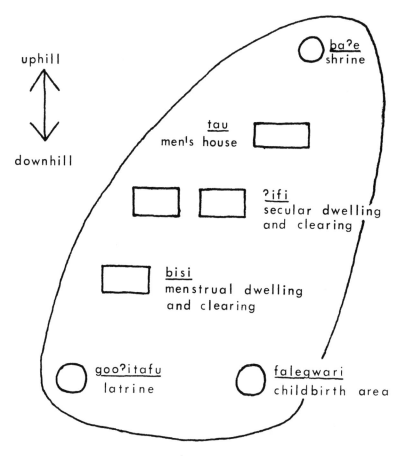

FIGURE 2   Diagram of a traditional Kwaio village.

that served as her companion after the birth. The two were secluded in the falegwari for 7 to 14 days following birth (the length of time differed from lineage to lineage). Visitors were allowed but could not touch the infant, mother, helper, or hut. The helper harvested all food, collected fuel and water, and cooked. When the seclusion in the falegwari ended, the three ritually ascended to the bisi. The next day, the infant was seen by men and held by women. In five days, the helper and infant ascended to the house and the helper returned to her home. At 12 to 19 days after the birth, the mother began to harvest her own food. She did not do any household work as she was still forbidden to visit the secular dwelling and gardens. She and her infant were restricted to the bisi for a minimum of 20 to 40 days.

Taboo violations included overstepping any of the above restrictions. If any of the women's bodily eliminations occurred in the house, it was torn down and pigs sacrificed to expiate the pollution.

Sukuru. The sukuru people were followers of three Christian denominations: South Seas Evangelical Church (SSEC), Seventh Day Adventist, and Roman Catholic. The overwhelming majority of sukuru people were SSEC or Adventist. These two sects were fundamentalist and evangelical. Missionaries introduced new housing styles, clothing, and new cultigens, among other things. Converts were encouraged to move into large mission villages on the coast (accessible to missionaries and government officials). The Adventists in particular emphasized Western health care, nutrition, and hygiene in their religious teachings. The Adventists also wrought a significant change in their converts' diets by forbidding pork, shellfish, and insects.

Women's Work and Duties: A Comparison Between Pagan and Sukuru

Pagan women, coastal or bush, were heavily burdened with work. Marriage and children increased that burden. A married woman helped garden for her in-laws and cognates, her own household, and for feasts. Women did the majority of garden work, including the men's work of clearing and fencing if the men were at the plantations.

All pagan women fed several pigs for ritual purposes, for relatives or their husbands, for sale, or for feasts. Many planted produce for trade. When at rest, most made shell money or crafts to sell for shell money. Because of taboo violations, houses were frequently replaced, which necessitated finding scarce thatch. When the entire hamlet moved, a new area was laboriously cleared.

The duties of coastal sukuru women were somewhat lighter. Certain chores were shared more efficiently in the large sukuru villages. As sukuru people did not observe the taboos, all work necessary in connection with their observance and expiation was eliminated. Sukuru men were encouraged by the churches to work in the gardens even at such female chores as harvesting. Furthermore, clearing was less strenuous as gardens were not left fallow for as long as in bush areas. More use of imported goods relieved women of food- and fuel-getting chores, as it was men who earned the cash. Fish was a more important part of the diet than in the bush, and fishing was a male job. In general, part of the subsistence burden had shifted from women to men. Additional duties of sukuru women included laundry, mending, dishwashing, and church work, which did not obligate them to be absent from home.

Nevertheless, even quite progressive families did not adopt the system in which the mother is exclusively a homemaker; women were essential to the subsistence economy, and women of both communities earned a great deal of prestige from this role.

METHODOLOGY

Data on feeding and child care practices were gathered betwen September, 1979, and March, 1983, in the Sinalagu Ward of East Kwaio, Malaita, while the author was employed as an anthropologist and community development worker at the Kwaio Cultural Centre. Supplementary data were collected from Uru and ?Oloburi Ward informants.

The major research technique was the standard ethnographic one of participant observation. Child care practices were observed over a period of $3\frac{1}{2}$ years, particularly among the coastal slope pagan communities, but also among pagan communities further inland, and in sukuru villages. In addition, 25 Kwaio mothers participated in semi-structured interviews. Data on socioeconomic background, household census, childhood illness, as well as infant feeding and care practices were collected. The interviews took from one to three hours to complete. When possible they were conducted privately, but more frequently at least one member of the interviewee's family was present and involved. All interviews and discussion took place in the Kwaio language, in which the author was fluent.

The sample was almost evenly divided between pagan (12) and sukuru (13) women. Of the total group, 64% (17) were from inland communities and the rest were from the coast. Forty-eight percent (12) were from Sinalagu proper, 28% (7) from the Sinalagu/Uru border area, 8% (2) from the Sinalagu/?Oloburi border area, 8% (2)

from Uru along the Sinalagu border, and 8% (2) from ?Oloburi proper.

Of the pagan interviewees, 8% (1) were 20-34 years old, 33% (4) were 35-49 years, 42% (5) were 50-64 years, and 17% (2) were 65-79 years. Two lived on the coast and ten lived inland. Of the sukuru women interviewed, 62% (8) were 20-34 years, 8% (1) were 35-49 years, 15% (2) were 50-64 years, and 15% (2) were 65-75 years. One lived in town, seven inland and five on the coast.

Two interviewees were originally from other language groups entirely. One was a sukuru woman from the north of the island, working as the health aide at Sinalagu Clinic at the time of the study, and the other was an older pagan woman from the southern part of the island. These two women both reported practices consistent with those of the inland pagan women.

Further information on beliefs and practices was offered spontaneously in informal discussion with mothers and caregivers (females and some males 14 to 85 years of age).

All verbal reports were cross-checked with other informants and verified by observation when possible. Nearly all aspects of infant and toddler care were observed, the major exceptions being the practices immediately following birth: use of colostrum and the coming in of milk. Only one case of insufficient lactation was actually observed due to its infrequency.

Uru practices were not actually observed. Data on ?Oloburi practices were gathered through the interviews and a very short period of observation. The observational data are thus biased toward the practices of younger (18 to 50 years old) pagan women of Sinalagu with whom I lived and worked most closely. Data on coastal sukuru practices are weaker because I lived inland; younger sukuru women were willing to visit for an interview, but the terrain made the trip difficult for older women. When I visited the coast, circumstances frequently made interviews difficult, so data on these women are primarily observational.

Traditional practices in all three wards differ mainly in the introduction of solid foods. This is true both synchronically (as evidenced by comparison of reports from women of different areas) and diachronically (as evidenced by comparison of reports from older and younger women). Newer sukuru practices, however, were found to differ from traditional practices on a number of points.

TRADITIONAL INFANT CARE

These practices were followed by bush pagans and sukuru families of traditional orientation. Coastal pagans also followed these practices, with a difference which will be noted. The stages of the child's growth used in the following section are those that the Kwaio themselves considered significant, and are correlated with age in months.

Mother's Diet

After birth, women reported increased appetite and desire for *kofila* "condiments". They ate as much as possible of the staples and searched for fish, mangrove pods, and shellfish. These foods were believed to stimulate lactation, as were greens. Women who lived far from the source of these craved edible insects and frogs.

Implicit in their statements was the belief that what the mother ate was passed to the child through breastmilk. Fish believed to cause digestive problems in the child were forbidden to the mother. In some lineages, women were not allowed to eat foods that were forbidden to her nursing child as part of a traditional cure, for example, foraged yams, tree fern (*Cyathea* sp.), papaya, pork, crab, and prawns.

The Breastfeeding Cycle

The newborn. The infant was called *biibiu* or *ʔita*. Women said that breastfeeding started within 12 hours after birth. The milk was first expressed by hand to make it "open" and "clean" and to remove "bad" milk at the surface. This practice was followed after every absence from the child. The breast was milked only a little, and the infant reportedly received most of the colostrum. The milk was reported to come in 24 hours after the commencement of breastfeeding. The colostrum was referred to as *malatarufaʔi* "like water". Breastmilk was called *susu*.

The newborn was nursed on demand. Its needs were immediately met, as it was believed that a crying infant attracted dangerous wild spirits and that crying caused the infant's mouth to become dry, resulting in the infant's death.

Feeding of solid food started early. About ⅓ of the pagans questioned first fed their children solids one to two weeks after birth, and another ⅓ fed solids three weeks to two months after birth. The first solid food introduced was premasticated sweet potato or taro, given mouth-to-mouth in small well-chewed amounts (*mariko ʔani* "premasticate for"). Afterward, the breast was given to make the infant swallow. Later, when the mother left the infant in the care of a non-lactating woman, this woman followed the same procedure.

Women said they gave solid food because the baby wanted it. They thought the infant was hungry when it was restless, fussy, and refused the breast. The infant appeared satisfied after being fed. Early introduction of solid food was believed essential for the child's growth. One woman explicitly stated that she fed her children solids early so that she could leave them to work as long as was necessary. Solid food was given by caretakers and other relatives as well.

Women in seclusion often reported boredom, and so foraged in addition to the little garden work required to meet their needs. The infant was not left longer than four hours at this point, nor did the mother leave every day. Occasionally an especially industrious mother left her child for six hours. The infant did not accompany the mother because, until it could sit alone, it was considered vulnerable to sorcery, the ill will of wild ancestors, and the rugged physical environment. The mother might take the infant to visit her natal hamlet, however, if not too distant.

Two to four months. After ending the birth isolation, the mother more or less resumed her normal duties. When the infant was deemed fat and strong (*ʔofuʔofu*) the mother left it for longer periods of time. It was still fed on demand when the mother was present. She gave the infant a good feed in the morning, emptying both breasts, then left for an average of four or five hours, although seven hour absences were not unusual. She did not fetch large amounts of produce at one time and finish her chores in one trip as women usually did, but rather arranged to return early to feed the child and then go to work again closer to home. Women who did leave for a greater length of time had started to feed solids to their infants. Infants were not taken to the gardens. Because the infant was fed frequently in the night, the mother did not attend night festivities.

Some coastal pagan lineages first gave premasticated food to their infants in this period, rather than during the ritual seclusion following birth. This seems to be a long-standing difference in practice, as reported by older informants.

Five to eleven months.   This period was called *langoniwela na ʔifi* "feed child of the house". The infant's increased strength and ability to survive was signalled first by sitting alone, and later by pulling itself up to stand. The infant did not breastfeed passively but played, about which mothers complained. The infant ate more at one time, so the mother spent more time apart from it. She left for about seven hours at a time, but not every day. It was reported that the infant required less nursing at night.

Food was premasticated less thoroughly, and was transferred to the infant's hand to eat, or given mouth-to-mouth. Water was given from a bamboo container. Some infants begged for premasticated betel nut when their families chewed. Among a very few groups the baby began to eat some premasticated fish; generally, fish and shellfish were denied to infants as they were reported to cause constipation. Other families did not give fish or betel nut as they did not want the child to cry for it when it was not available. When the child teethed it received its first pork (if available) and was given other, soft unpremasticated foods to hold and gnaw on. Edible insects were not given because the outer shell was considered "too hard". Despite the child's growing independence, it was often seen actively nursing when the mother was available. In all, this was a transitional period.

Twelve to twenty-four months.   The child was called *wela ʔabala noʔo* when it learned to walk and to eat by itself. It now began to take part in the ritual life of the community, which required restrictions on food and contact with the mother. As part of one ritual, *ribanga*, the participants received a portion of taro and pork, which was to be completely eaten by only that person. Young children participated as early as 1½ years. Boys first ate with ritually mature men at this point.

The child nibbled constantly on staples kept aside for that purpose. Food was occasionally premasticated if it was too hard to chew. By two years the child could eat a diet largely similar to an adult's, but most high protein foods, greens, and foraged foods were not introduced until nearly four years of age, as they were believed to be indigestible for a young child.

TABLE I

Acceptable time limits of absence from their infants[a] for Kwaio mothers (central Malaita, S.I.)

Age of infant (months)	Time limits for absence of mother
0-3	Infrequent absences of not longer than 4 hours
3-5	4 to 5 hours every other day
6-11	Up to 7 hours, consecutive days if necessary
12-17	Up to 10 hours regularly
18-24	14 hours or overnight
>24[b]	2 or 3 days

[a]These are acceptable upper limits only. The mother should not find it necessary to be absent from the child for these time periods every day. Mothers also, on occasion, are absent for periods of time longer than this.

[b]Or next pregnancy.

The mother was allowed to leave for two or three days at a time, especially after the child reached 18 months. She was less likely to take a heavy child on her visits to her natal hamlet. She returned to a full ten hour day of work and could leave her child to attend night activities, a separation of at least 14 hours. This is what was possible, but it did not occur every night, or even every week. Table I summarizes the allowable upper limits of the mother's absence from the nursing child according to the child's age.

The child spent more time with caretakers even when the mother was present. The breast was largely replaced by solid food by the time two years was reached; breastmilk was more an important condiment than a staple.

Breastfeeding continued in this vein until the child was three or four years old (or occasionally ten) unless, as was usual, the mother became pregnant.

Weaning

Except in cases of distant birth spacing or the last-born (*susuburi*), weaning (sevrage) was completed at the next pregnancy or birth.

Women spoke disparagingly of those who were highly fertile, as it deprived the previous child of breastmilk. Ideally, the child was weaned before pregnancy. No one, including the older women, seemed to have regularly achieved that goal.

Birth control through abstinence seems to have been achieved through numerous ritual and social obligations which in total substantially limited sexual activity. In addition, the practice of marrying late limited the reproductive time of each woman. No specific postpartum sex taboo existed.

Based on ten birth intervals observed in the course of fieldwork, the previous child, on the average, was 25.2 months old at the birth of the next child. This is based on only $3\frac{1}{2}$ years of observation of younger women with no health or marital difficulties, and is therefore biased toward shorter intervals. Reproductive histories suggest a longer average birth interval of three to four years. Other reported birth intervals were as long as six years.

When a woman became aware of her pregnancy, she discouraged breastfeeding. Sevrage was brought about by a combination of transferring the mother's care of the child to others, absence of the mother and instruction of the child. Gentle teasing took place when the child sought to breastfeed. If a male child had not yet eaten with the ritually mature men, he did so at this point. This was the time to convince the child that it was too grown-up to need to breastfeed. It was explained that as a new child was coming there was no milk left. By the fourth or fifth month of pregnancy, the flow of milk had substantially decreased and women gave this as a reason for weaning. Most women said that they could not withhold the breast. The child had to reject it alone, and so the caretakers tried to convince the child that it did not want the breast.

There were no restrictions on the mother's activities at this stage. The child spent a great deal more time with the caretakers and slept with them throughout the process of withdrawal from the breast. If the grandmother dry-suckled the child, it was not weaned from her breast at that time.

The final break came when the mother gave birth, and the child did not see its mother for one to two weeks due to the ritual regulations. During the separation, the child was constantly reminded that there was a new baby and that he could not breastfeed anymore. When the mother ascended to the house, the first thing she did was to sit with the older child and have a serious talk with him about his obligation to give up breastfeeding in the newborn's favor.

There was a difference between the weaning of the first and last children and the middle ones. The last-born, of course, was not weaned until much later than was average. When the mother did not become pregnant again, the child was usually weaned at three or four years by rubbing tobacco, ginger, coleus, or ants on the breast. The mother and others also scolded the child for nursing.

The first child was sometimes more difficult to wean. Mothers claimed that their first-born did not want food. This may have been a reflection of the mother's inexperience in weaning a child. If sevrage was not accomplished when the second child was born, the older child was sometimes allowed to breastfeed a little. In such cases, the mother complained and others shamed the child. The situation did not continue for long.

Child Caretaking

This was called *nana?iwelanga*. The child was cared for by all occupants of the parents' hamlet. The parents were not exclusively responsible for their children; a good part of the care was rendered and decisions made by other adults and even adolescents of the hamlet. No child was left alone for even a short period.

As can be seen, the mother was not relieved of her subsistence duties. Only if her child was seriously ill for a long period did her in-laws take over the gardening chores so that the mother could nurse the child.

An older woman was the preferred caretaker because of her experience and because she could use her breasts as pacifiers or in giving premasticated food. One woman still menstruating resumed lactation under this stimulus.

In infancy, the paternal grandmother was, after the mother, the most common caretaker. Maternal grandmothers could look after the infants nearly as frequently because of occasional uxorilocal residence, the tendency to marry as close to home as possible, the daughter's constant visits to her natal place to garden or to help her own family, the maternal grandmother's residence in her daughter's settlement, or the recent tendency to marry relatives. Many women stated a decided preference for having their own mothers watch their children.

Unmarried or widowed childless sisters of the child's father living at their natal settlement, if any, cared for "their" child frequently. At weaning, they took over a great deal of the child's care, sometimes informally adopting it. The same applied to childless couples related to the father or to elderly relatives of the second ascending generation who lived apart from the child's family.

At five years, female children began to care for the younger children under the supervision of an adult. Infants and toddlers were not put under the care of unsupervised children as it was believed that the older child was afraid without an adult to give instructions, and that an infant preferred to be cared for by an adult. In the absence of a daughter, a male child (not the eldest) was given these duties.

The child's grandfather and father also minded the children when other duties allowed. They were restricted by ritual regulations which prohibited contact with female urine or feces. Therefore, most men watched over male rather than female children, waited until the child was a toddler and easier to control, or supervised a young girl minding the infant or toddler. However, once the child was old enough, the father spent a great deal of time caring for him. It was considered one of the father's duties, and men showed considerable knowledge about the proper care of a child.

In the absence of a grandmother or other suitable caretaker in the hamlet, a young family experienced conflict between subsistence, ritual regulations, and child care duties. It was not possible to take the young child to the gardens nor could the mother neglect her subsistence work lest she and her family go hungry. In such cases, a woman either convinced a sister (classificatory or biological) or niece to stay with her temporarily as caretaker, or the father became the primary child caretaker. If the father did so, he risked ancestral sanctions of illness and death if expiatory sacrifices were not made for contact with feces or urine.

To quiet a crying infant, caretakers lay with it on their stomachs, cradled it in their arms, patted it, droned or sang to it, walked around with it, washed it, slept with it, gave a little water or food, and gave a dry breast as pacifier.

Wetnursing was acceptable, services being exchanged primarily among sisters of the child's mother and father. It was always acceptable for the father's sister to nurse the child and the mother to nurse the infant of her sister-in-law, as long as the nurse was not ill. Similarly, mother's sisters or the wife of one's husband's brother nursed each other's children. Some may never have exchanged services if no suitable nurse was nearby, however. Occasionally, more distant female relatives breastfed each other's children when the two were together. It was done much as solid food was given, to acquaint the child with its relative as well as to satisfy it. Asked why she wetnursed a child, one woman replied, "It's my child, we're related." It also indicated a close friendship between the mother and the nurse.

Unrelated women never exchanged these services for fear of transmission of illness. It was thought best for the child of the wetnurse to
be of the same age as the infant she nursed to ensure that the breastmilk was of the appropriate consistency and to the child's taste. Milk
for a newborn was said to be watery, that for an older child to be
thick.

CHANGING INFANT CARE PRACTICES

Most changes were observed in sukuru communities, both coastal and
bush. Some were also observed among coastal pagans. Only those
practices which differed from traditional ones will be discussed.
Introduced practices were more variable than traditional ones.
Families who were highly acculturated in one area, such as in
dependence on cash, were traditional in another, such as use of
traditional medicine. Modernization occurred in a patchwork fashion
(Igun, 1982). These data did not disclose any correlations or patterns.

Birth

Young sukuru women reported that they preferred to give birth at the
Adventist Hospital, and did so when possible. Some pagan women
also went for the first birth to the hospital, knowing the labor was
reputed to be long and difficult. The sukuru mother stayed eight to
ten days at the hospital, or stayed in her house for a week if she gave
birth at home. After birth, the mother was supposed to rest for three
months and be fed by in-laws, but the actual resting period was closer
to one month. They performed lighter duties for another one to two
months, depending on her health and the availability of relatives to
help her. Sukuru women were kept from going out or travelling
widely sooner than this by fear of pagan taboos. If she violated those
prohibiting contact between a woman who had recently given birth
and men, she was liable to pay compensation to those so violated.

Some women milked the colostrum from their breasts because they
"disliked" the color, but the common practice was to feed the child
the colostrum, as was traditional.

During pregnancy and lactation, sukuru women desired storebought foods such as rice, biscuits and tinned foods, as well as local
foods. These foods were believed to be particularly nutritious. In the

vernacular, the preferred and nutritious traditional foods were called *maasi?a* "sweet"; by extension; since imported foods were "sweet", they were thought to be nutritious whether they were or not.

Absence from the Child

Sukuru women, in interview, gave the same acceptable time limits of absence from the child for each period as did pagan women (Table I), yet they also claimed, without exception, that pagan women left their children too early and for too long. This is indicative of tense relations between the two interdependent communities, and not necessarily of a real difference in practice. However, due to the nature of their duties and settlement pattern, most sukuru women left the child only once or twice a week.

Child Caretaking

In a big village, it was easier to find people to mind the child, but the quality of care may have deteriorated. There were fewer restrictions on men minding small children, so elderly male relatives served as main caretakers more often than in pagan communities. Because there were more children in one place, an adult supervised a larger number of children. However, in a large village it was less easy to keep track of the children and the sense of responsibility was diffused. Thus, there were accidents because everyone assumed that the child was with someone else. Drowning was the greatest danger.

Weaning

Women weaned their children at the same age as did pagans, but depended less on the child's own rejection of the breast. Techniques used in weaning included wearing tight clothing so that the child was not able to reach the breast, going or sending the child away for four or five days, and rubbing chili or chloroquine on the breast.

Introduction of Solid Foods

The major difference between pagan and sukuru infant feeding practices was in this area. Only approximately $\frac{1}{4}$ of the sukuru mothers admitted to following the traditional practice of introducing premasticated food mouth-to-mouth well before the child was crawling.

Giving of premasticated food mouth-to-mouth was held in disrepute by most sukuru people, and by some modern pagans, for a number of reasons. Expatriates, mostly missionaries, taught that premastication is a filthy habit. This was based on personal repulsion and on their concern that respiratory illnesses so common in Kwaio, especially tuberculosis, not be passed to the infant. It also reflected the Western belief that solid food was inappropriate in the first semester.

Reasons given by Kwaio informants for discontinuing premastication were that it was a dirty practice and transmitted sickness to the infant. Reasons for not introducing food in the first semester were that a child who ate that early would crawl around on the ground looking for food and put dirty items in its mouth and that it distended the child's stomach. Some also said that their children did not like premasticated or mashed foods, and could not eat unprocessed solids before they began teething.

Solid food was introduced between six and nine months of age. Solids were introduced in one of three fashions: Food was premasticated into a spoon or the child's hand, boiled foods were mashed and fed to the child with a spoon, or soft foods were given to the child to hold by itself.

"Soup-soup", boiled sweet potato in coconut milk, was the preferred introductory food as it was believed to be as nutritious as breastmilk. Boiled sweet potato was sometimes cooked especially for the child, as most adults preferred it roasted. A few wealthier parents bought imported foods for their children, believing them to be nutritious. Most did not because, as with fish in bush communities, they would have been unable to satisfy the child's desire for that item when it was unavailable and unable to explain to the child.

Despite these reports, the theory did not always stand up to necessity. Observation showed that some women who disparaged early feeding and premastication actually gave premasticated food mouth-to-mouth to their charges when the mother was gone and the infant was hungry. Nor were the parents' preferences consistently observed by elderly caretakers, who were set in their ways. Generally, however, modern practices differed from traditional ones.

Wetnursing

Although many sukuru women allowed wetnursing, some definitely did not, even by biological sisters. This was reported to be because of fear of transmission of illness. Sukuru women who did allow wetnursing did not share services with as wide a range of kinswomen as did pagan women, as they did not keep track of kinship in the same depth as did pagans. Only the parents' sisters and the father's brother's wife (if the two families lived together) normally exchanged wetnursing services.

EXTRAORDINARY SITUATIONS

In the course of the breastfeeding cycle, any number of things could have gone wrong. What happened in cases of death, abandonment, or insufficient lactation?

Traditional Responses

Fosterage. Up until 20 years ago, if the mother died in childbirth, the child was buried with her. Then a new type of magic became available which allowed the still living infant to be taken to the menstrual area immediately, where it could be fostered. If the mother died in the menstrual area or afterward, there were no ritual difficulties, and the infant's survival depended on finding a suitable wetnurse.

In cases in which the child had to be breastfed, a wetnurse was sought among the infant's relatives. Biological sisters of the deceased mother were most willing to take on the added burden. Barring that, a sister of the father took the child. In the event that a close female relative was not available, a male relative of the infant's parents whose wife was lactating sometimes offered to take the child. The further the kin relationship and geographical separation, the less chance there was of quality care. The foster mother had a child of her own to care for, and may have neglected the other. A good foster mother was said to think of her foster child before her own children, because she pitied it and feared that the dead mother would cause it to die if she heard it crying. The child stayed with its foster mother until well after weaning.

If there was no lactating woman available, the infant was fed with premasticated taro and water. Survival of an infant under six months was unusual in these cases. A child who was able to eat when its mother died (at six months or later) was simply weaned early.

Insufficient lactation.   This was an uncommon problem. According to informants, the child breastfed a little, then jerked away and cried. When squeezed, the breast did not drip milk. It was believed to be caused by ancestors, so the appropriate ritual action was taken, after which the breastmilk was reported to return. The mother did not stop putting the infant to the breast. A customary cure was effected by rubbing the milky looking sap of the *suala* (*Alstonia scholaris*) on the breast, although none of my numerous informants have ever seen it used, or could tell me of someone who had seen it used. If the mother's milk did not return, another lactating woman in the same hamlet wetnursed the child, but the usual response was to feed solids to the child.

Insufficient lactation was most common in cases of multiple birth. One informant bore triplets and was dry within a week; two others bore twins which suffered from malnutrition at nine months. It appeared that at nine months or earlier the mother began the weaning process because of absences from the home, with dangerous results for the infants.

Modern Responses

The availability of substitutes for breastmilk changed the responses to problems that arose in breastfeeding the infant. These substitutes were powdered milk and soup-soup. The latter only recently became available because it is made of sweet potato, an introduced crop.

The Kwaio first became aware of a substitute for breastmilk in the late 1920s when a mother was taken from her two month old infant. A Kwaio man working on a ship obtained tinned milk for the infant and it lived. Ever since then, reliance has been placed on tinned or powdered milk in situations in which the infant could not be breastfed and milk was available.

In the case of mother's death in a sukuru community, a foster mother was sought, but if not readily available it was easier to give the child to a clinic or get powdered milk on the coast than in the bush. Another alternative, sometimes preferred, was to feed soup-soup, especially if the child had begun to eat already. If a foster mother was not available in a pagan community, the child was given to the sukuru, where tinned milk and medical care were available. Feeding soup-soup was not an alternative in many inland communities because of the scarcity of coconuts past the coastal slope.

In cases of insufficient lactation, the family sought help from the clinic or hospital more often at the time of the study than in the past. They received a bottle and powdered milk for the infant, or milk to be mixed with the child's food. Some parents did not follow the doctor's orders to give milk supplements to malnourished children because they claimed the children disliked the milk. (These cases of malnutrition were all situations of multiple birth.) In the Sinalagu coastal sukuru community of 500 people, there were four reported cases of insufficient lactation not due to multiple births or severe illness of the mother. These all occurred within the last six years, but before the beginning of the research period. Three of the four children were of the same nuclear family.

Bottlefeeding

In the eleven cases in which bottle use was noted, five were reportedly due to insufficient lactation, as noted above. In one other case, a Kwaio woman living in the capital city at the time of the interview was feeding her child with a bottle because she had been recently cured of tuberculosis and feared "giving sickness" to her child. The remaining five women (four coastal and one inland) used bottles to leave breastmilk for their children while working. Sometimes water, limeade, or (rarely) milo drink mix was put in the bottle. In these cases, the caretaker gave the bottle to the child. Only one of these five families, having a steady cash income, occasionally bought powdered milk for their child. Although coastal and inland informants had heard of powdered milk through cases of fosterage and considered it an excellent substitute for breastmilk, they did not buy it. At U.S.$1.70 per 300 gm container, powdered milk was too expensive for all but the wealthiest sukuru families. Infant formula was unheard of in this area.

In only one case during the course of this study was an infant removed from the breast and put on powdered milk mix in a bottle. The father did this because he had worked as a houseboy for an expatriate woman and had learned to prepare a bottle for his employer's child. When his own child became ill and lost weight on this regimen, he removed the child from the bottle. The mother had ceased lactating, so the child was put on a diet of solid food, although only nine months old. The father had prepared the bottle by boiling the bottle and nipple, and used boiled, cooled water to prepare the milk. All other people who used bottles, cups, or spoons to feed the child washed these utensils as they washed all household utensils: with cold water and no soap.

Attitudes Toward Bottlefeeding

Use of powdered milk will not be discussed here because the Kwaio do not perceive it to be a viable alternative to breastmilk, due to its cost.

Bottles, however, were considered a viable alternative — not to breastfeeding, but to other forms of supplementary feeding. Women frequently left their infants to work. Perhaps in response to this necessity, premasticated food was introduced to the infant early in the first trimester. This allowed caretakers to satisfy the child's hunger in the absence of the mother. However, the sukuru women believed it to be unhygienic. This belief was penetrating the inland areas, and younger pagan mothers were beginning to reject the practice. If it was believed (as Solomon Islands medical personnel and Social Services officers taught) that a child should not receive solid food until six months of age, the mother was left with a serious conflict between her duties as food provider and as mother. Attitudes toward bottle use were positive because women, old and young, traditional and modern, pagan and sukuru, believed a bottle filled with expressed milk would relieve the tension between their duties to their infant and to the rest of their families. With a bottle, they could provide breastmilk to their children even when absent.

There was one functional radio in the bush at the time of study. There were more radios on the coast, but there were no advertisements for infant formula or baby bottles on the radio. Nor were there any in the national newspaper, which rarely got to East Kwaio. On the other hand, there were advertisements on the radio advocating breastfeeding, local clinic and mission hospital staff advocated it, and officers from the Ministry of Health and Social Services came around on government ships to teach proper infant care — breastmilk only until six months.

DISCUSSION

In this study, two different, although interdependent and related, communities were examined. Infant care and feeding practices were described in the sociocultural context. The mother-infant dyad was shown not to interact in isolation, but to be affected by the needs of the rest of their family, economic constraints, and sociocultural expectations.

Given that a woman's prestige in the traditional Kwaio system derived as much from her role as provider as from her role as mother, it was important that she continue to work despite having a dependent infant. Child care was accommodated to the economic needs of the family. Prestige was a factor, but not in terms of acquiring prestige by imitation of Western practices. Pagan Kwaio gained no prestige by such imitation. In relation to modernization, the values of the society lay in being largely self-sufficient and selectively adopting items of Western technology which they found useful. The values of the sukuru Kwaio differed from those of the pagan Kwaio and the factor of prestige through identification with Western culture can not be entirely discounted in an explanation of sukuru feeding practices.

It was found that traditional bush practice was to introduce solid food to the infant in the first trimester of life. Coastal practice was to introduce solids in the second trimester. Modern practice, found mostly on the coast, was to introduce solids in a different fashion, at about six months of age. Given the importance of premasticated foods in these infants' diets, information on the health benefits or dangers of the practice would be useful. Premastication may be a good way of introducing the starchy local diet, as ptyalin, an enzyme in saliva, breaks carbohydrates down into sugars, perhaps making it easier for the infant to digest these foods (Jelliffe, 1968). Emanuel and Biddulph (1969) noted the low occurrence of gastrointestinal disease among inland pagan children and speculated that this was related to feeding premasticated food mouth-to-mouth. Introduction of premasticated foods in the first semester of life was apparently common in the Pacific Islands (Jelliffe, 1968) and throughout Mongo-Malaya (Jelliffe, 1962).

The changes that have occurred in the introduction of solid foods were not possible until the recent pre-eminence of sweet potato, which can be boiled and mashed (taro must be premasticated to give to an infant); the wide availability of pots, dishes, and other utensils; and the cash to buy these items. Older women did not cook foods in a pot for their infants, or even use sweet potato to such an extent, until the 1950s. These changes have occurred among sukuru people first and foremost because they have had more contact with agents of change than have the Kwaio living in the inaccessible hills.

Kwaio women were concerned about the health and well-being of their infants. It is this that makes them susceptible to bottle use, if enough cash should ever become available to the community. Told that premastication and solids in the first semester were bad for the infant, mothers have been left in a quandry as to how to provide for their infants while at work. Breast is best, but a woman can not detach part of her body to provide food for her infant while she is getting food for the rest of her family — or can she? Bottles filled with breastmilk seem to offer this option.

Even should bottle use increase in frequency and infant morbidity and mortality increase with it, women might still want to use bottles. Kwaio medical beliefs are not the same as Western medical beliefs, so they would not associate disease with food. (Most illness is believed to be supernaturally caused.) Women are unlikely to give up the early supplementation which allows them to return to their productive activities, even if there are health risks to the infant, unless substantial changes in belief systems and community organization occur. This leaves policy-makers in a quandry: How can the needs of both infants and their mothers be met? This discussion raises more questions than it answers, but hopefully these will be helpful questions for policy-makers, for mothers, and for infants.

ACKNOWLEDGEMENTS

The author wishes to thank Leslie Marshall for her advice and comments on this paper as organizer of the ASAO symposium on Infant Care and Feeding in Oceania; Helen Patton, Florence Sofield, Brigit Obrist, David Akin, and an unidentified reviewer for *Ecology of Food and Nutrition* for their comments; and especially the people of Sinalagu who so graciously submitted to observation and interview. This research was made possible by employment by United States Peace Corps as a community development worker at the Kwaio Cultural Centre, Ngarinaasuru, Sinalagu, Malaita. The author alone is responsible for any errors and omissions in this paper.

REFERENCES

Burt, B. (1982). Kastom, Christianity and the First Ancestor of the Kwara?ae of Malaita. *Mankind* **13**(4), 374-399.

Department of Community Medicine, Papua New Guinea (1979). *Malaita Nutrition Survey: Preliminary Report.* University of Papua New Guinea Faculty of Medicine, Boroko.

Emanuel, I. and J. Biddulph (1969). Pediatric field survey of the Nasioi and Kwaio of the Solomon Islands. *J. Trop. Pediatr.* **15**(2), 56-69.

Gaull, G.E. (1979). What is biochemically special about human milk? In D. Raphael (Ed.), *Breastfeeding and Food Policy in a Hungry World.* Academic Press, New York, pp. 217-227.

Gude, R. (1979). *Assessment of Nutrition in Solomon Islands.* Ministry of Health and Social Services, Solomon Islands, Honiara.

Huffman, S.L., A.K.M. Alauddin Chowdhury, J. Chakraborty, and N.K. Simpson (1980). Breastfeeding patterns in rural Bangladesh. *Amer. J. Clin. Nutr.* **33**, 144-154.

Igun, U.A. (1982). Child-feeding habits in a situation of social change: The case of Maiduguri, Nigeria. *So9c. Sci. Med.* **16**, 769-781.

Jelliffe, D.B. (1962). Culture, social change and infant feeding. *Amer. J. Clin. Nutr.* **10**, 19-43.

Jelliffe, D.B. (1968). *Infant Nutrition in the Subtropics and Tropics,* Second Edition. World Health Organisation, Geneva, pp. 58-60.

Jelliffe, D.B. and E.F.P. Jelliffe (1978). *Human Milk in the Modern World.* Oxford University Press, Oxford.

Keesing, R.M. (1967). Christians and pagans in Kwaio, Malaita. *J. Polynes. Soc.* **76**(1), 82-100.

Keesing, R.M. (1982). Kastom and anticolonialism on Malaita: "Culture" as political symbol. *Mankind* **13**(4), 357-373.

Morse, J.M. (1982). Infant feeding in the Third World: A critique of the literature. *Adv. Nurs. Sci.* **5**(1), 77-88.

Nerlove, S.B. (1974). Women's workload and infant feeding practices: A relationship with demographic implications. *Ethnol.* **13**(2), 207-214.

Pitt, J. (1979). Immunologic aspects of human milk. In D. Raphael (Ed.), *Breastfeeding and Food Policy in a Hungry World.* Academic Press, New York, pp. 229-232.

Popkin, B. (1980). Time allocation of the mother and child nutrition. *Ecol. Food Nutr.* **9**, 1-13.

Popkin, B. and F. Solon (1976). Income, time, the working mother and child nutriture. *J. Trop. Pediatr.* **22**(4), 156-166.

Thompson, B. and A.K. Rahman (1967). Infant feeding and care in a West African village. *J. Trop. Pediatr.* **57**, 124-128.

Van Esterik, P. and T. Greiner (1981). Breastfeeding and women's work: Constraints and opportunities. *Stud. Fam. Plann.* **12**(4), 184-197.

CHAPTER 13

KWARA'AE MOTHERS AND INFANTS: CHANGING FAMILY PRACTICES IN HEALTH, WORK, AND CHILDREARING

DAVID WELCHMAN GEGEO AND KAREN ANN WATSON-GEGEO

INTRODUCTION

Very few studies of health and nutrition have as yet been conducted in the Solomon Islands. There is a particular need to look carefully at infant and maternal health in light of the social changes in the Solomons over the past few decades, including the introduction of wage labor economy with its concomitant diet, the malaria eradication program, better medical care, and migration from mountain to coastal plains and from rural to urban areas.

Within the Solomons, Malaita island provides an interesting case because of its relatively high infant mortality rate, low birth rate, and low life expectancy rate for women.[†] These rates may be influenced by the fact that the malaria eradication program did not arrive on Malaita until 1970, and then has been hampered by the upsurge of malaria that began in 1976 when mosquitoes started showing an increasing resistance to DDT spray. It has also been suggested, however, that cultural factors such as local childbirth customs and women's heavy work also influence these figures.

This chapter summarizes findings from initial observations and interviews on maternal health, childbirth, and infant care in Anglican

---

[†]According to the 1976 census, the birth rate for the Solomons as a whole is 3.7 %, and for Malaita 3.0%. The infant mortality rate for the Solomons as a whole is 46/1000; for Malaita it is 49/1000 — which nevertheless represents a sharp decrease from the 1970 census (Statistics Division, 1980). Whereas men's life expectancy at birth is 52.8 years on Malaita, women's is only 50.6 years (Statistics Division, 1980). Both of these figures are lower than the over-all Solomon Islands figure of 54 years, and Malaita is the only island where men live longer than women.

J

villages along the coastal plain of West Kwara'ae, north Malaita. The
data, which were collected in association with the authors' ongoing
study of children's socialization, are strictly preliminary in character;
details of the sample are reported below. The purpose is twofold.
First, the descriptive information presented on Kwara'ae mothers and
infants invites comparison with health, nutrition and work organization reported in other chapters of this volume. Secondly, the data
indicate how Kwara'ae parents 25 to 45 years old have adapted to the
pressures of social and economic change. The adaptations that these
Kwara'ae parents have made in family organization, work, and infant
care reflect their determination to maintain central values in the traditional culture while participating in a modernizing society. One
implication is that the Kwara'ae have had to decide which values are
the most important to them. For the families interviewed in this
study, this decision-making process was conscious.

METHODS

The data consist of observation and interviews conducted during a
total of 14 months of fieldwork in West Kwara'ae. For three months
each in 1978 and 1979, background cultural studies and spot observations were made on caregivers interacting with infants and young
children, in two villages. In 1981, a sample of five families in three
villages were observed intensively over a seven month period, and an 80-
hour tape-recorded sample of caregivers interacting with their
charges 6 months to 3.6 years old, during babysitting and routine
household activities, was compiled. The caregivers (all of whom were
25 to 45 years old) in these and five other families were twice interviewed in depth on health, pregnancy, childbirth, infant care, childrearing, and family history. Three *gwaunga'i ki* (elders) — two
women and a man, all over 70 years of age — who were locally
regarded as experts in health, child care, and tradition, were also
interviewed. Occasional observations were made among five other
families in these villages, and in one family in a fourth village.[†] The

†See Watson-Gegeo and Gegeo (1985a) for a full discussion of the sample. Our
analysis also draws on Gegeo's knowledge as a member of Kwara'ae culture, whose family
has lived on the coastal plain of West Kwara'ae for the past 50 years. Our field research in
1978 was supported by small grants from the National Institute of Mental Health, the
Milton Fund, and the Spencer Foundation. Field trips in 1979 and 1981 were supported by
seed grants from the Spencer Foundation.

data reported here were collected in preparation for a future systematic study.

SOCIAL CHANGES IN KWARA'AE

A Melanesian subsistence-gardening people, the Kwara'ae are the largest cultural/linguistic group in the Solomons (over 14,000 native speakers).[†] They occupy a large cross-sectional slice of north-central Malaita, just to the northwest of the Kwaio region. (See Figure 1.)

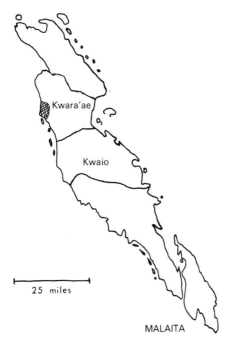

FIGURE 1   Map of Malaita with fieldwork area in West Kwara'ae indicated by shading.

[†]The 1978 census recognizes 85 local languages and dialects (exclusive of non-Solomons languages and Pijin). After Kwara'ae, the two languages/dialects with the most native speakers are Nggela (5300, of Gela island), and Maringe (5100, of Santa Isabel) (Statistics Division, 1980).

238   D.W. GEGEO AND K.A. WATSON-GEGEO

This study pertains primarily to Church of Melanesia (Anglican) villages on the coastal plain.† Kwara'ae society is essentially patrilineal and patrilocal: Most rights to land and inheritance pass through the male line, and married couples are expected to live in the husband's village. Thus Anglican villages, like traditional villages, are usually made up of members of a single descent group within a tribe. The Kwara'ae are strictly exogamous: Relatives within the descent group, or who are connected by close marriage ties, cannot marry. In the past, high brideprices often meant that men reached their thirties in age before assembling enough shell money and pigs to get married. The Anglican and Catholic churches have greatly limited brideprice amounts, while the South Seas Evangelical Church has abolished them altogether. These moves may have contributed to the lowering of age at first marriage over the past two generations. Kwara'ae men are usually in their mid- to late twenties at marriage, and women in their early to mid-twenties,‡ although teenage marriages are becoming increasingly common.

Relationships within the Kwara'ae family are organized hierarchically. Parents and other adults are senior in rank to children, and older siblings senior to younger. The oldest son is head of the sibling group and will continue to be so throughout life. Often in childhood, however, the oldest daughter has a great deal of power in the household because of her supervisory role in family work. Parents supervise older children who then supervise their younger siblings in household work and child caregiving, as is the case in Hawaii (Boggs, 1984) and Samoa (Ochs, 1982). The oldest daughter is usually the head sibling

---

†The Anglican and Catholic churches have embraced more of the traditional Kwara'ae culture than has the South Seas Evangelical Church. Kwara'ae people's behavior, speech, and marriage practices are greatly influenced by which church they have joined. Anglican and Catholic villagers are conservative, expressing a desire to balance the traditional with the introduced in all spheres of life. In 1981, the Seventh Day Adventists had just begun a strong missionary push in northwest Kwara'ae, but had not yet reached the areas where this research was conducted.

‡Mean age of marriage on Malaita is 24.5 years for men and 22.3 years for women, the lowest in the Solomons for males and the highest for females. High female age at marriage may be due to the rate of outmigration of men to Honiara. For the Solomons as a whole, the mean age of marriage is 22.5 for men and 21.1 for women (Statistics Division, 1980). The Kwara'ae cultural expectation is that men be in their late twenties or early thirties and women in their mid- to late twenties when they marry; the present generation is marrying younger than their parents. See Gegeo and Watson-Gegeo (1981) for a further discussion of marriage.

in this regard, although an oldest son may take the role if all the older children are boys.

Several factors have contributed to rapid social change in West Kwara'ae over the past two generations. First, the forced and voluntary migration of West Kwara'ae people from the *tolo* (mountain) areas to the coastal plain as they were converted to Christianity has left relatively few, unChristianized villages in the mountains. Secondly, due to land pressures and other factors, many West Kwara'ae have migrated to other islands in search of work, especially to Honiara (the capital town of the Solomons) where there are large Kwara'ae settlements at Gilbert Camp and Kaibia.[†] Most of the migrants are young men who eventually return home to marry and settle down in the village, maintaining their land rights and kin ties while they are away with regular visits home and contributions to marriages. The money, household goods, and gifts that these migrants send or bring home, along with the new ideas and behaviors they pick up when away, are an important factor inducing social change. Malaita's proximity to Honiara (half hour by airplane, or five to six hours by ferry) makes visits and even shopping trips essentially easy for the West Kwara'ae. The location of Malaita's capital town of Auki and the governmental airstrip at Gwaunaru'u in West Kwara'ae are other contributing factors, as are the increasing number of children going to local schools, the firm establishment of the Christian churches in coastal and lower tolo villages, and the introduction of a wage labor economy.

The quality of life on the densely populated coastal slope of West Kwara'ae has begun to deteriorate in the past ten years. Building materials — wood, sago palm for leaf, rattan — are now scarce, and families live in smaller, more crowded houses than a generation ago. The carrying capacity of the land has almost been reached in some areas, resulting in continuously planted gardens, soil exhaustion, lower productivity, and less diversity in garden crops. Kwara'ae gardeners grow several varieties of Pacific sweet potato and yam, *ba'era* (shrubs whose leaves are eaten as vegetable), coconut, cassava, swamp taro, and such introduced plants as papaya, pineapple, sugar cane, and Chinese cabbage. Dry land taro, formerly a staple, was destroyed by a taro blight several years ago, and today must be sought from relatives in the tolo or purchased at the Auki market. Streams

---

[†]Nearly 3000 Kwara'ae have migrated to other islands, according to the 1976 census (Statistics Division, 1980).

and rivers have been nearly fished out, though coastal villagers fish at sea. Chickens and pigs are kept, but usually eaten only at feasts.

Traditionally, food exchanges among households ensured sharing in times of heavy production, famine or flood, or availability of highly valued foods such as fish. Exchanges may help provide a balance of nutrients in the diet, a question that needs to be explored. One social change in the past 20 years has been a reduction in food exchanges, influenced by lower garden productivity and the growth of the cash economy. The Kwara'ae meet their income needs (taxes, church offering, household goods, school tuition) in ways similar to Fijians (Katz, ch. 15): sale of garden produce at the market, sale of small amounts of copra, and contributions from relatives working in Auki or elsewhere. Marriage expenses (brideprices, feasts, debts) are met cooperatively within the extended family and descent group. The local economy is depressed, however, providing little regular income for most families.

The agricultural and economic problems sketched above are factors in the growing number of cases of mild malnutrition occurring in West Kwara'ae villages.[†] There is thus a need for in-depth study of health and nutrition in this region.

PREGNANCY AND CHILDBIRTH

By the time a Kwara'ae woman reaches the end of her childbearing years, she will have usually given birth to between five and six children, exclusive of stillborns and miscarriages.[‡] However, families of 10 to 13 children are not uncommon. The fertility rate and family size have greatly increased in the past two generations as the result of several processes. Better health care, control of malaria, and the less strenuous lifestyle on the coastal plain have no doubt increased fer-

---

[†]Cases of malnutrition in the villages studied have been observed by the authors and by medical personnel in the islands (Helen Patton, personal communication). The 1976 census also reported mild malnutrition and anemia as health problems on Malaita.

[‡]The mean number of children born alive per woman on Malaita, for those women who were 45 to 49 years old in 1976, was 5.81. For the Solomons as a whole, the same aged women had given birth to 6.23 offspring (Statistics Division, 1980).

tility and decreased miscarriages and stillbirths.[†] Under the traditional *tabu* system, women did not resume sexual relations with their husbands until at least a year after the birth of a child, and breastfeeding continued until the infant was as much as three years old. These practices contributed to spacing births well apart. In fact, *falafala* (tradition or culture) prescribed that births be spaced apart by at least two to four years. The ritual separation of women from men, especially during menstruation and the child's first year, but also in connection with many ritual events, probably helped to keep births well-spaced. In situations of famine, war, exile from the village, adultery, and menopausal pregnancies, the Kwara'ae also practiced infanticide.

The abolition of the tabu system and the ascendance of Christianity has meant that West Kwara'ae men and women live in the same house, and ritual separation and infanticide are no longer practiced. Today, couples vary in when they resume sexual intercourse after childbirth. Interviews and unsolicited comments indicate that some refrain for a year, others until the baby can sit up, and still others for only a few days or weeks after the birth (until the women's bleeding ceases). Women today also vary in how long they continue to breast-feed their infants.

Some Kwara'ae women say that they still use traditional methods of contraception, though a few are beginning to respond to the Solomons' campaign for smaller families by taking contraceptive injections. However, most of these women become interested in

---

[†]In *tolo* regions, gardens and drinking water are often very far from the village. This may mean hiking over extremely rough trails, up and down steep ridges, and fording streams to arrive at the gardens or water source; then gardening for several more hours before returning, and carrying back heavy bamboo containers of water, bags of potato, or loads of firewood. Dwellers on the coastal plain no longer have to clear primary (virgin) forest to plant gardens; at the most, scrubby secondary forest must be cleared. Often gardens are continuously planted. In the past, the heavy work of the tolo had to be done with stone tools and digging sticks, whereas today gardeners can use steel axes, spades, machetes, and hoes. In the past, all food was cooked in the *gwa'abi*, necessitating lifting and carrying of wood, stones, and leaves. Although the gwa'abi is frequently made today, families also have cooking pots which greatly relieve the effort of making family meals. (See also footnote on p. 250.) The main crops planted in the past were taro and yam, which required more labor than the introduced potato crops relied on today. Life today is still very strenuous, particularly in the tolo regions. Tolo women on Malaita suffer a continuous weight loss from the time they reach adulthood until death, whereas women on the coast are able to maintain their body weight throughout adulthood (Friedlander and Rhoads, 1982).

contraception only after having had several children, or when advised by medical doctors that another pregnancy would endanger their health. Others take the injections only to space their births. Large families are still desired, although in villages where garden land is scarce couples may be gossiped about for having too many children, especially if they are mostly boys (who grow up and raise their own families in the village). Couples also express a strong desire to have approximately equal numbers of boys and girls, and may continue having children if they have produced a string of only one sex. The women interviewed knew the traditional herbal mixtures that induce abortion, but because the churches oppose abortion they did not see it as an alternative way to limit the size of the family. The few abortions that occur are usually among women who have become pregnant outside of marriage. Premarital sex and adultery are both strictly tabu today as they were in traditional times; this is one set of tabus on which falafala and the churches agree.

During the first and second trimesters of pregnancy, Kwara'ae men and women say that a great deal of heat is generated in the abdomen by the pregnancy. The heat can be counteracted by eating certain foods, such as slightly green pineapple and banana. Otherwise the expectant mother is supposed to eat foods that will promote healthy blood, which is seen as the source of the fetus' nourishment. These foods are taro leaves and young 'amau (Ficus copiosa) leaves boiled or baked in coconut cream, and other leafy green vegetables (ba'era).

The pregnant mother's nutrition may be affected by systematic restriction of foods from her diet due to food tabus and family theories of health. There are three kinds of food tabus that may affect a woman's diet. First, there are traditional tabus general to the whole of Kwara'ae in which certain foods cannot be eaten by pregnant women or children because they are thought to affect physical characteristics in the infant. Very few women in West Kwara'ae heed these tabus today, according to those interviewed, and it is mostly the grandparental generation who knows what the tabus are and what they mean.

A second kind of food tabu consists of those particular to a descent group or extended family. Some are part of the larger category of name tabus: A food included in the name of an ancestor worshipped by the descent group might be tabu for his/her descendants to eat. Other tabus have to do with experiences of an important ancestor with that food, leading to a tabu being placed on it. The men and women interviewed said that West Kwara'ae people today may or may not observe such descent group tabus. A woman marrying into a family that observes tabus may or may not adopt those of her husband's family. However, even if she does not, foods that her spouse's family abstains from may not be readily available to her. A married woman may also choose to continue her own family's tabus or to drop them at marriage. Depending on the nature of the tabus, a woman who both continues her own family's tabus and adopts those of her husband may be restricting her diet seriously. A nutrition study of individual families is needed to examine this problem.

A third set of tabus are those observed only by pregnant women in a descent group or extended family — a woman's own or those of the family into which she marries. These tabus are formed when one (or more) pregnant woman has repeatedly reacted negatively to a particular food in the past. Some say that the food is therefore tabu to pregnant women of that descent group, while others say that they will refrain from it "just in case." Most women interviewed said that they observe this kind of tabu.

What the pregnant woman eats is also determined by family theories of health. A woman will adjust her diet based on her previous experience with pregnancy, her cravings while pregnant, and her dislikes especially during the first trimester. Interviews and observations indicated that there is a strong emphasis on eating foods that promote "good blood," eating clean and freshly cooked foods, and restoring the body to its proper temperature. Except for the emphasis on dark green, leafy vegetables and fruit, however, there is no clear agreement in West Kwara'ae on what pregnant women should eat.

The shifts that occur in family workload during pregnancy and childbirth reflect important changes in family organization for West Kwara'ae Christians. Traditionally, a woman's support system during pregnancy and childbirth consisted of paid and unpaid women of her and her husband's descent groups, who assisted her with heavy work and giving birth. Today the West Kwara'ae, like the Ponam (Carrier, ch. 11), regard these tasks as primarily the responsibility of the nuclear family. Interviews and observations showed that husbands take over many duties traditionally prescribed as women's work during pregnancy and the first month after birth, including collecting drinking water, chopping firewood, and carrying firewood and sacks of potato home from the gardens — assisted by their older children (if any). Sometimes an unmarried pre-teen or teenage female relative of either spouse may come to stay with the family to help out, and occasionally a family may hire a relative for a short period. Hiring a female relative has its precedence in tradition, where three different women served as paid *kini kwate* (woman who gives help) assistants during birth, purification, and the new mother's return to the family household.[†]

Another important change from traditional practice is the father's role in childbirth. Today many West Kwara'ae babies are born at Kiluufi Hospital, located on the main road between Auki and Gwaunaru'u. Because of difficulties in getting transportation to the hospital, and women's complaints about their treatment there, many women still prefer to give birth at home in the village. Among those who do give birth at home, the fathers exhibit and express pride in helping to deliver the baby themselves. Some of the male interviewees boasted that they had delivered all of their own children. In the home births we observed, an experienced woman was also sometimes present or called in, to tie the umbilical cord and make sure the afterbirth was expelled. Certainly fathers' role in childbirth is a radical departure from falafala, in which women were secluded in birth houses, and birthing blood, like menstrual blood, was seen as polluting for men.

INFANT FEEDING AND THE LACTATING MOTHER

Traditionally, the colostrum was expressed and the first nourishment given to a baby immediately after birth was a mouthful of roasted taro, premasticated to a liquid and passed orally from mother to infant. This *ma'e meme* (classifier for dished things + premasticated food) was given to the infant again after three weeks, and thereafter the amount was doubled or tripled, depending on the infant's age. Once the infant had two teeth, the mother chewed the taro and sweet potato to a solid mass rather than a liquid, and transferred it by hand to the baby's mouth; this *meme* was called *meme sa'e tutu* (pre-

---

[†]Two other shifts in a woman's life are important at pregnancy. Heat is believed to endanger the fetus, so pregnant women usually refrain from making the traditional gwa'abi (heated stone oven). Sexual intercourse is also believed to be too taxing to a pregnant woman's energy, and possibly endangering to the fetus. Interviewees stated that they refrain from sexual intercourse during pregnancy, at least in the final trimester.

masticated food + classifier for exposed flesh = freshwater clam, hence meme that looks like clam flesh). The degree of premastication thus depended on the infant's age, and chewing and digesting ability. The meme offered to the newborn may have symbolized its entrance into the human world, where taro was the main staple of the diet.

Nine of the women interviewed expressed all or part of their colostrum postpartum. Some thought the yellow milk was an infection of the breasts called *susu ngwata'a* (milk/breast + infected?). They argued that the colostrum would make an infant ill; vomiting was said to indicate that the baby had contracted the mother's infection. Others called the colostrum "watery milk," saying that during pregnancy excess water from the woman's body accumulates in the breasts and turns yellow. Some who held this view said they express all of the colostrum, and others said they express only some of it. The one mother who fed all of her colostrum to her infants argued that colostrum cleans the infant's intestinal system of fluids ingested in the womb; and said that she had arrived at this view through observation and experience.

In traditional times, a relative who had already given birth would be asked to wetnurse an infant if its mother died, or if her milk production was inadequate. If the infant was old enough to consume solid foods, it might be fed exclusively on roasted taro or sweet potato. Interviewees said that today, if a woman's milk production is low, she ingests copious amounts of water and food. Some women said that an infant may be breastfed by a relative if the mother is sick. But if the newborn's mother dies, the infant may be taken to an orphanage hospital where it will be cared for until it can walk. If an older infant's mother dies, it will be fed papaya, green coconut, and sweet potato.

Most families cannot afford to purchase powdered milk to supplement or replace the mother's milk; only the most acculturated and affluent use bottles and nipples. In the villages studied, cow's milk was said to be nutritionally equivalent to mother's milk in a general way, but inferior because it was produced outside the mother's body, and did not contain her *alafe'anga* (love). This view is related to the deeply held traditional value of women as the source of food and food-givers in the family, a belief shared by the Kwaio (Akin, ch. 12). Nevertheless, cow's milk is preferred over that of another woman, due to fears of illness and a sense of privacy, no doubt related to lingering beliefs about pollution. Even when an infant seems completely satisfied with cow's milk, caregivers say that it is dissatisfied;

any signs of discomfort are attributed to the absence of maternal love associated with breastfeeding.

The caregivers interviewed followed two basic patterns in introducing solid foods to young infants. All expressed a preference for taro being introduced first, as it was traditionally; but in practice, sweet potato is usually the infant's first food. One pattern is to introduce cooked potato mixed with water at one month of age. The other pattern, which a majority of the caregivers follow, is to wait until the infant's first two teeth have emerged (between the fourth and eighth month). All those interviewed said that they proceeded according to the infant's development, spacing attempts apart by a few weeks if the infant vomited or rejected the offered food.

Traditionally, taro was baked or roasted over a fire or steamed in bamboo, and then premasticated by the mother. Today the taro and/ or potato is boiled in water, mashed, and fed to the infant by spoon. Observations and interviews showed that early solid foods given to infants include boiled *kumara* (potato) with or without ripe papaya or ba'era; papaya boiled alone; and fresh banana and pineapple. The infant's diet, therefore, appears to be rich in Vitamin C, complex carbohydrates, and iron (from the dark green ba'era leaves), but affected by seasonal availability. Fish and meat are introduced last. In families that observe food tabus, moreover, children adopt their father's tabus at birth, but may or may not adopt their mother's — a decision made by the parents.

The main family meal, which is eaten in the evening, is based on the Kwara'ae theory of health embodied in the concept of *ādami'anga* (eating a variety of foods at the same sitting). Foods are classified as causing either heat or cold in the body. For example, pork causes heat, which leads to diarrhea if consumed in large amounts or unaccompanied by a cooling food such as potato. Ideally, a meal should contain potato, ba'era or papaya stewed in coconut cream and a meat. In practice, we observed that the poorest families often have only potato in the evening, or at the most, potato with ba'era cooked in water. The latter is the most common meal among other families as well, since coconut cream, papaya, and especially meat are often not available.

Women in the *nguda*, or lactating period, are encouraged to eat heartily. According to falafala, the lactating mother is supposed to eat a *sibolo* (animal protein) every day because it is thought to produce good milk. Normally, families in the villages studied had meat or fish about once a week, depending on the season. Men with lactating wives are expected to provide sibolo every day; those with cash can purchase tinned meats and fish, but others face a protein-poor hunting/gathering environment. It was observed that men with lactating wives did spend more time than usual fishing or hunting, and were assisted by sons and close male relatives. Sibolo in the environment include lizard, tree frog, flying fox, fish, eel, prawn, shellfish, freshwater clam, crab, locusts, and larvae of sago palm. (Lactating mothers and infants do not eat opossum.)

Certain foods are believed to have special benefits for lactating mothers and infants. Interviewees said that taro, Pacific yam, and prickly yam make the mother and infant strong; that fish cooked in coconut cream increases a woman's milk and improves her blood; that kumara and water clam stewed in ba'era strengthen the walls of the uterus, the intestines, and the stomach, and produce good milk. Women were also observed to eat cooked *samo* fern; its bitterness is said to protect the milk and infant from infection, and prevent infant constipation. Lactating mothers and children of all ages are to avoid eating dirty food, food touched by flies, and cold food unless reboiled. All of these are recognized as sources of illness, and were strictly avoided in the families observed.

The couples interviewed abstained from sexual intercourse anywhere from one month to one year after the birth of a child. Opinions varied about what length of time is still absolutely tabu, and the interviewees were aware of different opinions on the matter. None resumed sexual relations until postpartum bleeding and discharge had completely subsided.

The caregivers interviewed and observed varied greatly in when they weaned their infants. Some said weaning occurs only at the onset of another pregnancy, while others said they weaned their infants when they could walk well and eat solid foods. Our observations indicated that women were getting pregnant at intervals of 12 months to 4 years, and that weaning was taking place with children 12 months to 3 years old. One couple with five children reported that they count the months after a birth and wean each child at 16 months when, they said, a child can walk and talk.

Once it is decided to wean an infant, the mother rubs hot chili or quinine on her breasts, tells the infant that the breast is bitter or bad, and firmly removes its hands from the breast when it tries to nurse. A crying infant is soothed by either parent, often by holding it and walking back and forth through the house, lullabying and telling it gently to hush. At other times, the infant may be carried outside by the father to distract its attention with the *lia* routine.[†] Interviewees said an infant cries constantly for the first three days of weaning, then has crying spells frequently after that for from one week to one month. Infants are also described as rejecting food, and as being listless for several days. An inconsolable infant, especially one younger than 18 months, may be briefly allowed back on the breast for one or two weeks, and then another attempt made later, as we have observed.

INFANT CARE, CAREGIVERS, AND SOCIALIZATION

Although in many Pacific societies the entire extended family may be involved in child caregiving, this is less true for Kwara'ae. The extended family takes a very personal interest in the Kwara'ae child and plays a significant role in socialization. However, infant caregiving is the responsibility of the child's biological parents, sisters, and brothers. Under the traditional tabu system, most caregiving responsibility for infants was borne by mothers. Women took their infants to the gardens with them, though men might temporarily tend an infant. The gwaunga'i men and women interviewed said that in the past, men's degree of interaction with infants was limited by beliefs that urine and feces were polluting. The infant's skin was also considered potentially polluting until it reached full pigmentation. Men did take over the instruction of their sons after a ritual initiation at about five years old.

Current observations, however, strongly supported interviewee statements that fathers in West Kwara'ae are expected to, and actually do spend as much time babysitting as mothers. Some spend more time at it than their wives. Moreover, men dispose of the infant's wastes and bathe it themselves, prepare food for and feed the infant, tend it all day at a stretch, and carry it to the hospital if it is ill. (Formerly, men handed a defecating infant to a female, did not prepare food for infants, and did not carry infants on journeys in order to keep their arms free in case of ambush.) A nursing infant may be carried by the father or a daughter to the garden when it is hungry, or

---

[†]*Lia* is a widely used routine in Kwara'ae to distract an infant when it is fussing, crying, or doing something the caregiver wishes to stop. The child is taken to a door or window in the caregiver's arms. The caregiver points to a butterfly, bird, flower, or other object while chanting lia (look) in a low to middle pitch, and then describes the object, encouraging the infant to repeat its name or label. Lia is a remarkably effective technique for calming infants, whose crying usually stops instantly when the routine is initiated. See Watson-Gegeo and Gegeo (1985a).

the father may send a child to fetch the mother home.

Second in importance to parents in childcare are usually older daughters or sons. Whenever parents are home, however, they are the primary caregivers — the mother first, and the father second. In families without daughters, the oldest son will be trained in women's work and raised like a daughter until late teenage, or until the younger children are capable of taking over household tasks. As mentioned earlier, a parent's unmarried sister or other relative may be asked to stay with the family for several months or even years as a live-in babysitter and household helper. Grandparents may also play this role, but to a lesser extent. Most older men and women have achieved the status of gwaunga'i. According to interviewees, it is inappropriate for such elders to regularly contend with an infant's biological needs and behaviors. Instead, grandparents serve as companions and teachers of children three years and older.

Infant caregiving is also seen as too strenuous for older individuals. Observations of ten families indicated that the Kwara'ae style of infant care is definitely labor intensive. Infants are frequently bathed throughout the day, protected from flies and mosquitoes by fanning, shaded from the sun, and constantly talked to and entertained by their caregivers. The movements of caregivers and infants are very synchronous and coordinated. Most caregiver movements are smooth and slow so as not to startle an infant; items are seldom snatched from an infant's hands, and infants are rarely slapped or harshly scolded. Redirection of an infant's attention is a major strategy for controlling behavior. Babies are not allowed to crawl on the sago palm floor or ground, but are expected to walk by ten months of age. From six months on they are taught to be responsible for items they drop, to give to and share with others, and to exhibit polite interactional behavior. An important emphasis is on teaching infants to decenter, that is, to detach themselves from their own emotions and focus instead on the group and social concerns. As in Hawaii and many other Pacific societies, affiliative values are extremely important in the adult culture (Howard, 1974). Kwara'ae caregivers use intensive verbal interactions to help the child achieve an early independence, and to demonstrate adult-like behavior by age three years. Both parents take responsibility for this early training.[†]

[†]Kwara'ae socialization and children's language acquisition are discussed at length in Watson-Gegeo and Gegeo (1985a, 1985b).

Interviewees gave the same list of qualities for mother and father when we asked what made a good mother and good father. The highest ranked quality for good mothers was providing food and care for the children, and for good fathers was teaching and disciplining them. The West Kwara'ae talk openly about parents as models for their children's behavior, especially in the important cultural values of love, peacefulness, sharing, stability, and being welcoming. Parents express unwillingness to relinquish their babysitting to others, for fear their children may be exposed to different values. In fact, a couple's reputation in the village and district rests to a considerable extent on what people think of how they are rearing their children. People observe each other very closely in this regard. Someone whose children are naked or ragged, dirty, fed left-over food, or ill-behaved quickly becomes the talk of the surrounding villages. Achieving gwaunga'i rank itself can be affected by a couple's reputation as child caregivers, and by how their children turn out as they grow up.

WOMEN AND MEN: ROLE, WORKLOAD, AND VALUES

Men and women agree in the important role accorded to women in Kwara'ae falafala. The Kwara'ae say that the woman is the source of life, center of the family, and food-giver. In the past her power was based on generation and control of most of her husband's wealth. Women were the peace-keepers in the family and descent group, took important roles in deciding and negotiating marriages, displayed shell money at brideprice payments, and gave speeches at meetings and feasts. Yet women also worked extremely hard under severe conditions, and their lives were restricted by men's fears of pollution from menstrual blood. The coming of the Christian churches and other Western social institutions freed women from most of their ritual segregation, and other social and economic changes mentioned above have alleviated some of the heavy workload for West Kwara'ae women.[†] In other ways, however, Christianity led to a loss in status

---

[†]In her discussion of Samoan women's workload, Nardi (ch. 16) suggests that washing dishes and pots has actually added to the Samoan woman's workload, since serving food on disposable leaves and cooking it in earth ovens was simpler and involved less labor. For the Kwara'ae who live along streams and rivers, stews are much easier to concoct than a gwa'abi, and washing dishes occupies little time. (Villagers far from a water source constitute a different matter, of course.) If West Kwara'ae women cooked exclusively the traditional way, given the size of the population they soon would denude the area of cooking leaves: banana, *rako* (species or relative of ginger), *la* (type of ginger), *kakamo* (swamp taro), and *tangafino* (*Macaranga aleuritoides*).

for women. For instance, the church sees women as subject to men, men run the church, and men do nearly all the talking at public meetings.

Nevertheless, West Kwara'ae women are beginning to reclaim some of the leadership roles they formerly exercised. One sign of change is men's taking over more and more of the household work and childrearing responsibility, formerly seen as strictly women's work. Today one hears couples say that although they still respect the traditional division of labor, they are teaching their children to do whatever task needs doing and to think less about whether it is a male or female task. Men are thus providing considerable assistance to their wives, helping them balance family needs with conserving energy and with taking a greater role in social activities such as feast preparation and village meetings. Women are able to attend the "mother's union" and other women's organizations sponsored by the church, although Kwara'ae women have not taken to these organizations in the way that Samoan women (Nardi, ch. 16) have. Kwara'ae women complain that most meetings involve all talk and little action, and that their gardening and food-preparing schedules are interrupted by them. Women with young infants often shun roles in church organizations if their breastfeeding schedule might be disrupted. They consider infant care to be their most important task.

As with Fijian women (Katz, ch. 15), while having children greatly increased the Kwara'ae women's workload, it also brings an increase in status. As in Fiji, there is a strong cultural emphasis on procreativity and continuing the descent line, seen from the patrilineal viewpoint. Having her first child accords a woman the status of a mature adult. We also observed that the behavior of new parents changes in culturally expected ways, and this change itself causes people to see the couple as mature. Having a child is also the first step to becoming a gwaunga'i, the basic qualifications for which are reaching middle age and having one child marry.

We have seen that in addition to the help of her husband, a woman may obtain paid or unpaid help from a female relative during her childbearing years. She also relies on her oldest child, especially a daughter, to be her main assistant. Today a first daughter is often kept out of school to help her mother, and to be raised as a strictly traditional woman. Families are especially judged by the behavior and skills of an oldest daughter. Daughters born later usually have more choice in deciding whether they want to go to school. This means that the daughters in a family may, among them, represent different stages of acculturation to Western ideas and behaviors. This differential treatment of daughters, and sometimes also of sons, illustrates one way that the Kwara'ae attempt to maintain falafala while adapting to the modern world.

In conducting this study we were particularly struck by three characteristics of the Kwara'ae. First, as indicated above, the West Kwara'ae are determined to retain certain important values and practices of traditional culture. Even without the worship of ancestors, the Kwara'ae regard falafala as holy and sacred, and as very much alive; this contrasts, for example, with the A'ara of Santa Isabel (White, unpublished observations, 1982),[†] another highly Christianized Solomon Islands society. Second, the level of consciousness with which the Kwara'ae approach discussions of values and childrearing was surprising. The husbands and wives interviewed all reported having discussed with each other how they were going to care for and rear their children, shortly after marriage. They seemed very clear about strategies for raising healthy children with proper Kwara'ae values, and were analytical in evaluating other people's caregiving. Third, the level of knowledge men showed on childbirth practices, nutrition, infant caregiving, and women's health issues, both in falafala and as practiced today was striking. There appeared to be no gap between women's and men's knowledge on these topics. This adds support to the contention that men have taken a greater interest in these areas over the past two generations.

Looking to the future, the very changes that have alleviated women's workload may become problematic if their husbands become increasingly involved in the wage labor economy. Since men must pay taxes whether they are employed or not, since the churches press for larger offerings, and since prices for clothing, tools, household goods, and tuition continue to rise, more and more men will need to seek jobs in Auki or elsewhere. This is likely to put new stresses on Kwara'ae family organization.

REFERENCES

Boggs, S.T. (1984). *Learning to Communicate Hawaiian Style.* Ablex, New York.
Friedlander, J.S. and J.G. Rhoads (1982). Patterns of adult weight and fat change in six Solomon Island societies: A semi-longitudinal study. *Soc. Sci. Med.* **16**, 205-215.

[†]White, G.M. (1982). Culture and social images: Constructions of identity and history in Santa Isabel. Unpublished manuscript, East-West Center, Honolulu.

Gegeo, D.W. and K. Watson-Gegeo (1981). Courtship among the Kuarafi of Malaita: An ethnography of communication approach. *Kroeber Anthropol. Soc. Papers*, **57/58**, 98-121.

Howard, A. (1974). *Ain't No Big Thing: Coping Strategies in a Hawaiian-American Community*. University Press of Hawaii, Honolulu.

Ochs, E. (1982). Talking to children in Western Samoa. *Language in Society*, **11**: 77-104.

Statistics Division (1980). *Report on the Census of Population, 1976*. Ministry of Finance, Solomon Islands, Honiara, Vol. 2, pp. 11, 14-15, 19, 73, 82, 101, 103, 114, 171-172.

Watson-Gegeo, K.A. and D.W. Gegeo (1985a). Some aspects of calling out and repeating routines in Kwara'ae children's language acquisition. In E. Ochs and B. Schieffelin (Eds.), *Language Acquisition and Socialization Across Cultures*. Cambridge University Press, Cambridge. Submitted for publication.

Watson-Gegeo, K.A. and D.W. Gegeo (1985b). The social world of Kwara'ae children: Acquisition of language and values. In J. Cook-Gumperz and W. Corsaro (Eds.), *Children's Worlds and Children's Language*. Cambridge University Press, Cambridge. Submitted for publication.

CHAPTER 14

THE CULTURAL CONTEXT OF INFANT FEEDING IN FIJI

JANICE M. MORSE

INTRODUCTION

Assessment of infant feeding practices cannot be isolated or removed from the underlying cultural beliefs and practices. Yet medical research on infant feeding has, with few exceptions, largely violated this principle. With the incorporation of Western health care systems and the availability of technological advances such as infant feeding bottles, nipples and milk substitutes, infant feeding methods and patterns change. However, that change occurs within the cultural context and must be interpreted within the cultural *milieu*. This paper examines the cultural beliefs and values pertaining to infant feeding and infant care in two groups: the Fijian and the Fiji-Indian.

SETTING

This research was conducted in the Ba District of Viti Levu (Figure 1) from October, 1980, to February, 1981. Viti Levu is a continental type island with narrow coastal plains and river deltas on which most of the population is located. It is a center for government, industry and education in Fiji, and over the past three decades has developed a major tourist industry. Nadi, an international airport, is only 40 miles from the research area.

Two main groups inhabit Fiji — the Fijians, the indigenous population, who are Melanesian, and the Fiji-Indians, who came to Fiji as indentured workers from East India between 1880 and 1920 to work on the sugar plantations (Naidu, 1980). The Fiji-Indians now comprise more than half the total population.

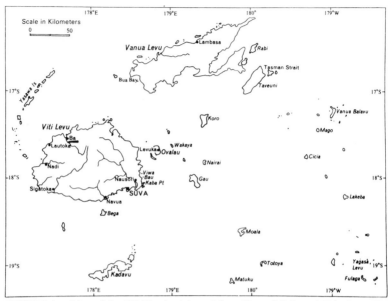

FIGURE 1   Map of the Fiji Islands. Adapted from Kerr, G.J.A. and T.A. Donnelly (1969). *Fiji in the Pacific.* Jacaranda Press, Milton, Queensland.

As Ba is the center for Indian culture in Fiji, the Fiji-Indians are the dominant ethnic group in the district. The population census of this region recorded 69.8% Indian (of which 77% are Hindu and 22% Moslem), 27.1% Fijian and 3.1% Chinese, European and other Pacific islanders (Parliament of Fiji, 1977). The Fijians reside in Ba town and in the traditional villages throughout the district. The Indians, on the other hand, reside in the town, dominating commercial activities (including the sugar mill), and in hamlets along the coastal plain, where they are employed as small crop farmers raising two crops of sugar cane per year on leased land.

The tropical environment provides an abundance of fish, fruits, vegetables and grazing land, but malnutrition is still a common public health problem. Intestinal parasites such as hookworm result in endemic anemia. Although malaria does not occur in Fiji, diseases such as tuberculosis, dengue fever and influenza are common.[†] Health care is

---

[†]The incidence of infantile diarrhea has improved from previous years with 8,700 reported cases resulting in three deaths in 1976 (Ministry of Health, 1979).

available to all at minimal cost, and public health nurses provide pre- and postnatal care throughout the district. A private hospital, Ba Methodist Mission Hospital, also provides prenatal and postnatal care (to the sixth week) and inpatient maternity services, as well as general and emergency services to the community.

DEFINITION OF TERMS

In this paper the "breastfed" infant is defined as an infant who does not receive a breastmilk substitute in a feeding bottle; a "mixed fed" infant is one that is both breastfed and receives supplementary or complementary feeds of breastmilk substitute from an infant bottle or a cup; and the "bottlefed" infant is one that receives a milk substitute from an infant feeding bottle or a cup.

METHODOLOGY

In this study, multiple methods were used to address the research question. To investigate traditional infant feeding practices, open-ended interviews were conducted (through an interpreter) with the Fijian and the Fiji-Indian traditional birth attendants and with elderly multigravida women. To assess current practices, interviews were conducted with mothers, with nurses and with some fathers; and participant observation in the postnatal ward and in the postnatal clinic of Ba Methodist Mission Hospital was conducted. To determine postnatal clinic utilization and feeding patterns, a survey of hospital records of all infants (and their siblings) born at the hospital for a 12-month period (from December 1, 1979, to November 30, 1980) was conducted.

RESULTS

The Fijian Culture: Traditional Infant Feeding Practices

Two cultural values, those of community and sharing, are pervasive in the Fijian way of life (Lester, 1946). The Fijian society is based on mutual services (rather than individual effort) for communal advantage. This attitude is reflected clearly in the perinatal period (Morse,

1981). The Fijian attitude toward the pregnant and parturient woman is one of interest, support, caring and assistance. Care of the pregnant woman is considered a community concern and the responsibility of all. As soon as the woman's pregnancy is revealed, she is relieved of many chores, receives a special diet, and is given massages and herbs to prepare her for her labor by the traditional birth attendant (*yalewa vuku*). The fetus is protected from the evil eye and other malevolent influences by the other women of the tribe, who accompany the expectant mother whenever she leaves the house.

Birth is conducted in a specially prepared *bure* (traditional dwelling) that has a *tapa*[†] cloth partitioning the room. Special support is given the mother by the yalewa vuku, female relatives and, if the labor is difficult, by her husband. The men sit awaiting the birth around a *yaqona* (kava)[‡] bowl on the other side of the tapa cloth.

After birth, the infant is passed to the female relatives, who care for the child. High ranking infants are not placed on the ground for ten days following birth but are constantly held in the arms of the attendants. According to informants, this pattern of group caring for the child continues for several years. If the infant naps, then someone will lie on the mat with the infant, and this responsibility is shared by all members of the family and elderly village women.

Following birth, Fijians withhold the infant from the breast for three or four days, until lactation is established, as it is believed that colostrum is impure. The colostrum is manually expressed from the breast and discarded. For the first few days the infant is given water or water and glucose, or *lolo* (coconut milk) to suck from a rag or a piece of cotton wool. Occasionally, if the child is very hungry, he will be given to a close relative who is lactating from whom he will nurse. A special bond and a sense of responsibility then develops between this surrogate mother and the infant. Fijians stated that they were reluctant to give the infant cow's milk, as this will make the baby "stubborn like a cow," but no such beliefs were attached to goat's milk or to formula.

The main problem concerning lactation was "poison milk" (*wai ni sucu gaga*). The main sign of poison milk was a change in the color of the milk to orange or yellow. It usually occurs in both breasts, but

---

[†]A bark cloth made from the inner bark of the *wauke* or paper mulberry tree (*Broussonetia papyrifera* (L.) Vent.).

[‡]Kava is used extensively throughout Oceania for ceremonial purposes. It is made from the dried root of the *Piper methysticum* forst.

may occur in one. The other main symptom is that the milk makes the baby sick, causing diarrhea or diarrhea and vomiting. A test for poison milk was reported as follows:

> Place a saucer of milk on the table. If the flies settle on it, then nothing is wrong with the milk. But if the flies don't settle on the milk, that means that it is poison.

No consensus could be obtained regarding the cause of poison milk. Most informants stated that they did not know; several stated that they thought it came from the influence of an unborn embryo or evil eye; and several others thought it came from eating bad shellfish. Treatment is to withhold the infant from the breast for four days and to give the mother *baka* (a herb tea).

The Fiji-Indian Culture: Traditional Infant Feeding Practices

In the Fiji-Indian culture female purity, innocence and virginity are highly valued, and childbirth is a very personal and private event. Childbirth is included in the category of subjects with a sexual connotation and is, therefore, not discussed with young women.

To maintain purity it is especially important to withhold information from young girls. Consequently, when they reach adolescence, they know nothing of menstruation until the bleeding begins. Even then some informants reported that they were too shy to tell their mothers, going instead to seek help from an older sister.

The sexual tensions of adolescence and courtship are removed by the careful supervision of teenage girls and the system of arranged marriages. At the time of marriage, Indian women usually do not know about sexual intercourse and are informed about this after their wedding by their husband. Nor are these women informed about the symptoms of pregnancy or the signs of labor and the physiology of delivery. Not surprisingly, the lifestyle of the pregnant woman continues without dietary supplement or change in the woman's work role.

At some time during the third trimester, the pregnant Indian woman will leave her mother-in-law's residence and move home to reside with her own mother. Following the birth of the infant, she will remain with her family until her husband comes to fetch her, usually six to eight weeks postpartum.

The *dai*, the traditional Indian birth attendant, does not care for the expectant mother until she is sent for after labor begins. The birth is conducted with only the dai present unless additional help is required, and the mother (and occasionally the mother-in-law) is then admitted.

The dai's role during labor is mostly one of passive non-intervention, watching and waiting. Comfort is provided verbally, rather than by touch, as it is believed that the birth can take place by itself. After the birth of the infant, the mother and her infant rest for ten days, although some informants reported that in the olden days mothers started to work as soon as they delivered. The dai visits the mother and her infant twice daily to give both a massage with coconut oil. The mother is fed "good food, with plenty of *ghee*."

As is the case with the Fijian, the Indian infant is withheld from the breast for three or four days. The infant is fed with water or water and glucose from a bottle, although a spoon was traditionally used. To initiate lactation, a hot ginger spice drink, *sont*,[†] is given to the mother three times daily.

The mother's diet does not contain any other nutritional supplements. However, special dispensation is given to lactating Moslem women during the Month of Ramzel (usually November), and they do not observe the fast.

Special privileges, such as relief from household duties, are given for approximately six weeks. Thereafter, the mother is expected to return to her duties, including working in the cane fields. The infant is the primary responsibility of the mother-in-law, who makes all the major decisions regarding infant care.

Fiji: Current Health Services

Maternal and infant health have been a priority in Fiji in recent years (Ministry of Health, 1979). Programs have focused on hospital delivery, pre- and postnatal care, infant immunizations and breast-feeding. Care is provided primarily by nurses, with physician support if required, and is free of charge.

The clinic at Ba Methodist Mission Hospital provided prenatal clinics one morning per week. Appointments were not required. The

---

[†]Contains milk, dry ginger (*Zingiber officinale* Rosc.), ghee (clarified butter), mangarel (*Nigella* L. spp.), harrae (*Glycyrrhiza* L. spp.), peeper (*Piper nigrum* L.), jeeine (*Cuminum cyminum* L.), and methi (*Trigonella Foenum-graecum* Linn.).

patients were booked for hospital delivery at that time, according to the expected date of the infant's arrival.

When the labor was well established, the mothers were expected to report to the hospital. Due to cramped conditions and little space for relatives, the women were cared for by the nurse and the relatives returned home.

Following delivery there was little opportunity for the establishment of maternal-infant bonding. Immediately after birth, the infant was wrapped to prevent chilling and shown to the mother by the nurse. If the infant's condition permitted, the nurse would give the infant to the mother to hold, but frequently the mother was uninterested and the infant removed to the nursery. The next opportunity for the mother to see the infant was at the next scheduled feeding time. The infant's father was not usually present at the birth and often did not remain at the hospital during his wife's labor. His first knowledge of the birth was frequently at the next visiting hour, when, to prevent infection, the infant was shown to the father and other relatives through a viewing window.

At this hospital, policy dictated that all mothers breastfeed their infants. Infant feeding bottles were not permitted in the wards, and infants who required bottlefeeds or supplemental feeding were fed by the nurses in the nursery. Because of space limitations in the hospital, a strict feeding regime had been established. This routine permitted scheduled doctors' rounds, visiting hours and meals to be conducted with maximum efficiency. The cramped conditions did not permit rooming-in or privacy for nursing mothers. Between feeds, the infants were separated from their mothers, sleeping in the ward nursery.

Both the nursing students and the staff nurses were responsible for teaching and assisting the mother with breastfeeding. In the hospital, cracked or sore nipples were not observed by the investigator, although the staff reported that they sometimes occurred. Mothers remained in the hospital until lactation was established on the third or fourth day postnatally, although occasionally lack of space forced relaxation of the rule.

Following discharge, the mothers and their infants were expected to return to the hospital's postnatal clinic weekly for six weeks and thereafter attend a government clinic at regular intervals.

The Fijian Culture: Current Infant Feeding Practices

The caring, concern and extended support system in the Fijian culture continues. Even with those who have shifted residence from the traditional village setting to the towns, relatives, usually adolescent female cousins, come to live with and help the pregnant woman. The pregnant woman is still expected to rest, to exercise, to take herbs to ease her labor and enhance lactation, and to eat a good diet. The support system is so strong that women reported that they would rather give birth at home where they felt "more loved" than in the hospital.

After birth, according to hospital procedures, the infant is removed to the nursery until it is feeding time. Fijian mothers occasionally resisted breastfeeding their infants. The cultural prohibition of feeding the infant colostrum was presented as "no one does that," but generally breastfeeding was not strongly resisted.

The relatives of the Fijian mothers brought traditional herbal remedies to the hospital. Most important was the hibiscus (*Hibiscus rosa-sinensis* L.) to clean the uterus and to initiate lactation. When I asked if the doctor knew of this practice, I was told, "No — unless they're Fijian and more understanding."

Infants in the Fijian culture are loved by all. It was observed that whenever they appeared, they would be fussed over and their cheeks pinched. They were indulged and petted, carried constantly and passed from one to another. One mother whose home was half a mile from the village said the old woman came to fetch her baby at dawn each day, for it is believed that the early morning rays are especially beneficial to the infant. They cared for the baby to the extent that "sometimes she did not even know where the baby was."

Lactating Fijian mothers fed their infants whenever the infants desired. One mother who was observed feeding her infant five times between 6:00 and 10:30 a.m., when asked about these frequent feedings, replied:

> The nurses tell us to feed the baby every four hours. They say, "Time your feeds." So when I go to see the nurse, they [sic] say, "Do you time your feeds?" and I say, "Yes," and they say, "How often?" and I say, "Every four hours." Then I do whatever I please.

Foods other than milk are given to the infant commencing at about three months of age. Pawpaw juice, and *bele* and *rorou* (dark green leafy vegetables) are given first, but I observed that infants are given "tastes" of food from a very early age. On the advice of the health clinic, the infants are given drinks of water, glucose and water or fruit juice in a feeding bottle daily.

This investigator did not see "poison milk," but it was frequently reported. Waqavonovona (1980) also noted the prevalence of poison milk and attributed this to mastitis. However, frequent occurrence of bilateral poison milk and absence of pain in the breast is not clinically consistent with mastitis and suggests that this condition should be investigated further.

In the weeks following discharge, breastfeeding is radically replaced by a pattern of mixed feeding but more slowly than the Fiji-Indian practice. By week six postnatally, 72.6% were breastfeeding, 11.9% bottlefeeding and 15.5% mixed feeding ($n = 125$). This pattern of nursing reduces the infant's dependence on the mother and permits a large number of caretakers to care for the infant.

The Fiji-Indian Culture: Current Infant Feeding Practices

The Fiji-Indian values that maintain purity and innocence for young women are still enforced, and the young woman may be ignorant of her pregnancy until the fourth or fifth month. The return of the Indian mother to her mother's home is still practiced, and this often results in a move of many miles. It frequently disrupts prenatal care, involving a change of clinic and physician.

Postnatally, breastfeeding was encouraged by the hospital, despite the frequent protest of the new mother who complained that, "It was no use, nothing is there." However, these mothers did feed their infants colostrum when instructed to do so. To initiate lactation, the relatives of Fiji-Indian patients were observed bringing sont into the hospital each day. The mothers kept this, without refrigeration, in their lockers. Sont was considered essential for successful lactation and was not provided by the hospital.

The staff recognized that rooming-in and breastfeeding on demand facilitated the establishment of lactation. However, with the strong modesty norm in the Fiji-Indian culture, the lack of privacy available in the open ward may be a very stress-inducing factor (perhaps even inhibiting lactation) should this be introduced. With fixed feeding times, the nurses could close the ward to visitors, thus ensuring privacy for the nursing mothers. Perhaps this is an example of a situation in which cultural assessment is essential prior to the imposition of change.

Following discharge, the return of the Indian mother to her mother-in-law's residence may disrupt postnatal care by requiring a transfer to another clinic in (and occasionally outside) the district. Examination of hospital records for 1981 ($N = 1610$) showed that 35% ($n = 563$) of the mothers who had delivered at the hospital did not return to the postnatal clinic, and it could not be determined if they attended other clinics or did not receive care.

In the following weeks, the breastfeeding norm established within the hospital was rapidly replaced by a pattern of mixed feeding. Between discharge and the first postnatal visit there was a 15.5% decrease in breastfeeding, and by the sixth week only 47.5% of the infants were breastfeeding (Figure 2). Fifteen percent were bottle-feeding, and 37.5% were receiving both breast- and bottlefeeds. The greatest decline in breastfeeding occurred between discharge and the first postnatal visit (15.5%) and between week two and week three (11.6%). The former decrease may be due to those not wishing to breastfeed and changing the method of feeding directly following dis-charge, and the latter to a misinterpretation of the normal course of

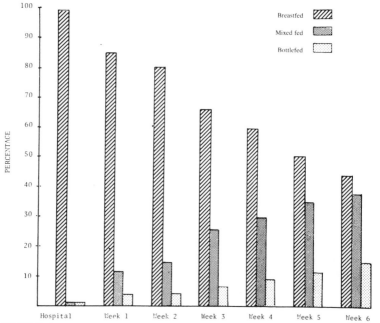

FIGURE 2   Percentage of Fiji-Indian infants attending postnatal clinic who are breastfed, bottlefed or mixed fed.

lactation. Initially, with lactogenesis, the breasts are extremely engorged; but as the milk supply adjusts to the infant's demands, this fullness subsides. As Indian mothers equate a soft breast with an empty breast, it is possible that the change may lead to the fear that they "do not have enough milk" expressed by many mothers. (See Gussler and Briesemeister, 1980.)

It is also important to consider other foods that the infants receive. On the advice of the nurses and physicians, all infants are given glucose powder and water or fruit juice, such as rose hip syrup, orange or mango juice, daily in a feeding bottle to prevent dehydration in the hot climate. Thus, theoretically, all infants are exposed to the dangers associated with bottlefeeding, such as exposure to diseases from contaminated water and unclean feeding bottles. Infants are also given tastes of many foods from an early age, usually partially masticated. These tastes are not reported to the nurses and counted as an infant food. When the infant is older, it is properly started on solids according to the advice of the nurse or the physician.

In the Fiji-Indian culture, several mothers reported to the investigator that Indian men preferred small breasts. Small breasts were associated with youth and purity, while large breasts were considered matronly. It was explained that the enlarged lactating breasts were not considered attractive, and it was believed that nursing further increased breast size.

Modern dress was also a deterrent to breastfeeding. For the Indian, the traditional Hindu *sari* or Moslem tunic scarf enabled the mother to nurse the infant discreetly. However, fashionable dress in Fiji at this time (and worn by approximately one third of young Indian women) was a full-length cotton dress with a back zipper. This clothing made nursing in public impossible.

In the hospital and postnatal clinics, the nurse's teaching is directed towards the patient. However, small examination rooms do not permit the patient's mother-in-law to be present. As the Indian mother has little control over the care of the infant at home, the nurse's advice on the care of the infant given to the mother instead of her mother-in-law is lost. The mother-in-law will continue to do as she thinks best, discrediting the nurse's advice to her daughter, as those nurses are young and inexperienced, and have not had children. Furthermore, it is in the mother-in-law's interest that the infant be bottlefed; only the mother can breastfeed the infant.

DISCUSSION

In both the Fijian and the Fiji-Indian cultures in the Ba area, cultural factors optimize neither breast- nor bottlefeeding, but a combination of both (mixed feeding) quickly develops as the norm. With supplemental mixed feeding, the infant is minimally breastfed[†] twice per day (in the morning and evening) and bottlefed during the day, rather than complementary mixed feeding (breastfed and bottlefed at each feeding). I am suggesting that supplemental mixed feeding is a physiologically adaptive strategy for both the mother and the infant, and is a compromise between the theoretical ideal and the real situation. In the case of the Fiji-Indians, the ideal total breastfeeding regime is not feasible, as the maternal work role demands a contribution to the household that separates the mother from her infant for lengthy periods. Poor maternal nutritional status resulting from suboptimal diet may adversely affect the quality and quantity of breastmilk (Hanafy et al., 1972; Jelliffe and Jelliffe, 1978; Lechtig et al., 1979; Miranda et al., 1983; Villar and Belizan, 1981). Furthermore, caloric and protein loss from the hookworm and caloric expenditure from physical labor (in the cane fields) may also contribute to maternal depletion. Breastfeeding the infant twice, rather than four times per day, reduces the nutritional demand on the mother and protects her depleted reserves. Although breastmilk is given in reduced quantities to the infant, the infant still receives some of the anti-infective immunological benefits of breastfeeding. In addition, the infant is receiving caloric supplements from breastmilk substitutes and weaning foods. Thus, while supplemental mixed feeding provides the infant with the benefits of breastfeeding, it also allows others to care for the infant, freeing the mother for other duties of the household.

The traditional practice of withholding of colostrum from the newborn infant by the delaying of breastfeeding until the third or fourth postpartum day[‡] has nutritional consequences for the infant (Jelliffe and Jelliffe, 1979). However, I am suggesting that this is a mechanism

---

[†]There is no evidence that lactation will cease using this pattern of breastfeeding (Morse and Harrison, unpublished observations).

[‡]At the symposium, "Infant Care and Feeding in Oceania," the infant feeding practices of 15 cultural groups were discussed. In her summary, Gussler noted that of these 15 groups, seven were reported to withhold colostrum from the infant, three were uncertain about this practice, and in five groups it was not known.

that may reduce maternal-infant bonding, preventing maternal attachment in areas of high infant morbidity and mortality. Most importantly, in both the Fijian and the Fiji-Indian cultures, the infant did not "belong to" the mother. In the Fijian culture, the infant was the responsibility of the entire community; and, in the Fiji-Indian culture, the mother-in-law was responsible for the care of the infant. However, this hypothesis needs to be tested further, and in other cultures.

The traditional beliefs and values regarding infant feeding are not discarded for Westernized technology; rather the advantages that infant feeding bottles and breastmilk substitutes provide are selectively used, and existing cultural practices are largely maintained. In the case of hospital births, the regulations pertaining to patient care are circumvented or adopted with or without the awareness of the staff, and traditional practices are maintained.

ACKNOWLEDGEMENTS

The author wishes to thank Dr. T. Bavadra, Fiji Ministry of Health; The Hon. Senator Ratu Vakalalabure; Matron S. Bali and the nurses of Ba Methodist Mission Hospital, Fiji; and Dr. Brij L. Kohli, Department of Botany, University of Alberta, for their assistance with this study. Funding was provided by the Isabel H. Robb Memorial Award, Nurses Educational Funds and Ross Laboratories, Columbus, Ohio.

REFERENCES

Gussler, J.S. and L.H. Briesemeister (1980). The insufficient milk syndrome: A bio-cultural explanation. *Med. Anthropol.* 4, 145-174.

Hanafy, M.M., M.R.A. Morsey, Y. Seddick, Y.A. Habib and M. el Lozy (1972). Maternal nutrition and lactation performance: A study in rural Alexandria. *J. Trop. Pediatr.* 18 187-191.

Jelliffe, D.B. and E.F.P. Jelliffe (1978). *Human Milk in the Modern world.* 2nd edition. Oxford Univ. Press, Oxford.

Lechtig, A., H. Delgado, R. Martorell and R.E. Klein (1979). Effect of maternal nutrition on the mother-child dyad. In L. Hambraeus and S. Sjolin (Eds.), *The Mother-Child Dyad: Nutritional Aspects.* Almquist and Wiksell, Stockholm, pp. 74-93

Lester, R.H. (1946). A few customs observed by Fijians in connection with birth, betrothal and marriage and death. *Fiji Soc. Sci. Indust.* 2, 113-129.

Ministry of Health (1979). *Annual Report of the Year 1976.* Parliamentary Paper No.53 for 1979, Government Printing Department, Suva, Fiji.

Miranda, R., N.G. Saravia, R. Ackerman, N. Murphy, S. Berman and D.N. McMurray (1983). Effect of maternal nutritional substances on immunological substances in human colostrum and milk. *Amer. J. Clin. Nutr.* 37, 632-640.

K

Morse, J.M. (1981). *Descriptive Analysis of Cultural Coping Mechanisms Utilized for the Reduction of Parturition Pain and Anxiety in Fiji.* Unpublished doctoral dissertation, The University of Utah, Salt Lake City. *Dissertation Abstracts International* **42**(2), B4363-B4364.

Naidu, V. (1980). *The Violence of Indenture in Fiji.* Fii Monograph Series No. 3., World University Service and University of the South Pacific, Suva, Fiji.

Parliament of Fiji (1977). *Report on the Census of the Population, 1976, Volume 1.* Parliamentary Paper No. 13 for 1977, Government Printing Department, Suva, Fiji.

Villar, J and J.M. Belizan (1981). Breastfeeding in developing countries. *Lancet* **2**, 621-623.

Waqavonovona, M. (1980). *A Case Study of Traditional Types of Healing Related to Women's Ailments and Conditions in Seven Villages and Settlements in Nadarivatu (Fiji).* The University of the South Pacific, Suva, Fiji.

CHAPTER 15

INFANT CARE IN A GROUP OF OUTER FIJI ISLANDS

MARY MAXWELL KATZ

INTRODUCTION

Recent surveys in Fiji directed by the National Food and Nutrition Development Programme found that breastfeeding is declining in urban areas (Lambert and Yee, 1981a), and that 8 % of Fijian children under age five are below 80 % of the Harvard Standard weight-for-age, with the highest percentage occurring between one and two years (Lambert and Yee, 1981b). In keeping with the concern to learn the causes of variations in infant care practices which has led to further studies at five sites in Fiji (Johnson, 1982), this report describes the care of infants aged six months to three years that was observed in a separate study in an outer-island district. While the study focused on cognitive development rather than nutrition, information is available on the relation of mother's work to infant care, and on variations among families in breastfeeding and spacing of recent births.

The Fiji Islands are located at 17-19° S. in the southwest Pacific Ocean, approximately 1300 km from Vanuatu and the Solomon Islands to the west, and 1000 km from Samoa and Tonga to the east. Australia is 2740 km southwest and New Zealand about 1770 km south. Mean monthly temperature ranges from 22 to 27°C. and rainfall varies 1780 to 5600 mm annually. The population of over 600,000 is mainly Fijian and Fiji-Indian and clusters on the two largest islands; the outer-islands are less populated, more dependent on traditional farming and fishing, and mainly Fijian in composition.

The families who participated in this study were all Fijian and lived in Kadavu, a group of volcanic outer islands that was accessible in a 6 to 12 hour journey by boat from the capital city of Suva.[†] Small cargo or fishing boats of 30 to 45 m length, and occasionally larger government boats, travelled to and from this area every one to three weeks, depending on cargo, passengers, and weather. The cost of the trip was expensive enough to prohibit frequent travel for many families, and visits to Suva among adults ranged from about twice a year to none over several years.

Nine villages whose total population was 945[‡] were surveyed during a 15-month period of residence in Nagone village[§] and in visits every three to five months to the other villages (June, 1977-October, 1978). These villages were distributed over an area of about 200 sq. miles, including land and sea. The data reported here were collected as part of a study of infant cognitive development (Katz, 1981), whose sample included all normal[‖] infants aged 13 to 36 months in the nine villages, as well as some of their three to four year-old siblings, for a total of 91 children in 70 households. Ages of children were determined from birthdates recorded in district medical records. It was clear from the fieldwork that these records were highly accurate and complete, and they were an invaluable aid in the study. Mothers of children in the younger age range (n=42) were interviewed for information about the birth, health and weaning of the child, and all mothers reported household membership, and parents' educational and residential history. Parental ages were determined from mothers' report of age or year of birth. Mothers were certain of their own year of birth, but could not always provide information about their husband. Systematic "spot" observations (Johnson, 1975; Munroe and Munroe, 1971) were made on a randomized schedule of daylight hours over a one-year period on all infants in Nagone and Tamana villages, noting who was present, their proximities to the infant, and ongoing activities. This group included ten infants who

---

[†]Recent studies of infant care and feeding in other parts of Fiji have been conducted by the UNDP/FAO Fiji National Food and Development Committee for five sites including one outer-island setting, and by Morse (ch. 14) for Fijians and Fiji-Indians in western Viti Levu, Fiji. The results of the National Food and Development Committee study were unavailable to the author at the writing of this article, but were learned to be in preparation in Johnson (1982).

[‡]Population figures were taken from district medical records.

[§]True village names are not used in this report.

[‖]Because the study was designed to assess normal infant cognitive development, two infants who showed clear physical or mental abnormality were eliminated from the study. No data are available on the families of these infants.

ranged in age from 6 to 17 months when observation began. Six of the infants were observed almost daily.[†]

The findings reported here include: a description and interpretation of the infant care system in relation to the nature of the subsistence economy; and an analysis of variation in duration of breastfeeding and the interval between mothers' most recent deliveries, focusing on their association with parental age, number of living children, education, urban residence, and proximity to maternal kin. These analyses show that, in addition to parental age, secondary education and proximity to the maternal kin group are both related to parental behavior in this setting.

ENVIRONMENT AND ECONOMY

There was a variety of terrain types in the district; some land areas were large and high enough to catch rain unequally, providing forested and less forested sections. Two small and low islands received less rain than other parts of the district, and were more dependent on fish and less on agriculture. The entire district was protected from ocean swells by a distant barrier reef, creating excellent fishing grounds in relatively protected waters. Nevertheless, high winds often made boat travel dangerous or impossible.

A village consisted of one or two groups of patrilineally-related families totalling 75 to 150 persons. Each village was located on a bay and near the mouth of a river or stream. Houses were clustered about 7 to 15 m apart in an open area; they were built with red, timber or cement block walls and had clean mat-covered floors. In most villages, water was piped into three or four village taps from dammed stream reservoirs high in the hillsides; otherwise it was caught by house roofs and directed into a central underground cistern. The first road in the area was being built during the time of fieldwork; as yet there were no motor vehicles in the district and travel between villages was by foot or by outboard motor boat.

Garden plots were located outside of the village in the surrounding hillsides, up to an hour's walk away. A field rotation system of agriculture was practiced, with tapioca (*Manihot spp*), taro (*Colocasia*

[†]Observations made on the random schedule for Nagone village only are included in the quantitative data reported in Table I; additional observations were used for general ethnographic purposes.

spp) and yams (*Dioscorea spp*) being the most common crops. A few vegetables were also cultivated — eggplant (*Solanum melongena*), cabbage, *bele* (leaves of *Hibiscus manihot Malvaceae*), and pumpkin (*Cucurbita spp*). Coconut trees were abundant, and papayas (*Carica papaya*), bananas (*Musa spp*), and breadfruit (*Artocarpus communis*) were common. Mangos (*Mangitera indica*) and oranges were available in some areas, as were two varieties of small nut. The exact food resources varied considerably from place to place due to variations in soil and rainfall. Barrier and fringing reefs, as well as coastal mangrove swamps, provided a variety of fish, shellfish and crabs. In addition, chicken and pigs were raised in all villages.

As in much of the rural Pacific, the principle component of daily meals was a root food (or occasionally breadfruit), complemented by smaller portions of green vegetables and/or seafood. Fish or shellfish were eaten about three to five days a week, depending on season and weather. Pork was eaten in quantity on ceremonial occasions, which were irregular but might average over a year to one to three days a month. Green vegetables and fish were often cooked in or served with sauce made from coconut. Fruit and nuts were seasonal snack foods eaten occasionally between meals, and with varying availability in different villages. Storebought foods — rice, Pacific biscuits, tinned fish or meat, sugar and tea — were a very small part of the diet. Consumption of these varied with family income. Rice was used only about once a week by an average family, and tinned fish or meat about two or three times a week.

Except for the few persons employed as teachers or nurses, villagers earned cash from goods produced through both individual and cooperative efforts for sale in Suva — mainly copra, fish, yams and kava.[†] Perhaps because of the difficulty and expense of transport to Suva, however, cash cropping consumed very little of villagers' work effort compared to production for local noncommercial use. The fundamental aspects of the economy were those which have been well-described for Fiji by others; they relied upon traditional co-operative work and customary presentations of goods, both formal and informal (Nayacakalou, 1978; Sahlins, 1962; Thompson, 1949). These mechanisms allowed the particular resources of one lineage group or village to be spread throughout the village and district, and

---

[†]Kava (*yaqona*) is the root of the shrub, *Piper methysticum*, which is processed into a beverage and drunk at social and ceremonial occasions throughout Fiji and other areas of the Pacific.

also exchanged between urban and rural areas. Cash was also subject to this kind of distribution; persons earning cash from employment in the towns shared cash, and purchased goods, with relatives in the villages through the same formal and informal means.

The daily activities of parents in the study district were very similar; they engaged in farming, fishing, domestic tasks, and communal events. Yet their educational, residential and employment histories varied (as well as their ages and size of families). Parental years of schooling varied from 4 to 12 years and was negatively correlated with age ($r = -0.41$, $n = 49$, $p < 0.01$ for mothers; $r = -0.52$, $n = 33$, $p < 0.01$ for fathers). The mean years of schooling for mothers was 7.2 years; for fathers, 6.8 years. Younger parents achieved more schooling mainly because the number of years of primary school available in the local area had gradually increased from four to six or eight years in the past. Secondary schooling was available only in boarding schools, most of which were located near urban areas on the 'main island.

Due to education and other factors, the extent of past residence in urban areas also varied: 60 % of mothers and 50 % of fathers had not lived continuously in an urban area for more than one year or travelled outside of Fiji. These parents had been exposed to urban life and values, and the customs of other ethnic groups, only in occasional visits to the city or in short-term work. In other families, parents had attended school on the main islands, or had worked there for several years. All families were able to hear national and international news and other programs through the Fijian language radio station. Radios were available in most households; since visiting between households was frequent, the presence of a radio in a household was not a measure of extent of radio listening.

The district was served by two nursing stations staffed by one doctor and two nurses. Care of infants was a high priority, and every infant received standard immunizations against diphtheria, tetanus, typhoid, polio, measles and smallpox. Sick infants (and adults) were treated with antibiotics.

MARRIAGE AND RESIDENCE PATTERN

Marriages in the study district followed the pattern described for Fiji more generally (Nayacakalou, 1971) in a preference for marrage outside of one's local lineage group,[†] and for patrilocal residence. Sixty-three percent of the married mothers originally came from a village other than the one in which they currently resided, which was the village of their husband: 21 % were from a neighboring village on the same island, 16 % were from a nearby island in same district, and 26 % were from a distant area. In only one case was a married mother living in her own parents' household, rather than near the husband's kin.

Marriages between neighboring villages were especially important because, in the Fijian kinship system, marriage bonds provide channels of political and material support (Becker, 1983; Nayacakalou, 1978; Sahlins, 1962; Thompson, 1949; and others). In this area, where natural resources varied noticeably between villages, the informal exchanges of goods that occurred between persons related by marriage in neighboring villages were clearly advantageous. Maintenance of good relations between neighboring peoples was very important and was reaffirmed in frequently-occurring ceremonial events between villages. Villagers prided themselves on their traditional displays of hospitality and generosity on these occasions, and also in sports and scholastic competition. Thus village pride and rivalry was a means by which these and other traditional Fijian values were maintained. (See Nayacakalou, 1978, p. 120; Thompson, 1940b, p. 40.) The results of this study imply that this applied to cultural values concerning parental behavior as well. Yet 71 % of fathers with more than six years of schooling had married a woman either from the same village or from a distant area, rather than from a neighboring area. Of the less educated fathers, only 35 % had done so ($\chi^2 = 4.8$, p < 0.05).

In 60 out of 70 households, the child in the study sample lived with both mother and father, and 13 of these 60 were three generation households. In these cases, a young couple was living in the husband's parents' household, or a widowed grandparent had joined the established household of their married child. As for the ten other households, three children had been adopted by grandparents and lived with neither mother nor father; in five, the mother was unmarried and lived with her parents; and in two the father was deceased or absent for longterm employment in town.

[†]Nayacakalou (1975) and Walter (1978) note that this group is defined by villagers in flexible ways, which has caused confusion to ethnographers.

INFANT BIRTHS AND MORTALITY

Female menarche was reported to occur at 14 to 17 years, and the mean age at first childbirth (of mothers whose first child was born within the previous ten years) was 20.9 years (n = 31, range = 17 to 28 years, S.D.= 2.5 years). The medical department recommended that first births and those later than fifth be delivered in the hospital in Suva, and most mothers followed this practice. They travelled to Suva in late pregnancy and stayed there for a few months. Mothers could also receive tubal ligation in Suva if they wished. Otherwise, births occurred in the local clinic or village, usually attended by a nurse or doctor, or if transport could not be arranged, by midwives. Even if the nurse was not able to attend the birth due to the difficulty of transport, she examined the infant within a day or two. The interval between recent births was calculated from medical records birthdates for those mothers who had more than one child under age six years, resulting in a mean birth interval of 25.3 months (n = 39, range = 15 to 36 months, S.D. = 5.6 months).

Infant mortality nationally for the Fijian population was reported in 1974 as 40/1000 for infants 0 to 12 months, 24/1000 in the second year, and 8/1000 in the third (Government of Fiji Bureau of Statistics, 1976). However, there were no deaths among the 77 infants in the study sample (born in 1975-77) during the 15 months of the field work, nor were any deaths of younger infants reported.

INFANT CARE

Women's Work and Infant Care

Adult women's work was culturally defined to include providing the green vegetable or seafood portion of the daily meal (*nai lava ni kakana*); preparing, serving and cleaning up after meals; gathering firewood and coconuts; washing clothes; carrying water from the tap; weaving mats; and making oil for the skin. Mats and oil were important goods that were presented in large quantities to other groups on ceremonial occasions. The food resources that women provided were an important part of the diet, and were perhaps even more essential in the past, before the availability of tinned fish.

Some of a woman's necessary food-providing work was incompatible with carrying an infant, raising the question of how women's subsistence role was combined with infant care — an issue of considerable discussion in crosscultural literature on infant care (Brown, 1970; Katz and Konner, 1981; Nardi, ch. 16; Nerlove, 1974; Van Esterik and Greiner, 1981; Whiting, 1981). In fishing and shellfishing, women waded or dove into the water; even the shallow water shellfish-gathering would have caused an infant being carried to become wet and chilled. Although plant- and firewood-gathering, as well as some of the domestic work, could perhaps be performed while carrying an infant in a sling carrier, this did not occur. Infant carrying devices were infrequently used, and apparently were not in much use in former times in Fiji (Thompson, 1940a).[†] Infants were tied to the back with a piece of cloth only when the mother had to walk between villages, or when a caregiver wished to "walk" the baby because it was especially fussy.

In this research setting, it appears that the availability of social resources offered a solution in which the mother was excused from work and could be near the infant continuously, which facilitated breastfeeding at frequent intervals. The typically female food resources — small fish, shellfish, and green vegetables — were provided to her by other women or her husband, compensating for her decreased foraging. Mothers were encouraged by several beliefs to remain in the village in the vicinity of the infant. It was said that going into the cold water would chill the milk and the nursing infant would become sick; neither should the mother leave the infant to go on long plant- or shellfish-gathering treks while nursing. A very similar system of material support and release from subsistence work was described by Thompson (1940a, p. 36) for a similar outer-island part of Fiji during the 1930s.

In addition to help received from her female kin, the husband's role in providing a variety of food resources was critical. While by cultural ideal he was responsible mainly for the root foods, that were the staple of the diet, the husband might provide the vegetable or seafood complement as well. He might grow and harvest some green vegetables, or provide fish by spear or line fishing. Men were also evaluated by women with respect to their ability to provide a variety of foods for the family.

Infant care in several other Pacific cultures where women engage in fishing follows a similar pattern. As in this part of Fiji, Iatmul

---

[†]Thompson (1940a) reports the use of a sling carrier in Lau, Fiji, a similar area, when the mother walked through the village.

mothers of E. Sepik, Papua New Guinea, are reportedly provisioned by other women in an exchange system while their infants are young (Hauser-Schaublin, cited in Obrist, unpublished observations, 1983).[†] on Ponam Island, Papua New Guinea, where women ordinarily do a smaller amount of the fishing, the husband takes over this activity (Carrier, ch. 11). In the Murik Lakes region, Papua New Guinea, where fish obtained by women is also an important part of the diet, daughters and their mothers cooperate in sharing childcare tasks to allow some women the opportunity for fishing (Barlow, ch. 8).

Although nursing mothers were freed from foraging by the provision of food resources by other women or their husbands, and were thus able to remain in the proximity of their infants, it was observed that mothers still received considerable help with infant care from both children and adults. While mothers usually bathed and dressed infants themselves, others often held, watched over, or engaged in playful social interaction with infants.

The distribution of caregiving observed at three age periods on the 13 preschool children living in Nagone village is shown in Table I. Although the mother was present in 70 % of observations of infants at 7 to 12 months, and in 60 % at 13 to 50 months, she was actually giving care (or was the only one available to do so), in half or less at all age points. Someone other than the mother was holding or touching the infant in 30 % of observations at 7 to 20 months. The mean number of other persons in the immediate environment of infants was four. The only times that infants were found alone was when they were sleeping, and then an adult was always nearby within hearing distance.

When the mothers were asked about who helps them care for their child, they mentioned grandmothers, adolescent and school-age (6 to 13 years) girls, their sisters and sometimes their husbands. In Nagone village, the second most frequent caregivers were adolescent or school-age girls. Children under age seven did not engage in significant infant or toddler care.

The allocation of care and attention can be explained utilizing two different assumptions: First, infant care is a chore; second, infants are persons with specific culturally-defined relationships to others. On

---

[†]Obrist, B. (1983). The study of infant feeding: Suggestions for further research. Paper presented at the twelfth annual meeting of the Association for Social Anthropology in Oceania, New Harmony, IN.

TABLE I
Infant care in Nagone village, Fiji

	Age (Months)[a]		
	7-12	13-20	38-50
Number of infants	6	6	6
Observations per infant	12-15	17-22	10-14
Total observations	87	103	71
Percent of observations (random schedule, daylight hours)			
Infant asleep or awake			
1. Infant is sleeping.	17	12	3
2. Mother is giving active care or is only person available for care.	51	39	27
3. Person other than mother is giving care or is only person available.	43	44	18
4. Mother and others are available, none actively giving care.	3	8	26
5. Mother absent, others available, none actively giving care.	3	9	10
6. Caregiver available only at a distance.	0	0	13
7. Child's whereabouts in the village not known by those at home.	0	0	6
Infant awake only			
8. Physical contact with anyone.	56	43	13
9. Physical contact with mother.	27	11	1.4
10. No one within one m.	18	31	43
11. Mother present in immediate vicinity.	69	62	60
12. Father present in immediate vicinity.[b]	18	18	19

[a]There was only one 2-year-old child in the village, so data for this age point are not presented.

[b]Fathers of only 5 infants age 7-20 mos., and of 4 at 38-50 mos., resided in the village.

the one hand, watching an infant was a simple chore that the mother could delegate to her juniors. On the other hand, attention and nurturance were culturally prescribed aspects of certain kin relationships, for example, grandfather to first grandson, mother's younger brother to her male child, father's sister to his female child, the child's namesake, and others. The social attention paid to an infant by various relatives, both male and female, constituted a form of culturally-prescribed help with infant care.

The nature of the infant's relationship with the mother and with others was different. Lively playful interactions with infants were more characteristic of non-parental interactions than parental. While mothers performed routine physical care of infants, they might engage occasionally in playful interactions with them. But the playful interactions of the *non*maternal caregivers in the household or relatives visiting from other households were much more notable. Various persons, of all ages and either sex, greeted infants enthusiastically and often initiated playful slaps or clapping games, or engaged the infant in reciprocal vocal or verbal routines. These more highly dramatized routines may have functioned to introduce infants to the significant others in the social matrix, as well as to confirm among the adults that the caregiver felt the appropriate concern for the infant, given their particular relationships.

The pattern of physical contact shows that the infant quickly became physically independent from the mother, while also experiencing contact with others. While infants 7 to 12 months were in physical contact with some person in 56% of observations, only about half of that was with the mother (Table I). At 13 to 20 months physical contact was 43% and only one quarter of that was with the mother. At age three years, it was 13%, and the proportion of maternal contact was one-tenth.

Examination of the kind of work that was occuring during the observations provides further evidence that the allocation of caregiving served purposes other than merely enabling the mother to perform necessary chores. As indicated, when the mother was in the infant's proximity, she was not always the one taking care of it. But the task that she pursued was often one that could have been performed appropriately, and equally well, by the person who was caring for the infant at that moment. For example, the mother might be peeling tapioca while an adolescent girl played with the infant. In other observations, the observed caretaker was a male, a higher status female, or guest, and it would have been inappropriate for that person to perform the mother's work. In most of these cases, however, there was no reason of health or safety for the infant to be held; the child would have been safe either lying on the infant mat or toddling around the house. Thus infant holding and social interactions in this setting appear to have a degree of social function, rather than deriving only from practical needs of mother's work, or of infant health or safety. A significant social basis for infant care patterns has also been proposed for several other Polynesian areas of the Pacific (Carroll, 1970; Firth, 1936; Levy, 1970; Martini and

Kirkpatrick, 1981; Rubenstein, unpublished observations,1978).[†]

Infant Feeding

It was the ideal that infants be breastfed until they were able to walk, preferably to run, and most mothers followed this practice. The six infants who were observed daily in Nagone village began walking between 9 and 14 months and discontinued breastfeeding between 10 and 16 months. Adding the mother interview data to this, the mean age of walking in the nine villages was 10 months (range=8 to 15, n=27), and mean age of discontinuing breastfeeding was 13 months (range=2 to 24, n=42). Among parents who were living together and where infant illness was not a cause for early weaning, the mean duration of breastfeeding was longer — 13.8 months (range=8 to 24 months, S.D.=4.1, n=34). Older women reported that they could remember times when infants were breastfed longer; they recalled seeing nursing infants who could talk.

Since infants under age six months were not a focus of this study, information on the initiation of supplementary feeding is not available. Mothers of six month-old infants reported feeding infants broth of cooked starchy or leafy vegetables, or mashed papaya. It appeared that the quantity and variety of infant foods was gradually increased, so that by age one year an infant was eating the foods being cooked for the rest of the family. Two complete meals were cooked each day, in addition to a breakfast of tea and tapioca, or re-cooked leftovers from the night before. Toddlers ate more frequently because they often requested and were given whatever recently cooked foods were available in the kitchen. They also snacked on small fruits and nuts that older children brought back from the bush.

Infant formula was not sold in village stores; although powdered milk was available, it was expensive and was given to an infant by bottle only in the unusual circumstances of adoption or severe illness in which the infant was weaned very early from the breast. Nor were fresh animal milks used; cows were not present, and, while goats were raised by a few villagers, these were sold for meat in the capital and were not milked.

[†]Rubenstein, D. (1978). Adoption on Fais Island: An ethnography of childhood. Paper presented at the 77th annual meetings of the American Anthropological Association, Los Angeles, CA.

Weaning (*kali*, also meaning to separate a thing from what it adheres to) was culturally defined as a four-day period during which the breast was completely denied to the infant and the infant's food was cooked specially in a separate pot. Traditionally the infant was cared for by one of the mother's female kin in another household during this period, but only 25 % of this sample had followed this practice. Of the others, some reported that they remained with the child in the same house but did not sleep together at night, while others slept with the child but did not nurse. After weaning, the infant might again sleep next to the mother, or with another household member. Weaning was observed to occur abruptly; for example, just prior to weaning, a 15-month old infant was observed to nurse five times during the day, and presumably also nursed at night. After weaning, a child was considered to be more socially mature.

There were differences between individual mothers in feeding practices, depending upon whether or not their husband or other male kin grew green vegetables in addition to the essential root crops, and on how much cash was available to the family to buy food. Family composition and personalities could also affect an individual child's diet, because animal foods were accorded a high value and were served first to the adult men in the family. All of these factors created variations in the diets of toddlers, whether weaned or soon-to-be weaned.

Maintenance of Birth Spacing

Birth spacing was apparently maintained through aspects of natural fertility and through customary abstinence. Although birth control devices were available to mothers at the nursing stations, informants reported that very few women used these.[†] Cultural values supported abstinence from sexual relations during pregnancy and lactation for the good of the infant's and mother's health. As noted, it was the ideal that infants be breastfed until they were able to walk well, preferably able to run; but if a nursing mother became pregnant, she was supposed to wean the nursling. If not weaned, the child could develop a life-threatening condition known as *save*, in which its legs

---

[†]Infertility may have contributed to birth spacing to some degree because it was reported that some women did not conceive readily.

and body would become weak and the child would not eat well.[†] It was also believed that if husband and wife begin intercourse too soon after childbirth or a miscarriage, the mother could become ill (*guruvou*). For example, the condition of a woman who miscarried and appeared to other village women to be ill was of great concern among them. Although she was physically active, they perceived her as pale and distracted in her behavior and considered her vulnerable to severe mental derangement; this condition was ascribed to the couple's early resumption of sexual relations. The blame for both of these types of illness was said to fall on the husband.[†] Reference to a man's child as save was an insult which, though it might be made in jest, was a definite comment on his lack of character. In general, however, close birth spacing was not a joking matter but a cause of shame and embarrassment.

Correlates of Duration of Breastfeeding and Spacing of Recent Births

Two analyses were made to examine further the influences on breast-feeding and birth spacing: study of cases of early weaning; and correlational analysis of selected demographic variables on duration of breastfeeding and recent birth intervals.

Early weaning, according to the current indigenous definition, would refer to children weaned before they are able to walk. In a sample of 27 infants whose mothers could report the age of both weaning and walking, 7 were weaned before walking. Examining their mother's subsequent reproductive history and other circumstances showed that three conditions — pregnancy, infant illness, and unmarried motherhood — might account for five of the seven. Two mothers were pregnant, two infants had been ill, one infant's mother was unmarried and the infant was being adopted by the grandmother. The sixth mother was a schoolteacher who wished to attend a teacher's meeting in another district; the seventh has no known special circumstance.

[†] The custom of sexual abstinence during the nursing period is described as the traditional pattern for Fiji in the country report for Fiji (1981), presented at the Nutrition Seminar of the Foundation for the Peoples of the South Pacific, May 11-15, 1981, in Suva.

[†] As in the present study, Thompson (1940a, p. 33) observed in Southern Lau, Fiji, that "if a woman becomes pregnant while she is nursing, people blame the husband ..."

These same conditions, however, were also found for infants weaned at later ages. Of 36 cases where the age of weaning and the mother's subsequent reproductive history are known, 8 mothers were certainly pregnant at weaning, 19 were not, and in 9 cases the mother's reproductive status was unclear. The average age of weaning of pregnant and non-pregnant mothers is the same — 13 months. Infant illness was reported by the mother to be the cause of weaning in four cases; there was no reason why this would have been under-reported. If an infant had severe diarrhea, the district nurse advised that the infant drink only boiled water for a few days, and several mothers had weaned at this time. Finally, of the four infants of unmarried mothers in the sample, one was weaned at two months and adopted by the grandmother. The other three remained with their mothers and were weaned slightly below the mean age — at 10, 11 and 12 months respectively. Thus pregnancy and infant illness, while they might have been sufficient causes for weaning, did not explain variation in the age of weaning in the entire sample, and the effect of mother's unmarried status is minor.

In order to explore further the factors that might typically affect duration of breastfeeding and birth intervals, correlations of these with three sets of variables were computed (see Table II): namely, parental age and sibling position of the child (among living children); extent of parents' exposure to urban life, through formal education and past residence in urban areas; and proximity to maternal kin, as indicated by the location of mother's natal village. All cases in which the mother was unmarried, or in which illness was reported as the cause of weaning, were deleted from this sample since they represented atypical circumstances. The mean duration of breast-feeding for this group was 13.8 months (range=8 to 24, S.D.=4.1, n=34). The birth interval data are those described above; that is, recent births of mothers living with their husband, and with more than one child under six years of age (mean=25.3 months, range=15 to 36, S.D.=5.6, n=39). These data therefore represent typical mothers in reproductive years who have demonstrated no significant infertility.

Among mothers who could report the age of weaning and who bore another child during the fieldwork (n=17), the correlation between the elder child's age at weaning and at the birth of the following child was 0.52 ($p<0.05$). Thus these two variables are not strongly related to each and may have different determinants. The recent birth interval was longer among older mothers and less

TABLE II

Correlation (Pearson r) of demographic variables with duration of breastfeeding and recent birth intervals[a] for selected Fijian families

	Duration of breastfeeding[b] (in months)	Recent birth interval[c] (in months)
1. Mother's age	0.21	0.46**
	(33)[d]	(37)
2. Father's age	0.43**	0.34
	(31)	(34)
3. Sibling position of child (1-9)	0.04	0.24
	(34)	(38)
4. Sib position later than 5	−0.09	0.33*
	(34)	(38)
5. Mother — years of schooling	−0.07	−0.30(*)
	(31)	(33)
6. Father — years of schooling	−0.09	−0.63***
	(27)	(32)
7. Mother has lived in urban areas one year or more (0=no/1=yes)	−0.06	−0.08
	(30)	(35)
8. Father has lived in urban areas one year or more (0=no/1=yes)	−0.05	−0.05
	(30)	(35)
10. Mother's kin live in the village (0=no/1=yes)	−0.39*	0.08
	(32)	(38)
11. Mother's kin live in a neighboring village (0=no/1=yes)	0.35*	0.32*
	(32)	(38)

[a]Both samples include only parents residing together.

[b]This sample excludes infants who were reported by the mother to be weaned on account of illness.

[c]This sample includes only mothers with more than one child under age six years.

[d](r/n).

(*) $p \le .10$
* $p \le .05$
** $p \le .01$
*** $p \le .001$

educated fathers, where the mother came from a neighboring village, and among children of sibling position greater than five. Duration of breastfeeding was longer among older fathers, and where the mother came from a neighboring village. Mothers from the *same* village as their husbands weaned earlier. There was no significant association between continuous past urban residence of more than one year, measured dichotomously, and either dependent variable.

Since parental age, education and marriage patterns were inter-

TABLE III

Multiple correlations showing incremental effects (ΔR^2) of parental age, education and proximity of mother's natal village on duration of breastfeeding and recent birth interval

	Duration of breastfeeding				Recent birth interval			
	R^2	F	ΔR^2	F_Δ	R^2	F	ΔR^2	F_Δ
Mothers	n=31				n=32			
1. Mother's age	0.03	0.74	0.03	0.74	0.14*	5.03	0.14*	5.03
2. Mother's education	0.03	0.36	0.00	0.00	0.15	2.42	0.01	0.75
3. Mother's kin in same village	0.20(*)	2.20	0.17*	5.73	0.15	1.57	0.00	0.05
4. Mother's kin in neighboring vill.	0.21	1.71	0.01	0.42	0.31*	2.45	0.16*	5.47
Fathers	n=26				n=30			
1. Father's age	0.24*	6.49	0.24*	7.21	0.07	2.05	0.07	2.05
2. Father's education	0.26*	3.78	0.02	0.50	0.40***	9.08	0.33***	15.07
3. Mother's kin in same village	0.31*	3.13	0.05	1.63	0.41**	5.92	0.01	0.18
4. Mother's kin in neighboring vill.	0.32(*)	2.36	0.01	0.38	0.45**	4.95	0.04	1.61

(*) p ≤ .10
* p ≤ .05
** p ≤ .01
*** p ≤ .001

correlated, multiple correlations using a hierarchical analysis were computed to determine their combined effects and whether education and proximity to maternal kin provided any incremental effects over and above parental age alone. Mothers and fathers were considered separately. (See Table III.) Parental age was entered first; the additional contributions of parental education and proximity to maternal kin (over their relation to variables previously entered) are indicated by the increment in R^2 (ΔR^2) as each is added to the multiple correlation.

The proportion of variance explained by these selected independent variables was significant and moderately strong, ranging from 0.21 to 0.45 (0.09 to 0.35 adjusted). Some of the unexplained variance may be due to error in the measurement of duration of breast-feeding, based mainly on mother's report, and to lack of information on infertility or miscarriages. The exclusion of mothers with only one child under age six years, however, helps to exclude cases of infertility. Father variables predicted more variance than mother variables in both duration of breastfeeding and recent birth interval. Father's age predicted a significant portion of variance in breastfeeding and his education did not add to this; father's education, however, was more important than his age in prediction of recent birth interval. Father's education predicted 33 % of the variance in birth interval over and above the 7 % predicted by father's age alone. Of the mother variables, the proximity of her natal village was equally or more important than her age, and much more important than her education. As noted, mothers who came from the same village weaned earlier, whereas mothers from neighboring villages weaned later and had longer birth intervals (Table II). When proximity of mother's village was added to father variables, its additional predictive effect was small; however, this variable was more highly correlated with father's age and education than mother's, and so its effect was partially included when they were entered.

DISCUSSION

Villagers in this outer-island setting practiced a high degree of utilization of local food resources, as opposed to reliance on imported manufactured foods. There was considerable local variation in food resources; their production and distribution depended upon cooperation and generosity, requiring the maintenance of social ties both within and between villages. Within the village, the infant caregiver system reflected the importance of nurturant social relationships. Although the mother was available, the infant was introduced early to a number of caregivers and social interaction partners; these were significant relationships that would continue throughout its life.

The mechanisms for sharing of food resources in the local group also allowed the mother and infant to be provisioned during the period of lactation, a period of over a year in which the mother did not perform some of her usual subsistence work. During this period, the woman's husband and her female kin were critical figures in determining the food that she and the infant received. The father affected the infant's care and feeding in three ways: through provision of food resources; through his own food consumption pattern within the family, since he was served the most valued foods first; and through his observance of abstinence, since it was believed that a mother should not continue nursing while pregnant.

Given that the mother and infant were to some extent a burden on others during the period of lactation, what conditions were associated with earlier weaning of the infant, permitting her return to subsistence work? In this study, mothers whose consanguineal kin resided in the same village ended breastfeeding earlier; either plentiful food supply or greater availability of infant caregivers could have allowed this. These women already had more channels for food supply than did women whose kin lived at more distance. The implication of this correlation is that mothers did not wean early in order to return to subsistence work and thus increase the food supply for the family. Rather, they were inclined to reduce the burden of lactation on themselves, and the need for provisioning by others, when a *more* secure food supply was available for the family, including the infant.

The availability of caregivers could also be a factor underlying the association of earlier weaning with mother's kin in the village. In discussions, village mothers indicated that their own mothers gave more help with infant care than did their husband's mothers, perhaps because of the more distant and respectful relationship that obtains between a woman and her mother-in-law compared to her own mother. Mothers with their own kin in the same village can also command their younger female siblings to care for toddlers. Thus mothers whose kin are not available in the village have fewer options available in arranging for the care of a recently weaned infant.

It is also possible that infants of mothers whose kin lived in the same village were developmentally stronger or healthier, enabling the parents to feel more confident about discontinuation of breastfeeding. The customary discontinuation of breastfeeding according to infant developmental stage — walking/running — was sensitive to the health and development of individual infants. Similarly, LeVine (1974) observed that mothers in parts of both East and West Africa adjusted age of weaning according to their infant's size for age. Since no measures were taken in the present study of infant physical development at the time of weaning, its possible effect as an additional determinant of parental behavior cannot be known.

Finally, the role of the father in determining duration of breast-feeding must also be considered since father's age explained 24 % (21 % adjusted) of its variance. As noted, villagers attributed blame for early resumption of sexual relations to the father, and pregnancy required the weaning of a nursing child. They also placed blame for close birthspacing on the father. This interpretation of the findings is consistent with villagers' own expressed concepts of good parenting, which were phrased in terms of the health of mother and child. Mackenzie (1976-77) has noted the need for health workers to attend to parents' own concepts of good parenting, drawing on studies in the Cook Islands and Vanuatu.

The decreasing length of recent birth interval with greater paternal education also points to the father's role. In this outer-island setting there appear to be two routes by which education may have affected parental behavior. First, while variations in family income appear minor, higher father education could have been associated with greater cash income which could have improved family health and led to production of children at closer intervals (either through enhanced fertility, or conscious decisions to increase the family); second, higher father education could have been associated with attitude change and shifting marriage patterns, causing decreased observance of trad-itional customs such as abstinence. Both of these routes have been indicated in studies of modernization and parental behavior in other parts of the Third World (Jelliffe, 1976; LeVine, 1980; Nag, 1980). However, parents in this sample differed from many such studies in that there were no differences in parental occupations and no wide socioeconomic variations, no differential use of bottlefeeding, no differential access to medical care, and no differential exposure to mass media (in this case, radio).

While it would bear more study, the first interpretation is *less* likely in this study for several reasons. A family's access to cash depended on its entire kin network, not only on the father. The education of fathers did not always affect their entry into small cash earning activities in the villages. Family use of cash for food versus other purchases varied, as did the distribution of foods within the family. Finally, it is not known whether increased use of purchased foods necessarily improved parent or infant health.

A more probable interpretation is that secondary education provided more exposure to urban life leading to changing patterns of marriage and to lesser observance of some traditional customs such as sexual abstinence. Secondary schooling was available mainly in urban areas. The more educated parents (both mothers and fathers) had lived for several years in urban areas and had observed the varying ways of life of both Europeans and Fiji-Indians, which may have lessened the feeling of obligation to Fijian customs in general. In addition, there was opportunity to meet and marry persons from distant parts of Fiji. As reported, 71 % of more educated fathers had married a woman from the same or a distant village; yet the presence of maternal kin in a *neighboring* village lengthened both breast-feeding and birth interval. Mother's education was not related to either duration of breastfeeding or recent birth interval, suggesting again the significance of the father's role. While no correlation was found between parental residence of more than one year in urban areas and duration of breastfeeding or recent birth intervals, this measure was only dichotomous, and did not indicate the full range of variation that was present.

Cochrane (1979) and LeVine (1980) have noted that the means by which formal education affects parental behavior are not under-stood, and the effects would be expected to vary markedly in different settings. In Fiji, the interrelations of education, marriage pattern, income and infant care appear complex. Ideally, a survey of foods eaten daily by the infant and family should be taken, as well as infant health histories and measures of health and nutrition. These were not attempted in the present study because it focused on cog-nitive development among normal infants, and infant health was, in general, good.

The view that decline in traditional infant care ideology affected parental behavior is most strongly supported by the finding that breastfeeding and birth interval lengthen if maternal kin are present in a neighboring village. As noted above, relations between neighboring villages are delicate — proud, respectful, and competitive. When parents are from neighboring villages, the father would wish to conform to the cultural ideal of longer breastfeeding

and birth intervals rather than risk either shame on himself and his
village, or insult to the mother's people. Within a village, however,
relations are more informal and tightknit, and the sanctions of pride
against violations of cultural ideals are less effective.

Finally, the Fijian outer-island pattern of infant care offers an
interesting contrast with other patterns of care among rural women
who make a high contribution to subsistence resources. For example,
!Kung San foraging mothers carry their infants with them while
gathering wild foods from the bush, and thus maintain prolonged
breastfeeding and physical contact with the infant (Konner, 1977).
On the other hand, East African mothers' farming work requires
them to be separated earlier from their infants and to rely on child
nurses and the early introduction of supplementary foods (LeVine
and LeVine, 1963; Whiting, 1974). Reports on other areas of the
Pacific where women are primarily horticulturalists, suggest a third
pattern — mothers may carry infants to the garden but have it tended
there by other family members (Akin, ch. 12; Conton, ch. 6; Gegeo
and Watson-Gegeo, ch. 13). In the outer Fijian Islands as on Ponam
(Carrier, ch. 11), there is a fourth pattern. Some of a woman's subsistence work cannot be performed while carrying an infant, but
social mechanisms exist to allow her to be excused from this work
and to remain with the infant during a nursing period of more than a
year. Differences in the role of fathers and others in the maternal support network, as well as differences in the nature of women's work
and the foods available to supplement breastfeeding, may be important determinants of parental behavior in these settings.

ACKNOWLEDGMENTS

This study was supported by a grant from the U.S. National Institutes of Mental Health
to Beatrice Whiting, Harvard Graduate School of Education. I wish to express my deep
thanks for this support. I am also grateful to the Government of Fiji, and especially to
the Ministries of Health, Education, and Fijian Affairs for their permission to conduct
research and to use medical records; to Asasela Ravuvu and Ron Crocombe of the
Institute of Pacific Studies of the University of the South Pacific for advice; to
Saiamone Vatu for his support and generous and thoughtful guidance; to Ifereimi
Naivota and Nurse Sereana Naivota for their support, advice and assistance; and to the
parents who participated in this study for their help and cooperation.

REFERENCES

Becker, A. (1983). Women's group formation and maintenance in rural Western Fiji.
    Unpublished senior thesis, Harvard University, Cambridge, MA.

Brown, J.K. (1970). A note on the division of labor. *Amer. Anthropol.* **72**, 1073-1078.
Carroll, V. (1970). What does "adoption" mean? In V. Carroll (Ed.), *Adoption in Eastern Oceania.* University of Hawaii Press, Honolulu, pp. 3-17.
Cochrane, S.J. (1979). *Fertility and Education: What Do We Really Know?* Johns Hopkins University Press, Baltimore, MD.
Fiji, country report for (1981). Breastfeeding in Fiji. In D.B. Jelliffe and E.F.P. Jelliffe, *Consultant Report for the South Pacific.* Prepared for the International Nutrition Communication Service, Newton, MA. Report of a Seminar of the Foundation for the Peoples of the South Pacific, May 11-15, Suva, Fiji.
Firth, R. (1963 [orig. 1936]). *We, the Tikopia: Kinship in Primitive Polynesia.* Beacon Press, Boston.
Government of Fiji Bureau of Statistics (1976). *Social Indicators for Fiji, No. 3.* Government Printing Department, Suva, Fiji.
Jelliffe, D.B. (1976). World trends in infant feeding. *Amer. J. Clin. Nutr.* **29**, 1227-1237.
Johnson, A. (1973). Time allocation in a Machiguenga community. *Ethnol.* **14**, 301-310.
Johnson, J.S. (1982). A national food and nutrition policy for Fiji. *Food and Nutr.* **8**, 19-26.
Katz, M.M. (1981). Gaining Sense at Age Two in the Outer Fiji Islands: A Study of Cognitive Development. Unpublished doctoral dissertation, Harvard Graduate School of Education, Cambridge, MA. *Dissertation Abstracts International* **42**(6), B2565.
Katz, M.M. and M.J. Konner (1981). The role of the father: An anthropological perspective. In M. Lamb (Ed.), *The Role of the Father in Child Development.* John Wiley and Sons, New York, pp. 155-186.
Konner, M.J. (1977). Infancy among the Kalahari Desert San. In P.H. Leiderman, S. Tulkin and A. Rosenfeld (Eds.), *Culture and Infancy.* Academic Press, New York, pp. 287-328.
Lambert, J. and V. Yee, (1981a). Suva infant feeding survey. *Fiji. Med. J.* **6**(5), 5-9.
Lambert, J. and V. Yee (1981b). Fiji national nutrition survey. *Fiji Med. J.* **6**(6), 79-85.
LeVine, R.A. (1974). Parental goals: A cross-cultural view. *Teacher's College Record* **76**(2), 226-239.
LeVine, R.A. (1980). Influences of women's schooling on maternal behavior in the Third World. *Comparative Education Review* supplement, 78-105.
LeVine, R.A. and B. LeVine (Lloyd) (1963). *Nyansongo: A Gusii Community in Kenya.* In B.B. Whiting (Ed.), *Six Cultures.* John Wiley and Sons, New York.
Levy, R. (1970). Tahitian adoption as a psychological message. In V. Carroll (Ed.), *Adoption in Eastern Oceania.* University of Hawaii Press, Honolulu, pp. 71-87.
Mackenzie, M. (1976/77). Who is a good mother? *Ethnomed.* **4**(1/2), 7-22.
Martini, M. and J. Kirkpatrick (1981). Early interactions in the Marquesas islands. In T.M. Field, A.M. Sostek, P. Vietze and P.H. Leiderman (Eds.), *Culture and Early Interactions.* Earlbaum, Hillsdale, NJ, pp. 189-213.
Munroe, R.H. and R.L. Munroe, (1971). Household density and infant care in an East African society. *J. Soc. Psych.* **83**, 3-13.
Nag. M. (1980). How modernization can also increase fertility. *Cur. Anthropol.* **21**, 571-587.
Nayacakalou, R.R. (1971). The Fijian system of kinship and marriage. In A. Howard (Ed.), *Polynesia: Readings on a Culture Area.* Chandler, Scranton, PA, pp. 133-161.

Nayacakalou, R.R. (1975). *Leadership in Fiji.* Oxford University Press, Melbourne.
Nayacakalou, R.R. (1978). *Tradition and Change in the Fijian Village.* Institute of Pacific Studies, University of the South Pacific, Suva, Fiji.
Nerlove, S.B. (1974). Women's workload and infant feeding practices: A relationship with demographic implications. *Ethnol.* **13**, 207-214.
Sahlins, M. (1962). *Moala.* University of Michigan Press, Ann Arbor, MI.
Thompson, L. (1940a). *Fijian Frontier.* American Council Institute of Pacific Relations, New York.
Thompson, L. (1940b). *Southern Lau: An Ethnography.* Bulletin 162. Bernice P. Bishop Museum, Honolulu.
Thompson, L. (1949). The relations of men, animals and plants in an island community (Fiji). *Amer. Anthropol.* **51**, 253-267.
Van Esterik, P. and T. Greiner (1981). Breastfeeding and women's work: Constraints and opportunities. *Stud. Fam. Plann.* **12**, 184-197.
Walter, M.A.H.B. (1978). Analysis of Fijian traditional social organization: The confusion of local and descent grouping. *Ethnol.* **17**(3), 351-366.
Whiting, B.B. (1974). Folk wisdom and child rearing. *Merrill-Palmer Quart.* **20**, 9-19.
Whiting, J.W.M. (1981). Environmental constraints on infant care practices. In R.H. Munroe, R.L. Munroe and B.B. Whiting (Eds.), *Handbook of Cross-cultural Human Development.* Garland, New York, pp. 155-179.

CHAPTER 16

INFANT FEEDING AND WOMEN'S WORK IN WESTERN SAMOA: A HYPOTHESIS, SOME EVIDENCE AND SUGGESTIONS FOR FUTURE RESEARCH[†]

BONNIE A. NARDI

INTRODUCTION

A key issue in breastfeeding research is the marked trend in underdeveloped countries toward earlier weaning from the breast, whether weaning to infant formula or non-milk foods (Jelliffe and Jelliffe, 1978; Greiner, 1979). Jelliffe and Jelliffe (1978, p. 213) have noted:

> ... there is increasing evidence of changing patterns of infant feedings, particularly moves toward early weaning, in most ... technically developing countries. The process of change has usually been rapid, with an infinitely shorter time-lag than in the Western world during the preceding 50 years.

The reasons for earlier weaning from the breast are not clear. Such factors as corporate advertising of infant formula,[‡] urbanization, female labor force participation,[§] insufficient lactation, medical intervention and breast eroticism have been suggested. (Hull, unpublished observations, 1982, reviews the literature on these factors.)[‖]

[†]This research was supported by a grant from the National Institutes of Health (#5F32 HDO5800).

[‡]Despite the co-occurrence of corporate advertising of infant formula and adoption of bottlefeeding, it is not clear from existing studies exactly why mothers choose bottlefeeding. Is it because they believe it will lead to the fat, healthy babies portrayed in the advertisements, or because bottlefed infants can be left in the care of others while the mother engages in other activities? Careful studies of the actual decision making process involved in feeding choices are needed to ascertain the real reasons for the adoption of bottlefeeding by Third World women.

[§]Labor force participation encompasses regular wage labor jobs outside the home.

[‖]Hull, V. (1982). *Breastfeeding and Fertility: The Sociocultural Context.* Paper prepared for WHO/NAS Workshop on Breastfeeding. Geneva.

This paper calls attention to another possible factor responsible for the decline in the age of weaning in Third World nations: *an increasing workload for women associated with increasing involvement in the cash economy.* It is hypothesized that, as women become more involved in the cash economy, their workload increases and they choose to breastfeed for shorter periods in order to have more time for their other activities.

There is a well-established correlation between duration of breast-feeding and women's workload (Whiting and Child, 1953; Jelliffe, 1962; Nerlove, 1974). This correlation is particularly important in understanding changing patterns of infant feeding in the Third World. In the last 50 to 100 years women in underdeveloped countries worldwide have experienced dramatic increases in the amount of work for which they are responsible. (See Minge-Klevana, 1980, for a review of time-allocation studies.) In pre-colonial times women and men toiled to provide themselves and their families with basic subsistence; now they work long hours in underproductive, often exploitative jobs which net small amounts of cash used to purchase food, clothing, etc. (Coontz, 1961; White, 1973). Or they may work at *both* subsistence and cash-producing jobs. In either case, there is more work to be done in the context of a cash economy, leaving less time for the prolonged, intense mother-child interaction of lengthy breastfeeding. It is the aim of this paper to show that, as participation in the cash economy increases, women may choose to breastfeed for shorter periods of time in order to turn their time and attention to other kinds of work.

It is important to consider the *total* burden of work for which women are responsible. Women's work may include income-producing activities, both within and outside of the home; subsistence production; and housework.

This paper analyzes the relationship between infant feeding and women's workload in rural Western Samoa. The data suggest a connection between the decline in the age of weaning in rural Western Samoa and an increasing workload for women resulting from their increasing participation in the cash economy. The data are suggestive, not conclusive, and are reported here to stimulate further research on the possible link between duration of breastfeeding and participation in the cash economy.

METHODOLOGY

Western Samoa is a small, independent island nation of about 160,000 people, most easily located on the map by finding the region of the intersection of the international Dateline and the equator. About 80% of the Western Samoan population lives in rural villages which are strung out along the coastline of the two main islands (Upolu and Savai'i), and two tiny off-shore islets. The remainder lives in the Apia urban area and a few rural inland settlements. All local and international commerce and trade take place in Apia; regional commercial centers are lacking.

Samoa was Christianized and colonized during the mid-nineteenth century. Involvement in the cash economy grew slowly throughout the nineteenth and early twentieth centuries. It increased rapidly after World War II when opportunities for migration to New Zealand and other countries opened up (Pitt, 1970). Many Samoans left (and still leave) the islands temporarily or permanently to obtain wage labor jobs abroad and were able to earn much more cash than ever before. Much of this cash found its way back to the rural villages in the form of remittances (Shankman, 1976). This migration had a profound effect not only on personal income, but also dramatically increased demand for imported Western goods which stimulated cash work at home in the islands (Shankman, 1976).

The research for this study was conducted in the rural village of Salamumu on the southwestern coast of the island of Upolu during April, 1980-February, 1981. At the time of the study, Salamumu had 409 residents with 50% under the age of 15 — a result of the high rate of population growth (about 3% per annum).

Salamumu is a typical Samoan village in having a mixed economy of subsistence agriculture, fishing, cash cropping and handicraft production. According to a survey done in Salamumu in 1978 (Orans, 1981, p. 136), cash income is derived from the following sources: 39% from agriculture, 23% from remittances from relatives with wage labor jobs in Apia or abroad; 18% from handicrafts and 4% from business. Copra is the main commodity sold for export, with some bananas, cocoa and taro.

The data in the study consist of a household survey, key informant interviews, informal interviews, and observations made during residence in the village. The survey contained fertility histories, questions on household economy and a breastfeeding questionnaire. Survey data were collected from every woman in Salamumu who has or has had children, natural or adopted ($N = 62$). Intensive in-depth interviews were conducted with 15 key informants who discussed their breastfeeding practices and ideas on infant care and feeding. These interviews were tape-recorded and translated into English for transcription. Informal interviews and observations made in the general round of participant observation contributed substantially to my understanding of the context of work and motherhood in the rural Samoan village.

RESULTS AND DISCUSSION

This section briefly describes breastfeeding patterns in Western Samoa and discusses the relationship between women's work and breastfeeding in Salamumu.

Style of Feeding

Village women give birth in their homes, travelling to the hospital in town only for complications of pregnancy or delivery. A new mother is surrounded by helpful kinswomen who take over her chores, keep an eye out for her other children and offer such comforts as a warm bath or a leisurely backrub. The newborn baby is held, rocked and fed while awake. It sleeps on a nest of soft pillows under a gracefully draped mosquito net during the day, and beside its mother at night.

The style of breastfeeding in rural Western Samoa may be described as "naturalistic." When a baby is born, some mothers feed the infant within an hour of birth. Others wait for as long as a day, giving herbs in water.[†] During the first three or four months of the infant's life, the mother does not leave him or her for more than an hour or two as she wishes to be available to breastfeed on demand. When the child is three or four months old, the mother may leave for a period of several hours in order to work in the garden, attend a funeral, go fishing or perform an important errand. When she departs, the child is left in the care of siblings, grandparents or other relatives. The mother may prepare a bottle of formula or leave some type of baby food to feed the infant.

Babies are breastfed until they are satisfied. They are fed on

---

[†]This herb is the moon-flower, or *fuefuesina* in Samoan. It is used for many childhood ailments and for some adult afflictions as well.

demand, when they cry. Young infants are fed several times throughout the night.

Most mothers do not bottlefeed in Salamumu; however, those who
do reported that bottles are given when they have to be away from
the child or if they feel they do not have enough milk. The survey
queried all mothers with children aged two years or younger.[†] ($N =$
32 mothers and 34 children.) Of the 32 mothers, five, or 16%,
reported bottlefeeding. In all cases save one, the bottle was used as a
supplement to breastfeeding. Only one child, an 18-month old who
lived with her grandmother while her mother worked in town, was fed
formula without breastmilk (in addition to a variety of other foods).

Breastfeeding women do not eat special foods, but they do eat
more than usual.

Wetnursing is not practiced.

Babies in Salamumu are usually weaned by about 15 months,
although there is some variation, and at the time of this study one
baby was still breastfed at 21 months.

Weaning foods in Western Samoa include taro, bananas, breadfruit, coconut milk, papaya, papaya juice, fish and sago. These foods
are given anywhere from two to six months, depending on the
mother's wishes. Usually the foods are mashed and boiled according
to special baby food recipes. A few mothers said that they premasticated food for their babies, which was common practice in the
nineteenth century (Pritchard, 1968, p. 24).

In key informant interviews, some women said that they used
breastfeeding as a form of birth control. Husbands could be coaxed to
abstain from sex in order to prevent the baby from getting sick via
germs passed through the breastmilk.

Informants stressed the nutritional benefits of breastmilk for
infants and very young children. This attitude is probably the result of
traditional knowledge underscored by the teachings of public health
nurses who visit the village and who favor breastfeeding.

Weaning and Women's Work

The age of weaning in Western Samoa appears to be earlier now than

---

[†]In many rural communities precise determination of ages is problematic. However,
in Western Samoa I was able to accurately determine the exact ages of the children
because their mothers had copies of the birth certificates.

in times past, showing a decline from about two years in the nineteenth century to about 15 months in the twentieth century. We do
not have survey data on breastfeeding practices for Western Samoa
before 1954 (Malcolm, 1954), but must rely on the statements of
observers which suggest a longer duration of breastfeeding than is
now found. W.T. Pritchard (1968, p. 141), a nineteenth century
British consul to Samoa, reported in 1866 that children were
breastfed until two years of age. Malcolm (1954, p. 19) also reported
that "It used to be customary to breastfeed the infants until they were
about two years old or even older."[†]

Malcolm's survey in 1954 ($N = 580$) found that 82% of the
children were breastfed at 0-9 months, 46% at 10-14 months and 8%
at 15 months and older. In the present study, 100% of the children
0-9 months were breastfed, 0% at 10-14 months and 12% at 15
months and older.

These data suggest that the vast majority of children are weaned by
15 months, with weaning taking place sometime between 10 and 14
months.

In the most recent large-scale survey of breastfeeding practices in
rural Western Samoa, King (1975)($N = 393$) found that 89% of all
children from ages 0-5 months were breastfed, 73% of all children
from 6-11 months and 52% from 12-17 months. In comparing the
three surveys, it is apparent that King's data suggest a higher average
age of weaning than Malcolm's data or my own. However, it is
possible that there is not really a discrepancy between the data sets
since King reported the frequency of breastfeeding only for the age
group 12-17 months — a rather large age interval which may contain
variable rates at its extremities. It is possible that many of the
breastfed infants in this age range actually occurred in the 12-14
month range, rather than the 15-17 month range, which would bring
King's estimates in closer conformity with Malcolm's and my own.
But overall, the three data sets strongly suggest that the age of weaning in Western Samoa has declined markedly between the nineteenth
century and the present.

[†]Other early observations, while not precise, suggest a lengthy period of breastfeeding. Wilkes, the head of the United States Exploring Expedition, visited Samoa in
1839, only three years after the arrival of the first missionary. He wrote, "The mothers
often suckle their children until they are six years old." (Wilkes, 1844). Brown, a
missionary who lived in Samoa for 14 years, 1860-1874, reported, "Suckling was
generally continued until the child was running about, or weaned itself." (Brown,
1910).

Why has this occurred? The hypothesis argued in this paper is that as women's workload increases due to participation in the cash economy, women choose to breastfeed for shorter periods in order to have more time for their increasing responsibilities. Rural Western Samoa is an apt test of the hypothesis because the other factors thought to account for a decline in the age of weaning are absent or insignificant there (corporate advertising of infant formula and widespread bottlefeeding, urbanization, female labor force participation, insufficient lactation, medical intervention and breast eroticism[†]).

Not only are these factors unimportant for rural Western Samoa, but data from interviews, observations and historical records suggest the plausibility of the hypothesis linking women's work and age of weaning.

In linking a decline in the age of weaning with increasing participation in the cash economy, this paper argues that women deliberately choose to reallocate the time they spend breastfeeding versus the time they spend working. Since this involves a conscious economic choice, it is important to ask women themselves if this is in fact what they are doing. (See Hull, 1982, for a discussion of the importance of studying breastfeeding decisions.) In key informant interviews and informal interviews, women were asked when and why they initiated weaning. These simple questions elicited detailed data on women's work schedules and strategies for successfully accomplishing the many activities they pursue. The following discussion summarizes what women said in the interviews, as well as what I observed while a resident in the village.

In key informant interviews and informal interviews in the present study, women reported that they generally initiated weaning because of their demanding work schedules. Women reported that they felt that they could accomplish more when assured the freedom of being able to leave their babies with other caretakers. Because babies are

---

[†]In the rural Samoan village infant formula is not advertised; nor is it advertised in town. Bottlefeeding is not widespread, as can be seen in the studies of Malcolm (1954) and King (1975), discussed previously. The present study, small though it is, confirms the findings of Malcolm and King. Women in rural villages have no opportunities for wage labor except for the occupation of village school teacher, a post often held by a man. There are no data on the amount of breastmilk produced by the rural Samoan women, but insufficient lactation seems unlikely as the vast majority of women breastfeed, and women do not complain of insufficient lactation. There is no medical intervention to inhibit breastfeeding; in fact, breastfeeding is strongly supported by the Public Health Service. Breast eroticism is not a feature of Samoan culture. In rural villages breasts are often exposed, with no sexual connotation.

L

weaned relatively early and given supplemental foods in the early
months, mothers said they felt less restricted in their activities than
they would be if they fully breastfed for long periods. Informant statements such as the following were typical:

> When I want to wean the baby I stop breastfeeding her. That can happen
> after eight months if there's a lot of work to be done, but if not I may wait till
> the baby's about a year old, or even older.

The women reported that they had many responsibilities to attend
to which required separation from the young child. These activities
included gardening, fishing and selling vegetables and fish in town.
Social and political activities were also important, including attendance at funerals, weddings, Women's Committee meetings and
church meetings. Although any one of these activities could be performed with a young child in attendance, the informants reported that
it is easier to leave the child with a caretaker. The mother can
accomplish more and with mimimum disruption.

Many informants described the conflict between work and breastfeeding:

> When I come back from the garden the baby's crying and wanting to be
> nursed and I don't want to pick him up because I'm so dirty. I have to feed
> the baby before I've had time to have a bath.

Another mother stated,

> If I'm at home doing some housework the baby gets breastfed about six or
> seven times a day because she cries for me all the time if she knows I'm here.
> If I have to leave, say to go fishing or to a funeral, then she gets more solid
> food and I only breastfeed her three times a day.

As an observer in the village it was easy to see that the activity
level of Samoan women is very high and that they are involved in a
large number of pursuits. Samoan women quickly resume their economic activities, social life and participation in community projects in
the early postpartum months. Their demanding schedules require
flexibility, freedom of movement and considerable time. Women in
Salamumu maintain a complex schedule of activities including
income-producing activities, subsistence agriculture and fishing, and
housework. Their tasks include cooking, sewing, laundry, ironing,
dishwashing, childcare, housecleaning, matweaving, animal tending,
food production, fishing, handicraft manufacture and the marketing
of food, fish and handicrafts. It seems likely that work time has
expanded throughout the colonial and post-colonial eras as women

(and men) have taken on a variety of new tasks. (See Minge-Klevana, 1980, for a review of the literature.)

In pre-colonial times Samoans produced primarily for subsistence; they were not, of course, connected to the world market economy in which production for cash is possible. Now all Western Samoans produce for *both subsistence and cash economies.* A woman's agricultural labor contributes to both feeding her family and supplying them with cash. In addition, women manufacture and sell handicrafts, catch and market fish and prepare cooked food for sale at the market. Data from the present study showed that 81% of the women in Salamumu make and sell handicrafts (for tourists) and/or *tauaga,* strainers used to make coconut cream. Seventy-six percent of the women sell homegrown vegetables and/or fresh or cooked fish in the market. Fifty-four percent sell *both* handmade articles and vegetables or fish. These items are sold by the women themselves, in town, which requires a 90-minute bus ride (each way) as often as twice a week.

Elderly informants (over age 65) reported that women (and men)† now have more work to do than in "old times" (that is, the days of their youth) because of the manufacture of handicrafts, selling food in town, cash cropping and the other cash-producing activities. As one informant put it,

> Samoans have too much to do because they want money to buy things other Samoans make, and to buy European food. They don't want to make things themselves — they just want to go to the market and buy things, like Europeans.

---

†Men's work has also increased with the advent of the cash economy. Men engage in cash work, especially growing cash crops. Some village men even hold wage labor jobs in addition to their farming and fishing. In Salamumu 16% of the men over age 18 had wage labor jobs during the present study. Most of these (86%) were in Apia and required commuting between Apia and the village, with weekends devoted to village life. The other jobs, such as school teacher and carpenter, allowed full-time residence in the village.

There are many other tasks for men associated with participation in the cash economy, such as maintaining the road to the village, which is the sole responsibility of the villagers themselves, building facilities for primary schools, and even corvee on church projects. For example, during my stay in the village about half of the adult men were absent from the village for three weeks as they labored on a new guest house for Piula College, a Methodist seminary in a village which is a two-hour drive from Salamumu. Not only do men contribute their labor to these projects, they also donate considerable cash which, of course, requires labor to obtain.

In asking elderly informants about work in former days I simply inquired, "Why don't Samoan women make tapa cloth and fine mats [ceremonial valuables] as much as they used to?" This question usually elicited a flood of responses about the new responsibilities of Samoan women, especially handicraft manufacture, which is seen as the direct replacement of the nearly-lost arts of tapa making and fine mat weaving. It used to be the task of women to produce painted tapa (bark cloth) and fine mats (mats of very fine fiber, densely woven) for use in ceremonial exchanges. It is time-consuming to execute an intricate tapa design, or a very finely woven mat of the texture of good linen. Now women feel that they do not have time for such work.

But if this work is so time-consuming, then what of the argument pursued here that women have more work now than before participation in the cash economy? Producing tapa and fine mats does take considerable time, but in earlier days there was no time pressure to produce quickly as there is with cash work where people feel the need of immediate cash for specific purposes. Informants in the present study recounted memories of the unhurried ceremonial aspect of tapa and fine mat making, telling how women gathered together to work on their art and were courteously served food by their admiring relatives.

Not only were tapa making and fine mat weaving accomplished at an unhurried pace, but these tasks were done in the home and were easily interrrupted by the demands of children (and others). Unlike daily housework which must be completed on schedule in order for the family to function, the production of tapa and fine mats could be easily put aside, if necessary, to tend to other chores, including the care of infants and toddlers.

There are a few categories of labor for which work in earlier times must have been more time-consuming. Before Europeans reached the islands, Samoans, of course, had no metal tools, no tin roofs (which last three or four times as long as thatched roofs) and no piped water.[†] Certainly water-fetching and roof building occupied more time. Metal tools reduce horticultural labor considerably, but since Samoans grow much more now as they produce for both subsistence and cash needs, horticultural labor is not reduced.

Throughout the twentieth century, the productive work of Samoan women has increased as they endeavor to find ways to supply their

---

[†]Most villages, including Salamumu, still do not have electricity or indoor plumbing.

families with money to buy coveted imported items such as tinned food, cloth, pots, lanterns and flashlights. But the acquisition of Western goods which is made possible by participation in the cash economy increases labor in another way, separate from the need to acquire cash to purchase items. For those who formerly lived in a "throw-away" culture, the acquisition of the more durable imported goods of metal, cloth, china and plastic opens up new vistas of maintenance and repair for which women assume much of the responsibility. Items such as clothes, pots, dishes, furniture, radios and lanterns must be cleaned, mended and carefully looked after if they are to provide continuing service. Previously for Samoans, throw-away materials were used in everyday life: plates were banana leaves, serving platters were woven mats lined with leaves which were discarded after becoming soiled. Modern Samoans wash dishes every day. In times past clothing was allowed to discreetly biodegrade and then cast away. Now it is washed, ironed, sewn and mended by women. Today cooking is done using metal pots and pans instead of the traditional *umu*, or ground oven, which sees service only on ceremonial occasions. Pots, pans, dishes and serving spoons are cleaned at least twice daily. Showers and latrines have replaced bathing pools and quiet spots away from the house for daily hygiene. Women take care of these just as they tidy the house, sweep the yard and collect the trash of tin cans, paper wrappers and cardboard boxes.

Another key factor which has increased women's work is that gardens are often farther away from the home and consist of more marginal land, because of increased population. Women spend more time walking to reach the gardens and cultivate more diligently to produce a crop from the less desirable land.

Not only have women's productive and household tasks increased; they have taken on expanded roles in church and village activities, *many of which are created by participation in the cash economy*, such as fund raising and maintaining property for the village and church. For example, new roles emerged with the development of the Women's Committees in the 1920s, a completely new institution in which women quickly became interested and involved. (See Schoeffel, 1977.) Women invest a great deal of time in Women's Committee activities including village inspections, educational programs and raising funds for new village facilities and special projects.

The church has also created new obligations of service for women, and they help to advance the fortunes of their local church through fund-raising, cleaning and decorating the church building, Sunday school programs and devotional activities.

The village primary school adds to women's work; they provide food for teachers, visiting officials and examination proctors as well as assisting in extra-curricular activities for the school children.

Samoan women have deliberately, actively chosen these extra-domestic roles. They enthusiastically acquire Western goods. Women relish their work in the church and Women's Committees, and they enjoy the Western goods they purchase, if not the chores associated with them. But the changes in income-producing activities, house-work and service have increased the daily burden of work for which women are responsible, making prolonged breastfeeding a less desirable choice for the ordinary Samoan woman.

SUMMARY AND CONCLUSIONS

This paper sets forth the hypothesis that a key factor in the decline in the age of weaning in Third World countries is an increasing work-load for women associated with increasing involvement in the cash economy. As women take on new tasks which increase their total workload, they choose to breastfeed for shorter periods of time because the prolonged, intense mother-child interaction of lengthy breastfeeding interferes with their other tasks.

In rural Western Samoa the age of weaning has declined since the nineteenth century. There has been increasing participation in the cash economy. That these two events are causally related in Western Samoa is suggested by two facts: Women themselves report that they initiate weaning in order to attend to other responsibilities; and other factors which might explain an earlier age at weaning are unimportant in rural Western Samoa.

This pattern is apparent elsewhere in the Pacific. Marshall and Marshall (1979) have shown that a key factor in the decline in breast-feeding in Truk is the increase in women's employment and edu-cational pursuits. They observed that a rising demand for cash stimu-lated women to take on the additional work of advanced schooling and employment outside the home. They concluded (1979, p. 248):

> We found a clearcut statistical association between mode of infant feeding and employment history and educational achievements of a mother ... These women were ... experiencing more demands on their time and energy incompatible with the demands of exclusive breastfeeding. Other ... female

relatives could always be found to take over childcare, *especially if by this means cash flow into the households would increase.* (Emphasis added.)

A rigorous test of the hypothesis that increasing involvement in the cash economy can lead to a decline in the age of weaning could be made by comparing quantitative estimates of women's workload and breastfeeding across regions which vary by level of involvement with the cash economy. There are many countries in the Third World (Mexico, Guatemala, Honduras) where regional participation in the cash economy varies greatly. Such countries would provide suitable settings for new research into the dynamics of changing infant feeding practices in the Third World.

Hull (1982) points out the need for studies which investigate how individual women make decisions about infant feeding practices. A productive research strategy would incorporate macro-level studies of quantitative estimates of women's workload as suggested above, along with micro-level decision making studies of women's perceptions of the new choices and constraints of a changing economy and society.

It is the author's hope that this paper has called attention to the possible link between women's participation in the cash economy which increases their total burden of work and the decision to breastfeed for shorter periods of time. Investigation of this relationship should prove fruitful in future research on infant feeding in today's world.

ACKNOWLEDGEMENTS

I would like to offer warm thanks to Mrs. Elizapeta Eteuati, who assisted me in the data collection process, and to the women of the village of Salamumu who graciously shared their thoughts and reflections with me. Valerie Hull, Leslie Marshall, and Maxwell Katz provided useful comments on earlier drafts of the paper.

REFERENCES

Brown, G. (1910). *Melanesians and Polynesians: Their Life Histories Described and Compared.* MacMillan and Co., London.
Coontz, S.H. (1961). *Population Theories and the Economic Interpretation.* Routledge and Kegan Paul, London.
Greiner, T. (1979). *The Economic Value of Breastfeeding with Results from Research Conducted in Ghana and the Ivory Coast,* Cornell International Nutrition Monograph Series No.6, Cornell University, Ithaca, NY.

306 B.A. NARDI

Jelliffe, D.B. (1962). Culture, social change and infant feeding. *Amer. J. Clin. Nutr.* **9**, 19-45.
Jelliffe, D.B. and E.F.P. Jelliffe (1978). *Human Milk in the Modern World: Psychological, Nutritional and Economic Significance.* Oxford University Press, Oxford.
King, C.M. (1975). *Nutritional Status of Preschool Children in Western Samoa.* Health Department of Western Samoa, Government of Western Samoa, Apia, Western Samoa.
Malcolm, S. (1954). *Diet and Nutrition in American Samoa.* South Pacific Commission Technical Paper No. 63, South Pacific Commission, Noumea, New Caledonia.
Marshall, L. and M. Marshall (1979). Breasts, bottles and babies: Historical changes in infant feeding practices in a Micronesian village. *Ecol. Food Nutr.* **8**, 241-249.
Minge-Klevana, W. (1980). Does labor time decrease with industrialization? A survey of time allocation studies. *Cur. Anthropol.* **21**, 279-287.
Nerlove, S.B. (1974). Women's workload and infant feeding practices: A relationship with demographic implications. *Ethnol.* **13**, 207-214.
Orans, M. (1981). Hierarchy and happiness in a Western Samoan community. In G. Berreman and K. Zaretsky (Eds.), *Social Inequality: Comparative and Developmental Approaches.* Academic Press, New York, pp. 123-147.
Pitt, D. (1970). Tradition and Economic Progress in Samoa. Clarendon Press, Oxford.
Pritchard, W.T. (1968). *Polynesian Reminiscences.* Dawson's of Pall Mall, London.
Schoeffel, P. (1977). The origin and development of women's associations in Western Samoa, 1830-1977. *J. Pacific Stud.* (University of the South Pacific, Suva) **3**, 1 – 21.
Shankman, P. (1976). *Migration and Underdevelopment: The Case of Western Samoa.* Westview Press, Boulder, CO.
White, B. (1973). Demand for labor and population growth in colonial Java. *Human Ecol.* **1**, 217-236.
Whiting, J. and I. Child (1953). *Child Training and Personality: A Cross-cultural Study.* Yale University Press, New Haven.
Wilkes, C. (1844). *Narrative of the United States Exploring Expedition During the Years 1838, 1839, 1840, 1841, 1842.* Volume 1. C. Sherman, Philadelphia.

CHAPTER 17

COMMENTARY:
A PEDIATRICIAN'S PERSPECTIVE

JOHN BIDDULPH

INTRODUCTION

The five million people who inhabit the South Pacific islands (excluding Australia and New Zealand) make up 0.1% of the world's population but live in an area of 29 million sq. km, about the size of the African continent land mass. Less than 2% of this region is land. Hence the populations are widely scattered with diverse cultures and languages. Papua New Guinea, the largest South Pacific island, whose population comprises about 60% of the total, has more than 700 distinct languages. Therefore, the 15 ethnographies described cover only a fraction of the multitude of cultural and linguistic groups in the South Pacific. Any generalizations must be made with great caution.

The methodology used in the ethnographies was that of participant observation. What people actually do rather than what they say they do is recorded. In this manner, Montague (ch. 5) found that wrong assumptions had been made by the Milne Bay Nutrition Monitoring Group who relied on brief survey interviews. Some of the ethnographic data recorded are anecdotal, and it is difficult to know how valid these data are for the rest of the group. However, more quantitative data, if carelessly collected, may be even more misleading. One suspects that the findings of the 1980 survey of the Milne Bay Nutrition Monitoring Group was seriously biased. I think that Montague quite rightly challenges the findings of the Monitoring Group on the extent and seriousness of malnutrition in Milne Bay. My brief nutrition surveys on Kiriwina in 1967 and 1975 confirm Montague's remarks. Only 17% of the under-five year old children in the surveyed villages had a weight-for-age less than 80% of standard.

Mortality data derived from the 1980 census also support the view that malnutrition in young children is not a major problem in Milne Bay. The infant mortality rate and one-to-four year mortality rate for the province was the third lowest in the country, being 50 per 1000 and 27 per 1000 respectively. This compares with the national figures of 72 per 1000 and 45 per 1000 respectively (Bakker, 1983). Only North Solomons Province and National Capital District had lower infant and one-to-four year mortality rates than Milne Bay Province.

The combined methods of ethnography and nutrition described by Jenkins, Orr-Ewing and Heywood (ch. 3) need to be used more often. Also, greater use should be made of conceptualizing growth stages emically, according to the informant, instead of etically, according to the observer, as a method of examining growth in relation to feeding practices. It would be a helpful achievement if ethnographers and nutritionists could agree on a standardized questionnaire relating to maternal and young child feeding practices. This could allow more valid comparisons between different groups.

BREASTFEEDING

Breastfeeding enhancing factors feature prominently in the ethnographies. Conton's account (ch. 6) of breastfeeding in Usino reads like a tract on "The Art of Successful Breastfeeding." Breastfeeding is seen as a natural, inevitable event and is publicly performed without embarrassment. There is no need for these mothers to go to court for their right to breastfeed in public (Lofton and Gotsch, 1983).

The new mother is often surrounded by women to support her (doulas). (By contrast, a characteristic feature of industrialized societies is the small nuclear family, removed from the supportive extended family.) The infant is fed on demand and the mother sleeps with her infant, thus allowing night feeds. The infant accompanies the mother everywhere in a netbag. Finally, the child weans itself gradually.

Counts (ch. 9) records how the Kaliai continue to think of the breast as *the* source of infant food despite the introduction of foreign attitudes toward the female breast.

Gegeo and Watson-Gegeo (ch. 13) recount that the Kwara'ae regard cow's milk as inferior because it is produced outside the mother's body and does not contain her love.

Akin (ch. 12) describes the use of protein foods such as fish, mangrove pods, shellfish and greens as galactagogues, while Tietjen (ch. 7) records use of a soup both as a galactagogue and sympathetic magic.

The length of breastfeeding is extremely variable. Conton (ch. 6), Counts (ch. 9), Marshall (ch. 2), Jenkins, Orr-Ewing and Heywood (ch. 3), Akin (ch. 12) and Chowning (ch. 10) all mention some children being suckled when aged four years or more. Becroft (1967) found in 1962 that 10% of a group of 125 Enga children were still on the breast at the age of 5 years 6 months. Berg (1973) has reminded us that earlier in this century, Chinese and Japanese mothers nursed their children for five or six years, Caroline Islanders up to ten years, and Eskimos up to 15 years. Use of the word "prolonged" in reference to the duration of breastfeeding is, therefore, best avoided.

The cultures described are strongly supportive of breastfeeding. Ironically, it is often health workers who inhibit breastfeeding in the hospital setting and in the rural clinic. An example of practices which interfere with breastfeeding is given by Katz (ch. 15) who found that several mothers weaned their babies after the district nurse had advised that their infants with severe diarrhea should drink only boiled water for a few days. Breastmilk, of course, contains water and food, both of which the infant with diarrhea needs. Therefore, it is important that the baby with diarrhea continues to be breastfed.

The commendable official policy in Fiji in favor of breastfeeding is undercut by the hospital practices described by Morse (ch. 14). Mothers are expected to feed their newborn babies at fixed times. Between feeds the babies are separated from their mothers and they sleep at night in a separate nursery. It is distressing that some hospitals still persist with such practices guaranteed to prevent the establishment of successful lactation. The claim that pressure of numbers and lack of space dictates these harmful practices is unwarranted. A baby sleeping beside its mother takes no extra space, and a mother looking after her own baby removes work from the nurse. The rules for successful breastfeeding — Breastfeed your baby whenever you want, as often as you want, for as long as you want — should be prominently displayed in all maternity wards.

Equally distressing is Morse's statement that "on the advice of the nurses and physicians, all infants are given glucose powder in water, or fruit juices such as rose hip syrup, orange or mango juice daily in a feeding bottle, to prevent dehydration in the hot climate". This is not reasonable. Breastmilk is a low solute fluid and provides sufficient free water for the baby's metabolic needs whether one is in a hot climate or not. The introduction of the bottle to the baby each day merely makes it less likely that lactation will be fully established.

In these circumstances, it is not the least surprising to learn that in the weeks following the mother's discharge from this hospital, breastfeeding is rapidly replaced by a pattern of mixed feeding.

The strong personal commitment to breastfeeding by a majority of public health nurses is highlighted in Marshall's chapter (ch. 2). This is an important report as it shows that when personal commitment, knowledge of infant feeding options and social support for breastfeeding from employers and coworkers combine, working mothers can continue to breastfeed their babies successfully. It also emphasizes the importance of providing creches, flexible work schedules and regular breastfeeding breaks if governments are serious about encouraging breastfeeding.

Papua New Guinea's legislation to protect breastfeeding by banning the advertising of feeding bottles and breastmilk substitutes, and by restricting the sale of feeding bottles has been successful in stemming the flow of feeding bottles to rural areas — Chowning (ch. 10) reports a decrease now in the amount of bottlefeeding in more remote villages. It has also been associated with a decrease in severe malnutrition and diarrheal disease and deaths in infants in urban areas (Biddulph, 1983). Before other countries rush to follow suit, it must be pointed out that its success in Papua New Guinea has been due to four important reasons: Breastfeeding is still very much the norm; there are no entrenched vested interests in the country to back artificial feeding; the medical and nursing professions are united on the issue; and the national government is strongly committed.

SUPPLEMENTARY FOODS

Growth-faltering has been recognized from around four months of age onwards in many breastfed children (Waterlow, Ackworth and Griffith, 1980). Chowning (ch. 10) quite rightly points out that most of the current infant feeding campaign centers on the importance of breastfeeding with too little emphasis on supplementary foods. From her experience among the Kove, which spans 18 years, she considers that the substitution of tradestore foods for traditionally grown foods is increasing. She foresees the desire for wealth leading to a neglect of food gardens, with children being weaned onto a limited range of foods of dubious quality, such as fried flour and sweet tea.

Conton (ch. 6) also sees the increased interest in and dependence on imported commodities as encouraging earlier weaning and alludes to the "suckling's dilemma" (Waterlow, 1981). If supplementary foods are introduced at the start of growth-faltering, diarrhea is likely to result under the existing conditions of poor hygiene. But if supplementary foods are not introduced early enough, the child may become malnourished. The answer to this dilemma is to start supplementary foods early (by four months) rather than later. Diarrhea is likely to occur at whatever age supplementary feeding starts, and it is better for this infection to occur while the baby is still in a well-nourished state.

Carrier (ch. 11) notes that the Ponam mothers feed their babies early with a variety of soft foods to keep them happy and comfortable.

The age of weaning (sevrage) is mostly declining. Nardi (ch. 16) suggests that in rural Western Samoa this is related to women's increasing work load due to their increasing participation in the cash economy. Carrier (ch. 11), on the other hand, points out that on Ponam Island, Manus, weaning follows conception, and, thus, any increase in sexual contact between husband and wife will contribute to a decrease in the age of weaning. Clearly many interrelated factors are involved, and the relative importance of each factor is likely to vary at different times in different places.

On a visit to Western Samoa in January, 1980, as part of a regional nutrition survey for the United States Agency for International Development, I observed that the prevalence of malnutrition in young children was greatest in areas where urban influences were strongest, and that it was a growing problem in the capital, Apia. The increase in bottlefeeding in Apia was an important factor in this malnutrition, as was the fact that many urbanized families no longer had access to plantations. With the change to a cash economy, traditional supplementary foods were being replaced by expensive and nutritionally inferior imported processed foods. Brewster and Brazill (unpublished observations, 1980)[†] have shown the high energy content of traditional Western Samoan weaning foods. *Fa'ausi*, for example, a mixture of taro and coconut cream, contains 280 Cals/100g. By comparison, tinned baby food labelled chicken and

---

[†]Brewster, D.R. and H. Brazill (1980). Social, economic and cultural factors in malnutrition in Western Samoa. Paper presented at the Centenary Scientific Meeting, Royal Alexandra Hospital for Children, Sydney.

vegetables supplies just over 60 Cals/100g.

The importance of fats in the diet has recently been stressed by Dearden, Harman and Morley (1980). Not only do fats increase the energy content of a weaning food, but the fat also decreases the viscosity and increases the palatability of the staple with which it is mixed. Thus a weaning food of given consistency, if it includes fat or oil, can be made with more staple and less water and thereby achieve a higher energy density. In this connection, one must dispute Lepowsky's comment (ch. 4) that fatty foods, which are prohibited to a young child on Vanatinai by custom, are not easily digested by young children in the first years of life and might cause diarrhea. In fact, modern treatment for toddler's diarrhea, a very common problem in the Western world, is to increase the amount of fat in the young child's diet.

The hypothesis put foward by Lepowsky (ch. 4) that the traditional taboos on the consumption of animal protein foods by young children in Vanatinai may be a cultural adaptation to an environment of endemic malaria, as there is evidence for an antagonistic relationship between host malnutrition and malaria, is intriguing but, I think, unlikely. Major social changes in the area, such as enforced settlement in large villages closer to the coast and swamps and the increased mobility of the population, which have probably increased the prevalence of malaria have only occurred in the past half century. But even if the hypothesis were correct, it would need to be balanced against the known fact that malnutrition is a major determinant of mortality risk in young children (Puffer and Serrano, 1973). On the average, child mortality is doubled between the ages of 1 to 36 months for each 10% decline below 80% of the Harvard weight-for-age median (Kielmann and McCord, 1978).

BIRTH SPACING AND FAMILY PLANNING

The case of Kuma and Aupai, so poignantly described by Barlow (ch. 8), well illustrates the problems of inadequate birth spacing. In a similar vein, Gegeo and Watson-Gegeo (ch. 13) show how the quality of life on the densely populated coastal slope of West Kwara'ae has begun to deteriorate in the past decade along with the great increases in fertility rate and family size in the past two generations.

The contraceptive effect of breastfeeding is now well recognized. Worldwide, breastfeeding is a more effective method of fertility regulation than the use of other forms of contraception (Anonymous, 1977). Unrestricted breastfeeding on demand, including night feeds, is important for delaying the return of ovulation in the mother (Davies, 1983) as well as for optimizing breastmilk output (Rowland, Paul and Whitehead, 1981). Basal prolactin levels in the mother correlate well with the frequency of suckling. Short's studies in Edinburgh (1983) suggest that a frequency of suckling of at least six times a day as well as the maintenance of nighttime feeds is important for lengthening the period of postpartum anovulation.

This helps explain the puzzle of the 44-month birth spacing of the !Kung. These traditional hunters and gatherers are pro-natalist, use no artificial forms of contraception and have no significant taboos on sexual intercourse during lactation. Their remarkable birth spacing is due to the frequency of suckling during both day and night (Konner and Worthman, 1980).

Short (1983) is emphatic that anything which reduces the infant's contact with the mother's nipple should be regarded as a potential conceptive (something that will stimulate fertility). In this category he includes feeding bottles, teats, dummies and the baby's own thumb. He regards thumb-sucking as abnormal behavior due to inadequate nipple contact and reminds us that young apes only suck their thumbs if kept in captivity and bottlefed. Likewise, !Kung babies do not normally suck their thumbs. Short suggests that the incidence of thumb-sucking could provide a useful clue as to the probable contraceptive effectiveness of breastfeeding in a particular community. None of the ethnographies mention thumb-sucking in any detail, but it would be interesting to have information on this.

Family planning services and advice are, therefore, an essential concomitant to changes in traditional life style.

A study of 1057 birth intervals from 12 centers in Papua New Guinea in 1975 (David Morley, personal communication) showed that the mean birth interval was 31 months and the fifth percentile was 13.5 months. The vulnerable month (the month by which time the first 5% of mothers had become pregnant again) was 13.5 minus 9 months (4.5 months). Thus the age of four months, at which time mothers are advised by the clinics to start giving mashed soft foods as well as breastmilk, is the time for appropriate family planning.

BELIEFS AND CUSTOMS

Health workers and nutritionists often portray a negative image. They tend to be more interested in disease than health, and focus more on taboos that may harm health than on beliefs and customs which may be beneficial. It is refreshing to read in these ethnographies of the many beliefs and practices that can positively contribute to good health and nutrition.

Gegeo and Watson-Gegeo (ch. 13), for example, point out that the Kwara'ae theory of health is embodied in the concept of eating a variety of foods at the same sitting. Pork, a hot food, should be eaten with a cooling food such as sweet potato or dark green leaves cooked in coconut cream — an ideal toddler's triple mix. According to Kwara'ae tradition, lactating mothers should eat animal protein each day to produce good milk. The environment provides a plentiful supply of animal protein such as lizards, tree frogs, fruit bats, fish, eel, prawns, shellfish, freshwater clam, crab, grasshoppers and sago grubs. Ferns, an excellent source of protein, are also traditionally eaten by women.

Certain foods are believed by many of these groups to have special benefits for lactating women and infants. Fish cooked in coconut cream, for example, is believed to increase lactation and improve the blood.

Tietjen (ch. 7) records that among the Maisin, pregnant women are considered to be highly susceptible to spirits and sorcery. It is recognized that pregnant women need special care, and so motivation is high for regular attendance at the antenatal clinic.

The Kwara'ae (ch. 13) also have an excellent appreciation of the need for food hygiene. Lactating mothers and children of all ages avoid eating dirty food, food touched by flies and cold food unless reboiled, as these foods are recognized as sources of illness.

Unfortunately, many health workers do not seize such splendid opportunities for health and nutrition education based on people's traditional beliefs and practices. Instead, there is an attempt to impose alien practices. Edible insects, such as grasshoppers, although widely eaten traditionally, do not usually feature on nutrition charts as a protein food. The contents of a health poster is more often related to the health worker's concepts, beliefs, and practices than to the rural villager's concepts of reality. A poster showing a huge fly hovering like a giant airplane over a village elicited the response from the village health committee that they were thankful such monsters did not occur in their area!

Premastication was a widely practiced custom in many traditional societies, and Akin (ch. 12) gives several reasons why it may have been beneficial. However, on Malaita, it has recently fallen into disrepute as "unhygienic". Health workers and missionaries have disliked the practice mainly for aesthetic reasons, and used hygiene as a convenient weapon to eradicate the practice. In fact, there is no evidence that premastication is harmful to the baby. Tuberculosis is often cited as being spread through premastication; but this is most unlikely. A mother or other caretaker with untreated tuberculosis is likely to transmit the infection to the baby through the normal close contact with the baby, and whether the baby is given premasticated food or not will make no difference to the baby's being infected.

Stopping premastication has, however, encouraged the use of bottles which are demonstrably unhygienic in the generally unsanitary environment of most rural villages and urban slums (Hendrickse, 1983). Moreover, as Akin (ch. 12) has pointed out, a Kwaio mother frequently has to leave her infant to work. Premasticated food allowed the caretaker to satisfy the infant's hunger in the absence of the mother. The active discouragement of premastication has left the mother with a serious conflict between her duties as a food provider and as a mother. In this situation the first principle of effective health education has been violated:

> Go to the people and learn from them;
> Start with what they know;
> And build on what they do.

Another important message to health workers from the ethnographies is that traditional beliefs and customs are not immutable. Indeed, the speed of change can be remarkable. Gegeo and Watson-Gegeo (ch. 13) record how Kwara'ae fathers sometimes now assist when their wives deliver at home. This is a radical change from the traditional seclusion of women in birth houses and men's fear of parturient women as a source of pollution. Men are now taking over more and more of the household work and childrearing responsibility formerly seen as strictly women's work. Lepowsky (ch. 4) also recounts how taboos against feeding animal protein foods to young children have nearly disappeared in the Misima language area due to pressure against them from health workers, missionaries and government personnel, although they are still in practice on neighboring Vanatinai.

These ethnographies provide some important insights into different cultural groups at different stages of transition. Although generalizations are risky due to wide differences between groups and the speed of change over time within the same group, all the cultural groups described are pro-child. Carrier (ch. 11) describes the parents on Ponam as anticipating the baby's wants. When the baby cries, the mother runs to pick it up. There are no "difficult infants", only careless, lazy or ignorant parents. Even the Kove women, whom Chowning (ch. 10) describes as being somewhat resentful of the social restrictions imposed by childcare and breastfeeding, all want to be married and to have children. The anti-child thrust of Western lifestyle (Greer, 1984) is not yet evident in the South Pacific.

Barlow (ch. 8) rightly emphasizes that infant care and feeding are embedded in a complex of social values, activities and expectations. Akin (ch. 12) reinforces this by the statement that infant care and feeding practices are best understood in the context of culture, economy and family. The mother-infant dyad is not an isolated set of interactions but dependent on the needs of the rest of their families, economic constraints and socio-cultural expectations.

Clearly, health workers need a more holistic approach and more flexible attitudes than they have customarily exhibited. The Western world, and the health profession in particular, should adopt a less arrogant stance. There is much that they can learn from these traditional societies.

REFERENCES

Anonymous (1977). Editorial: Lactation, fertility and the working woman. *I.P.P.F. Med. Bull.* **11**, 1-2.
Bakker, M.L. (1983). *Spatial Differentiation of Mortality in PNG: A Classification Based on the Results of the 1980 Census.* Working Paper No. 4. National Statistical Office, Port Moresby.
Becroft, T.C. (1967). Child-rearing practices in the Highlands of New Guinea: A longitudinal study of breastfeeding. *Med. J. Aust.* **2**, 598-601.
Berg, A. (1973). *The Nutrition Factor.* The Brookings Institution, Washington, DC.
Biddulph, J. (1983). Legislation to protect breastfeeding. *PNG Med. J.* **26**, 9-12.
Davies, D.P. (1983). The suckling child. In J.A. Dodge (Ed.), *Topics in Paediatric Nutrition*, Pitman Books Ltd., London, pp. 21-29.
Dearden, C., P. Harman, and D. Morley (1980). Eating more fats and oils as a step towards overcoming malnutrition. *Tropical Doctor* **10**, 137-142.

Greer, G. (1984). *Sex and Destiny: The Politics of Human Fertility.* Harper and Row, New York.

Hendrickse, R.G. (1983). Some thoughts about infant feeding. *Annals Trop. Pediat.* **3**, 163-168.

Kielmann, A.A. and C. McCord (1978). Weight-for-age as an index of risk of death in children. *Lancet* **1**, 1247-1250.

Konner, M. and C. Worthman (1980). Nursing frequency, gonadal function, and birth spacing among !Kung hunter-gatherers. *Science* **207**, 788-791.

Lofton, M. and G. Gotsch (1983). Legal rights of breastfeeding mothers: USA scene. In D.B. Jelliffe and E.F.P. Jelliffe (Eds.), *Advances in International Maternal and Child Health.* Oxford University Press, Oxford, Vol. 3, pp. 40-55.

Puffer, R.R. and C.V. Serrano (1973). *Patterns of Mortality in Childhood.* Scientific Publication No. 262, PAHO/WHO, Washington, D.C.

Rowland, M.G.M., A.A. Paul and R.G. Whitehead (1981). Lactation and infant nutrition. *Brit. Med. Bull.* **37**, 77-82.

Short, R.V. (1983). The biological basis for the contraceptive effects of breastfeeding. In D.B. Jelliffe and E.F.P. Jelliffe (Eds.), *Advances in International Maternal and Child Health.* Oxford University Press, Oxford, Vol. 3, pp. 27-39.

Waterlow, J.C. (1981). Observations on the suckling's dilemma — a personal view. *J. Human Nutr.* **35**, 85-98.

Waterlow, J.C., A. Ackworth and M. Griffith (1980). Faltering in infant growth in less developed countries. *Lancet* **2**, 1176-1178.

CHAPTER 18

COMMENTARY: WOMEN'S WORK AND INFANT FEEDING IN OCEANIA

JUDITH D. GUSSLER

INTRODUCTION

The body of breastfeeding literature has grown tremendously over the past decade. Some of the materials are ethnographic, but recently much of it has come from nutrition surveys and longitudinal studies which have isolated and analyzed breastfeeding and other feeding behaviors. As useful and important as these recent studies are, this book reminds us of the special value of ethnographic research, which reveals the context of isolated behaviors, and how they are shaped by underlying values and ideologies and the everyday needs of the real life actors. In this fashion, these chapters describe infant feeding practices in various Oceanic communities and demonstrate that these practices are determined by an interplay of traditional beliefs and values and the immediate work and social needs of individual mothers.

Mother-infant interaction and child care in traditional societies were important social science topics before the recent spate of breast-feeding studies. In 1967, for example, Mead and Newton described the breastfeeding stage of an infant's life as the "transitional period," during which "physiological separateness" is established. The concept implies that the period of breastfeeding is, in a sense, a continuation of pregnancy, the stage during which nutritional dependency on the mother is total. In traditional societies, where breastfeeding is universal and of long duration, and adequate breastmilk substitutes and artificial feeding paraphernalia are rare, nutritional dependency of an infant on its mother may continue for many months postpartum.

Sooner or later postpartum, a mother resumes at least some of her usual activities and chores, while also caring for the dependent infant.

The social and individual strategies that facilitate the conduct of these activities differ from place to place and, sometimes, woman to woman. The tone of earlier literature on women's arrangement of reproductive and productive roles suggests something akin to biological destiny; that is, that all women's work roles are shaped by the biological facts of childbearing and nurturing. As Newland says in a 1979 publication (p. 132):

> The unremitting requirements of what has been dubbed, rather inelegantly, the 'breeder-feeder role' provide the backdrop for all other work that women do.

In essence, this view suggests that the strategies by which women balance infant care and other economic activities must result in accommodation of economic roles to child care and nurturing roles, of production to reproduction.

A different view is proposed by Friedl (1975), who suggests that such a perspective is "sterile" and affords fewer insights than that which looks at the ways in which child care and nurturing roles are accommodated to women's other activities. The articles in this volume suggest that, in fact, adaptation may occur in both directions. Even within one delimited culture area, a great many strategies exist and coexist, and accommodation may be made in all the social arenas in which women act. In the same hamlet in Oceania, one sees women adjusting infant care and feeding to allow themselves to resume some subsistence activities, and foregoing or adjusting economic activities to allow for infant care and breastfeeding. In some communities, women are taught that the best expression of their female role is in nurturing and nourishing activities (Barlow, ch. 8; Counts, ch. 9). In others, women perceive the nutritional dependency of infants as a burden, and look forward to the end of exclusive breastfeeding and a return to other activities (Conton, ch. 6; Chowning, ch. 10). In most, where women conduct what they and their communities perceive as economically important and nondiscretionary activities outside the household, strategies are developed allowing the woman to balance infant care responsibilities with other work (Lepowsky, ch. 4; Morse, ch. 14; Chowning, ch. 10; Akin, ch. 12; Nardi, ch. 16). A major focus of these infant feeding ethnographies is the interaction of the mother and nursling with one another and their social environment and the adjustments made in that environment to accommodate the mother and dependent infant. (Specific strategies are discussed in greater detail below.)

GLOBAL TRENDS IN INFANT FEEDING

Much of the recent infant feeding literature has expressed concern for the so-called "decline of breastfeeding" in the developing world. In many urban areas of the developing world, some women are choosing not to breastfeed, some are using bottle feeds to supplement breast-milk, and some initiate breastfeeding but stop within a few weeks. One factor that is repeatedly cited as contributing to change in infant feeding is change in work roles of women. (Marshall, ch. 2; Nardi, ch. 16; and Katz, ch. 15, provide discussion and references.)

In developing, urbanizing, and industrializing communities, women are entering the cash economy, either pushed by social and economic necessity or pulled by increasing desire for consumer goods. They are assuming new non-domestic work roles, usually in addition to and not in lieu of traditional domestic tasks. Frequently, the new work roles seem to be incompatible with concurrent child-care and breastfeeding. Increasing distance from home and transpor-tation costs make breastfeeding breaks during the work day proble-matic or impossible. Child care facilities at work sites are rare and legislative protection (for example, paid maternity leave) is ineffective or nonexistent, especially for the large proportion of women outside the formal work force. Furthermore, the social context of these women's lives is changing. Many find themselves heads of house-holds, with no male breadwinners. Extended kin may not be near or able to help out if they are. Nontraditional infant feeding strategies frequently are developed by these women.

Despite these important trends in economic, technologic and social structures and organizations, recent research is showing us that the vast majority of infants born in the developing world still are being breastfed at birth (Kent, 1981; Gussler and Mock, 1983; WHO, 1981). Table I summarizes some of the major patterns of infant feed-ing in lesser developed and developing countries. There are, of course, many exceptions to the major patterns. However, in most tra-ditional and remote communities, prevalence of breastfeeding is high and duration tends to be long. In many of the developing areas experiencing some of the changes outlined above, the prevalence remains high but duration may be shorter, and concurrent breast- and bottlefeeding is common. In fact, in areas like the Caribbean, mixed feeding is the norm. There, where the majority of babies are born to single women, mothers introduce very young infants to the bottle and a variety of readily available commercial and non-commercial pro-

TABLE I
Major patterns of infant feeding in developing countries

Rural traditional	Urban poor	Urban elite
Prevalence of breastfeeding approaches 100%, especially in Africa and Third World Asia.[a]	Prevalence still high, but varies from region to region. (For example, nearly 100% in Kinshasa, Zaire; 85% in Cebu City, P.I.; less in urban Latin America).	Prevalence of breastfeeding quite low; in some communities, virtually all are exclusive bottlefeeders.[b]
Duration still long. Mean ages of sevrage of 1-2 years not uncommon.	Duration of breastfeeding tends to be shorter than in rural areas. (For example, in Guatemala, mean age of sevrage in rural areas is about 12 months; among urban poor of Guatemala City, it is 3 months.)	
Breastfeeding may be supplemented early, but usually not with bottle feeds. Traditional weaning foods generally are used.	Bottle use is common as a supplement to breastfeeding. Mixed breast- and bottlefeeding the norm in some urban areas. (For example, in St. Kitts-Nevis, W.I., over 90% of mothers employ the mixed feeding pattern.) A variety of products are used in bottles, but usually not commercial infant formulas. (Bush teas, evaporated and powdered milks, broths, paps, etc., all are common.)	Commercial infant formulas commonly are purchased and used.

[a] Most data are taken from the WHO Collaborative Study on Breastfeeding (WHO, 1981) and the World Fertility Survey (Kent, 1981).
[b] There is evidence of a resurgence of breastfeeding among professional classes in some areas, echoing the trends in infant feeding in developed countries.

ducts. Then they can leave them with other caretakers while they work away from the household, look for work, or participate in other extra-household activities (Gussler, 1979; Marchione, 1980; Johnston, unpublished observations, 1977).[†]

Some areas of Oceania have been touched relatively little by these changes. Others, such as the Port Moresby area of Papua New Guinea, already have experienced many of them. By and large, however, the communities represented in this volume are relatively remote and traditional, and feeding patterns reflect the lack of available cash and consumer products, such as commercial infant foods, bottles and nipples.

INFANT FEEDING TRENDS IN OCEANIA

Breastfeeding data in this volume are generally consistent with the summary in Table I. One is struck by the preponderance of breastfeeding in the rural areas, both the high prevalence and the long duration. Children as old as 12 years are reported to be still nursing. Exclusive bottle use seems to be very rare, limited to exceptional cases, such as the death or serious illness of the mother, twins, or a mother working off-island. That is not to say that no change is occuring in the infant feeding practices in Oceania. A mixed feeding pattern of breast- and bottlefeeding is not unusual in some communities (Morse, ch. 14; Nardi, ch. 16; Marshall, ch. 2), and two researchers (Nardi, ch. 16; Conton, ch. 6) report that duration of breastfeeding is shorter than in earlier times. But, consistent with the patterns described in Table I, these infant feeding practices are primarily seen in the more urban and developed parts of the region. (See middle column of table.)

Bottlefeeding is not supplanting breastfeeding anywhere in the areas of Oceania discussed here. Thus, a number of specific characteristics of these communities, even in those that are less isolated and less traditional, obviously support and reinforce the practice of breastfeeding. There are, for example, both *de facto* and *de jure* restrictions on feeding options in the region. Breastmilk substitutes generally are unavailable or very expensive (Akin, ch. 12; Counts, ch.

[†]Johnston, J. (1977). The household context of infant feeding practices in South Trinidad. Paper presented at the 76th annual meeting of the American Anthropological Association, Houston, TX.

9; Gegeo and Watson-Gegeo, ch. 13; Katz, ch. 15; Conton, ch. 6), and one government (PNG) has declared infant feeding paraphernalia like bottles and nipples "prescription-only" items.

Observance of several traditional practices also helps produce a good social environment for breastfeeding. A period of postpartum seclusion and a postpartum sex taboo are very nearly universal in this area. These practices serve to give the mother ample time with her infant to successfully initiate breastfeeding without having other duties interfere. Furthermore, her attention is not divided by wifely sexual roles, since she will not be sleeping with her husband for several months. This postpartum sex taboo also ensures that the new mother will not become pregnant again for some time. Pregnancy, in this cultural setting, necessitates sevrage.

Most of these communities also possess a system of beliefs which rationalizes and reinforces the accepted patterns of infant care and feeding (Montague, ch. 5; Tietjen, ch. 7; Conton, ch. 6; Katz, ch. 15). In some communities, women are, by tradition, socialized to view childcare and provision of food to children and other household members as their most important roles (Carrier, ch. 11; Gegeo and Watson-Gegeo, ch. 13). The belief that infants are particularly vulnerable to attacks from capricious spirits (Conton, ch. 6), keeps mothers and their youngsters close together in the protective environment of the home in the first weeks after birth. In such communities, separation of the nursing dyad is minimized, as is the need for breastmilk substitutes. The virtual absence of the "insufficient milk syndrome" (Gussler and Briesemeister, 1980) in the more traditional communities probably is due in large measure to the system of beliefs that maximizes mother-infant contact and minimizes the time the dyad is separated.

Extended family relationships, strong and ubiquitous in Oceania, may be as important as any other single factor in maintaining the high prevalence of breastfeeding. Most researchers report that friends and relatives care for the mother and newborn and assume some of the mother's usual duties for period of time after the birth. Residence rules and settlement patterns in several communities result in availability of both consanguineal and affinal female kin to provide aid and comfort. In three (Carrier, ch. 11; Katz, ch. 15; Tietjen, ch. 7), the husband serves as "doula" and later helps with child care. Community support for infant care is common and adoption of extra or unwanted babies is frequent (Barlow, ch. 8; Lepowsky, ch. 4; Counts, ch. 9; Jenkins, Orr-Ewing and Heywood, ch. 3; Chowning, ch. 10).

WOMEN, WORK, AND INFANT FEEDING IN OCEANIA

While these chapters describe a social environment generally conducive to breastfeeding, they also demonstrate that combining non-household economic activities and infant care may be problematic even in traditional societies. This point rarely is made in accounts of infant feeding. In fact, many of us have formed a picture of the traditional rural breastfeeder carrying the infant in a sling or other device, on front, back, or hip, offering the breast nearly continuously day and night. And we assume that she is able to do this while conducting her usual daily activities. This is the picture that Newland describes (1980, p. 23):

> Time-budget studies show that child care occupies little of rural women's time, usually 10% or less. The reason may be that the supervision of children is not seen as a separate activity. *It occupies the same time and space as do other tasks.* (emphasis added)

This picture of the traditional caretaker-breastfeeder probably was drawn primarily from accounts of hunters and gatherers. Katz (ch. 15) notes, however, that in some African horticultural communities, a nursling may be separated from its working mother for several hours each day. In a cross-cultural survey of traditional societies (excluding hunters and gatherers), Nerlove (1974) found a statistically significant association between early supplementation of breastmilk and mothers' importance in the subsistence economy. Women, that is, will leave even young infants with other caretakers if economically important, non-discretionary tasks take them from the household. Their caretakers will feed them a variety of semi-solids, mashed, or pre-masticated foods. Most of these chapters provide confirmation of Nerlove's thesis and reveal some of the ways in which mothers and societies resolve this apparent incompatibility between two types of potentially non-discretionary tasks — extra-household work and breastfeeding.

The articles also demonstrate that the concept of discretionary activity is shaped by underlying cultural values and beliefs about the proper nature and importance of women's work. It is emically, not etically, defined and varies from community to community. Thus, work outside the household may be foregone if it is considered un-important — to the woman, her family, or the community — relative to infant care and feeding. In this case, extra-household work is considered discretionary. On the other hand, if great value is placed on women's work activities and breastmilk substitutes are available, breastfeeding may be foregone, either wholly or partially. In this case, it is breastfeeding that is considered discretionary.

Economists (Butz, 1978; Popkin, 1980) who have studied the relationship between breastfeeding and out-of-household work (especially wage work), view them as two activities that have time and opportunity costs. This economic framework has proven to be a useful and productive way of looking at breastfeeding trends in developing countries. However, the tendency has been to focus on the impact of work opportunities on breastfeeding when the two activities are incompatible; that is, when they cannot "fill the same time and space." Thus, opportunities for extra earnings are expected to reduce the probability that a woman will breastfeed or breastfeed for a long duration. The impact of breastfeeding on work has received less attention. These data from Oceania suggest that the impact may be great, especially when alternatives to breastfeeding are inadequate or unavailable.

The following are some of the strategies suggested by these chapters by which women combine infant care and feeding with other activities. In each community, prevailing social and economic structures and organization, as well as a variety of beliefs and values, shape the form and content of the strategies.

Little or No Extra-Household Work

In some communities, women have few or no extra-household tasks in their normal repertoire of roles. Thus, no accommodation of extra-household work and child care is necessary when the woman has a dependent infant; she and the infant remain together in the home. Carrier (ch. 11) points out that Ponam women have very few activities that take them from home, a work situation compatible with their view that "... a mother should always be free to nurse her infant at any time and therefore should not leave the child's side for the first few months. ..."

Delayed Return to Extra-Household Work

In other social contexts, women normally conduct tasks outside the household, but they are considered discretionary and can be forgone until the period of nutritional dependency is over. Counts (ch. 9), for example, says that "... the value placed on the defining act of /Lusi/ motherhood — giving food to a hungry child — is so important that there are few things that could compete successfully with it for mother's attention." And village women "... do not engage in non-domestic labor that separates them from their children." They may return to their gardens when their infants are older and not totally dependent upon their mothers. Nardi (ch. 16) suggests that women may make a conscious decision to shorten this period of nutritional dependency by terminating breastfeeding earlier (or initiating supplementary bottle feeds?), in order to resume work more quickly.

Shared Responsibility for Extra-Household Work

In some communities, women normally conduct non-discretionary tasks, but husbands, other kin or friends assume responsibility for them during the period of the infant's nutritional dependency. Katz (ch. 15) indicates that in the outer Fiji Islands a number of beliefs "encourage" a woman to stay in the village and near her infant. However, "... social resources offered a solution in which the mother was excused from work and could be near the infant continuously ..." Husbands and other kin take over her important chores and provide her and her family with food for as long as a year.

Concurrent Extra-Household Work and Child Care

In a number of Oceanic communities, women normally conduct non-discretionary activities, which they continue by taking their infants with them. They may also take caretakers, such as older siblings, with them to watch the infants and keep them nearby as they work. Lepowsky (ch. 4) describes a "wall-less" garden house on Vanatinai constructed for this purpose. Conton (ch. 6) says that Usino mothers will not travel great distances with a young infant, but they will take them to the gardens and nearby villages, carrying them in netbags and feeding them whenever they begin to fuss. Although this is the common pattern, some mothers will leave infants to be fed sugar cane juice or water by caretakers. This alternative strategy attests to the existence of intra-community variation and individual decision making. (Also see Tietjen, ch. 7, and Gegeo and Watson-Gegeo, ch. 13.)

Non-Concurrent Extra-Household Work

In some settings, women with young infants resume important extra-household work that is not compatible with concurrent infant care and feeding, or is believed to be incompatible. (In some communities, such as described by Akin, ch. 12, for the Solomon Islands Kwaio, women's work is not physically incompatible with breastfeeding. However, infants are not taken to the gardens because they are believed to be vulnerable to supernatural harm.) This incompatibility of roles seems to be more common in developing and urbanizing areas, with cash economies, where women work under conditions either inappropriate or dangerous for infants. (See, especially, Morse, ch. 14, and Marshall, ch. 2.) Infants are left with caretakers and fed premasticated food, green coconut water, commercial milk products, etc., or wetnursed (Chowning, ch. 10).

Only the last strategy, employed when mother and infant must be separated, might require alternative feeding practices. The others do not interfere with breastfeeding, even exclusive breastfeeding, since social accommodations allow the mother and nursling to remain together, throughout the day and night. The high prevalence and long duration of breastfeeding in these Oceanic communities are explained by the existence of old and new social solutions available to mothers who must balance work and child care roles and responsibilities.[†]

CONCLUSION

This volume reaffirms the value of intensive ethnographic research to provide useful information on issues of international concern. These chapters, for example, demonstrate that a variety of solutions to a common human problem exist even within a relatively small area of the world. More specifically, they reveal that balancing out-of-household work with infant care and breastfeeding is not just a problem of the developed world. Indeed, it is an old problem to which human societies and individual mothers have responded in a variety of cultural and idiosyncratic ways. These discussions show us

---

[†]Age of infant and perceived adequacy of available breastmilk substitutes affect decisions about infant feeding and work. A mother may be comfortable leaving a four month old infant with a caretaker who provides mashed fruit and coconut water, but not a two month old. Thus, work-childcare strategies also will vary by age of infant.

once again that feeding and weaning an infant is a complex process of social, as well as nutritional, significance, and the decision to provide an infant with foods other than breastmilk may be based more on social and economic factors than nutritional ones.

REFERENCES

Butz, W. (1978). Economic aspects of breast-feeding. In W.H. Mosley (Ed.), *Nutrition and Reproduction*. Plenum Press, New York, pp. 231-255.

Friedl, E. (1975). *Women and Men: An Anthropologist's View*. Holt, Rinehart and Winston, New York.

Gussler, J. (1979). Village women of St. Kitts. In D. Raphael (Ed.), *Breastfeeding and Food Policy in a Hungry World*. Academic Press, New York, pp. 59-65.

Gussler, J.D. and L.H. Briesemeister (1980). The insufficient milk syndrome: A bio-cultural explanation. *Med. Anthropol.* **4** (2), 145-174.

Gussler, J.D. and N. Mock (1983). A comparative description of infant feeding practices in Zaire, the Philippines and St. Kitts-Nevis. *Ecol. Food Nutr.* **13**, 75-85.

Kent, M.M. (1981). *Breast-feeding in the Developing World: Current Patterns and Implications for Future Trends. Reports on the World Fertility Survey* (No. 2). Population Reference Bureau, Inc., Washington, D.C.

Marchione, T. (1980). A history of breast-feeding practices in the English-speaking Caribbean in the twentieth century. *Food. Nutr. Bull* **2** (2), 9-18.

Mead, M. and N. Newton (1967). Cultural patterning of perinatal behavior. In S.A. Richardson and A.F. Guttmacher (Eds.), *Childbearing — Its Social and Psychological Aspects*. The Williams and Wilkins Company, Baltimore, pp. 142-243.

Nerlove, S.B. (1974). Women's workload and infant feeding practices: A relationship with demographic implications. *Ethnol.* **13**, 207-214.

Newland, K. (1979). *The Sisterhood of Man*. W.W. Norton and Company, New York.

Newland, K. (1980). *Women, Men, and the Division of Labor*. Worldwatch Paper 37. Worldwatch Institute, Washington, D.C.

Popkin, B. (1980). Time allocation of the mother and child nutrition. *Ecol. Food Nutr.* **9**, 1-14.

World Health Organization (1981), *Contemporary Patterns of Breastfeeding. Report on the WHO Collaborative Study on Breastfeeding*. WHO, Geneva.

CHAPTER 19

COMMENTARY:
AN ANTHROPOLOGICAL PERSPECTIVE
ON INFANT FEEDING IN OCEANIA

PENNY VAN ESTERIK

INTRODUCTION

Everyone reading this monograph will approach the topic from a distinctive perspective. Let me identify my biases. First, I am not a specialist on Oceania but on Southeast Asia. Therefore, I cannot comment on the accuracy of the descriptions and interpretations. Second, I am a cultural anthropologist with a strong generalist background in the field and an optimistic enthusiasm about the importance of anthropology as an intellectual tradition. Third, for the last few years, I have been privileged to work as a nutritional anthropologist with a research consortium studying the determinants of infant feeding patterns among the urban poor of Bogota, Nairobi, Bangkok, and Semarang, Indonesia. Finally, I support and applaud the advocacy efforts of the groups working to limit the promotion of infant formula in Third World countries. Thus, my comments are coming from the direction of a Southeast Asian anthropologist with recent experience in comparing infant feeding patterns in urban contexts. In this capacity, it is an honor to comment on the papers collected here.

In these papers on infant feeding in Oceania, the authors all agree that ethnographic fieldwork and attention to cultural factors can enrich our understanding of infant feeding and potentially improve the health and nutritional status of infants in Oceania. I am delighted that anthropologists can serve in this capacity. But I am selfish. I want to see how these ethnographic insights into infant feeding can enrich cultural anthropology and improve anthropologists' theoretical models for interpreting human behavior in general.

331

M

It is a two-way street. If anthropologists provide technical assistance to other disciplines and work on interdisciplinary applied projects, it is important that insights gained from this experience feed back into the field as a whole.

It is important not just for the health profession but for the field of anthropology that anthropologists take on projects with application potential in the field of health and nutrition. The distinction between applied and basic research cannot be sustained in cultural anthropology. Most of the observations made by the ethnographers in these papers come from a sound methodological and theoretical basis in cultural anthropology. But this background is not always apparent to non-anthropologists who may value ethnographic observations as fortuitous anecdotes. On the other hand, anthropologists see the same observation as a very grounded piece of a puzzle. In applied projects — particularly interdisciplinary projects — there is more interest in having anthropologists show what they can do to help solve a problem than to explain why they approach the problem in a certain way. But we cannot keep working as anthropologists without occasionally making some connections between how we conduct research and why we approach problems in the way we do. Or one day we may reach back into our tradition to seek guidance, and come up empty handed. Personally, I have more confidence in anthropology as an intellectual endeavor than to believe that day will come in the near future. But, for the sake of argument, let us try to link some of the particulars of infant feeding in Oceania to broad issues in anthropology generally.

These papers raise many important questions and could be related to any number of arguments in cultural anthropology. I would like to raise three issues here:

- biocultural adaptation and utilitarian theory;
- time, work and the status of women; and
- infant feeding style.

BIOCULTURAL ADAPTATION

Infant feeding in general and lactation in particular is an excellent example of biocultural adaptation. Human milk has evolved as a product ideally suited to meet the needs of human infants; but lactation itself is a more complex process for, in primates, lactation is partly instinctual and partly learned. If it were strictly a biological process, we would not expect variation across cultures. Since breastfeeding is also a culturally constructed act, we see great variation in the way breastfeeding is managed, and even in maternal competence. The very fact that breastfeeding can fail suggests that our evolutionary development favored a more inclusive selection for good mothering rather than simply good lactating.

These studies demonstrate how different communities encourage good mothering. The seclusion of mothers and newborns in the Murik Lake district of Papua New Guinea (Barlow, ch. 8), and on Malaita, Solomon Islands (Akin, ch. 12) provide women with an opportunity for intense interaction with their newborns in the supportive environment of maternal kin.

In most of the communities described, the bonds between women appear exceptionally strong. This supportive network would be valuable for a mother in a number of ways, supplying her with food both before and after birth, relieving her of household and subsistence tasks, providing occasional care for the infant, and even wet-nursing her infant when necessary. Although this is not discussed in these papers, one could imagine that new mothers could receive breastfeeding advice and instruction from a group of supportive women. As several authors point out, breastfeeding may be viewed as so natural and unworthy of comment or discussion that some mothers may fail to bring up problems which others might view as insignificant. Conton (ch. 6) notes that in Madang, advice on breastfeeding is rarely sought or offered. Carrier (ch. 11) in particular illustrates how very rule-governed Ponam mothers are and how they insure compliance to their standards of child care through criticism and advice. Explicit rules for infant care and feeding must increase a woman's confidence and self esteem. It is of interest to note how few women in these communities complain of insufficient milk. The pattern of frequent feeds on demand and close body contact mitigate against these problems.

Another issue related to biocultural adaptation concerns weaning (or more precisely, sevrage, the cessation of breastfeeding). All papers discuss the mother's need to wean an infant as soon as she recognizes her subsequent pregnancy. While there are a number of reasons given for insuring this is done, most focus around the ideas of pollution.

From a biocultural perspective, we can understand the trade-offs between the mother's own fitness which is best served by weaning her infant, the fitness of her nursing infant which is best served by maximizing the amount of human milk it ingests, and the fitness of the next offspring best served by mother's food energy going to support that pregnancy (Daly and Wilson, 1978). The mother's decision to terminate breastfeeding needs to be examined from the perspective of this "weanling dilemma". Beliefs that intercourse and the semen produced contaminates breastmilk may delay pregnancy. Yet as Chowning (ch. 10) points out, pregnancy may be considered more polluting than intercourse.

One additional value of these intensive case studies is that they may throw into question some widely held assumptions basic to the dominant biomedical model of infant feeding as it is promulgated in health centers in many parts of the world. It is particularly refreshing to see induced lactation and relactation taken for granted, and insufficient milk rarely recognized.

Biocultural adaptation is most directly addressed by Lepowsky (ch. 4), who points out the fit between cultural rules surrounding infant feeding and broader ecological adaptation. She has cited evidence suggesting that the restriction on young children's consumption of animal protein may be an adaptation to an environment of endemic malaria.

The suggestion that poorly nourished individuals are more resistant to malaria would make sense of a wide range of reports of restrictions for young children worldwide. These arguments raise three questions which anthropologists have struggled with for decades. First, are the arguments about food restrictions for young children based on reports of normative rules or on observed behavior? So many reports of dietary rules break down under the weight of dietary records. I remember a Thai village woman telling me that toddlers her son's age should not eat cold foods. The toddler then walked by with a cucumber in his mouth. The mother, when confronted with her son's consumption of a cold food, argued that his gums were sore, and thus he enjoyed cold foods.

A second question concerns how functionalist and utilitarian theory should be applied to infant feeding. At one level, all infant feeding practices must be adaptive or the group would not have survived. Thus breastfeeding and giving soft foods are adaptive for infants in all societies. But if all infant feeding practices are adaptive, how do we account for shifts to poorer quality staples, processed commerical foods, and bottlefeeding with poor quality breastmilk substitutes in many parts of the world? These problems with functional explanations have been actively argued in the social sciences for the last 40 years. We must be wary of circular arguments which fail to account for differences and changes in patterns of adaptation.

Finally, we must ask what are the policy implications following from the theory that malnourished children are more resistant to malaria? Hopefully programs treating both malaria and malnutrition, as suggested by Lepowsky (ch. 4), would be undertaken. Let us hope that these approaches are not used as a rationalization for or acceptance of past and present malnutrition of infants and children.

TIME AND WOMEN'S ACTIVITIES

Several papers stress the relation between women's activities and infant feeding decisions. The communities differ in the division of labor between men and women for subsistence activities, and in the opportunities for women to participate in the cash economy. Nevertheless, women in all these communities have a wide range of responsibilities for food production, preparation and distribution, and household maintenance tasks.

None of the ethnographers provided detailed time allocation records, but from the observations made, it is clear that women must make a number of trade-offs regarding both formal and informal work and breastfeeding. Opportunities for women's participation in the cash economy are increasing in some communities. Although Nardi (ch. 16) argues that an increase in women's work load has been a chief reason for the decline in weaning age in Western Samoa, this does not apply in Ponam where there has been no significant increase in women's work load (Carrier, ch. 11). Women in Madang (Conton, ch. 6) return to work in their gardens a month after delivery, bringing their infants with them. Other communities relieve women of almost all subsistence and domestic responsibilities for a period after childbirth (Katz, ch. 15). Yet as Chowning (ch. 10) points out, women who stay around home to care for their infants may be called upon to be a wetnurse for other infants whose mothers have returned to subsistence activities.

It is interesting to note that it is not only subsistence activities that pull women away from child care. Social clubs, church groups, and women's meetings may also encourage the separation of mothers and infants (see Tietjen, ch. 7, and Nardi, ch. 16). It would be a mistake to focus exclusively on subsistence activities or paid employment as a factor influencing duration of breastfeeding without examining other activities which compete for women's time.

As these papers demonstrate, women's activities in rural communities can be incompatible with breastfeeding, but mechanisms exist to enable women to adjust their schedules in these rural settings. The strategies women develop to combine formal employment and infant feeding are different in urban settings. Marshall (ch. 2) describes how public health nurses with relatively predictable schedules maintain lactation by strategies such as wetnursing, shifts in breastfeeding schedule, providing early supplementary foods, and shifting the place of work closer to home. Nevertheless, return to full-time work usually requires part-time bottlefeeding.

Future work on this topic will require special attention to the amount of time a mother is physically separated from her infant, the regularity and predictability of hours of separation, and the nature of simultaneous tasks (that is, what activities can be performed while breastfeeding or caring for an infant).

The recent work of Mary Douglas on periodicity and division of labor may be instructive here (1979). Douglas argues that high frequency tasks are nonpostponable, menial, ordinary, and consequently of low value, and are therefore assigned to low rank individuals and groups. Generally, these are the kinds of tasks assigned to women: tasks associated with servicing bodily functions including food preparation, care of the sick, infant care and breastfeeding. She argues that people responsible for these high frequency nonpostponable tasks are constrained by periodicity. High frequency work is not compatible with being available for sudden windfalls and unpredictable tasks. Women can undertake tasks which would provide them with more prestige only when they can be relieved of high frequency tasks such as daily meal preparation and breastfeeding. What differs in these societies is the relative prestige value of mothering.

These ethnographies provide a valuable source of information for testing Douglas' hypotheses and for linking research on infant feeding more directly with research on the status of women.

INFANT FEEDING STYLE

It is paradoxical that the broadest and most general questions about infant feeding are best approached through understanding particular foods and particular people. The great complexity of infant feeding patterns illustrated in these papers underscores the need for new conceptual tools to deal with this diversity. It is difficult to describe and explain behavior as complex and personal as infant feeding. Neither women's activities, family income, nor mother's education accounts for all variability in infant feeding practices within and between communities.

Part of the understandings we accumulate about other groups is intuitive; it builds from sudden understandings coming unbidden and sometimes in spite of our preferred interpretations and models. These moments of insight can bring order to bits of disorganized observations and make sense out of disparate data. Although this mode of understanding of what is going on is seldom articulated, it colors our interpretation and analysis of social institutions and social interaction. Carrier (ch. 11), however, has made explicit how some of her personal experiences affected her analysis of infant care on Ponam Island. These insights, which are often lost in the task of writing ethnographies and in the analysis of specific hypotheses on infant feeding, are captured and made explicit in these papers.

To make full use of these insights and intuitions, I have found it helpful to reconstruct the old concept of style developed by Kroeber (1923) and Redfield (1960) and used by archaeologists and art historians to explain certain regularities, often in aesthetic expression. Style is not, nor perhaps can it be, a precisely defined analytical concept. Its value lies partly in the fact that it makes sense, and makes variables which can be more precisely defined, more meaningful. More significantly for explaining infant feeding, style combines both actual practices and the meaning of those practices to the participants.

Style refers to the manner of expression characteristic of an individual, a period of time, and a place. Infant feeding style refers to the manner of feeding an infant in a given community, and encompasses both the way the task is accomplished, shared images of the sanctioned way of feeding an infant, and the values, attitudes and beliefs associated with that behavior. It is a self-consistent way of performing certain acts requiring deliberate and coherent choices in the way something is done. In Kroeber's view, style is a way of achieving definiteness and effectiveness in human relations by choosing or evolving one line of procedure out of several possible ones, and sticking to it (Kroeber, 1923).

We speak of shared styles and personal styles. Since the locus of personal style is a single individual, then behavior and meaning are integrated and consistent for that person. In this collection of papers, Barlow (ch. 8) provides us with a good example of the interaction between infant feeding style in the Murik Lakes of Papua New Guinea and personal style interacting with the contingencies of every-day life.

These papers provide hints of what a South Pacific infant feeding style might look like. As these authors make abundantly clear, there is no single South Pacific infant feeding style, but rather a number of substyles reflecting the diversity of the region. In some areas differences in practices appear to be due to differences in personal style. In others, it might be possible to define substyles based on ecological context. However, perhaps the concept of style is particularly valuable in this region where there is so much diversity both in ideology and practice.

In order to define infant feeding style, we need to understand both breastfeeding style, the preferred foods and breastmilk substitutes introduced to the infant, and their patterns of combination. In these communities, coconut milk appears to be more common as a breast-milk substitute than commercial infant formula. Infant feeding style, however, is also part of broader patterns of mother-infant interaction and family eating style.

I have defined several dimensions of style which have been useful for examining infant feeding in our four country study mentioned at the beginning of this chapter. These include interaction style, eating style, breastfeeding style and feeding "in style." For each topic the meaning according to the participant is an important part of the definition.

Interaction Style

The ethnographers recorded many observations relevant to inter-action style: different modes of carrying infants (net bags, slings, or no carriers), responses to infant crying, mother-infant interaction such as cuddling and fondling. We learn, for example, that on Ponam, infant care is considered difficult, but "there is no such thing as a difficult infant, only careless, lazy, or ignorant parents" (Carrier, ch. 11). From the patterns of interaction, we can identify anxieties about infants and children, assumptions about when children need less care and attention (see Katz, ch. 15), and household priorities affecting their feeding.

Eating Style

Eating style was also well defined, with excellent analyses of how food exchanges structure social interaction, and how social structure affects food patterns (particularly Gegeo and Watson-Gegeo, çh. 13). Categories of food and the symbolic meaning of specific food items were also carefully analyzed. Other dimensions of eating style which would elucidate infant feeding practices include observations on meal formats, meal cycles, favored recipes and preferred combinations of foods. The number of times and hours when food is cooked and meals served is very relevant for young children. It is also useful to note how much snacking and gathering of food young children do. This information would be particularly valuable for relating infant feeding to intrafamily food distribution.

Breastfeeding Style

If breastfeeding were simply a physiological process unaffected by emotional or psychosocial factors, we would expect that breastfeeding style in the South Pacific would not differ from breastfeeding style in Southeast Asia or East Africa, for example. These articles demonstrate the diversity of practices and meanings surrounding breastfeeding just within the South Pacific region. To understand these differences in breastfeeding style it may be useful to distinguish between breastfeeding as a process and breastmilk as a product. Either process or product interpretations may be emphasized in different contexts, and in different cultures. For example, I found that breastfeeding style in Bangkok could best be understood by considering breastmilk as a product, while in urban Kenya a process interpretation was closer to the way mothers perceived breastfeeding.

Process versus product. The process/product contrast needs to be examined in historical perspective. For it is likely that the more traditional interpretations are primarily process models in societies where women breastfeed successfully. The biomedical model built on accumulated scientific evidence about breastmilk composition and the function of specific nutrients in breastmilk is a product oriented model. The imposition of this model through the spread of Western style health care and professionals has influenced women's interpretation of breastfeeding in the urban contexts in the four countries we studied. Several authors hint that these new interpretations may also be associated with health care institutions in the South Pacific.

It is this emphasis on breastmilk as a product which has been particularly advantageous to the expansion of the market for breastmilk substitutes. Infant formula is an item to be sold, and as such, can be compared with another product, breastmilk. This analogy is stressed in advertising copy. It would be difficult to imagine a process oriented campaign for breastmilk substitutes. In areas of the South Pacific where breastmilk as a product is emphasized, I would predict that the demand for breastmilk substitutes would be greatest. (Of course, this depends on the marketing structures and import regulations as well.)

There are certain dimensions of breastfeeding style that may affect the management of lactation. For example, the value of colostrum as a product and how soon women begin breastfeeding after childbirth; the social and economic support women receive during pregnancy, childbirth, and lactation; wetnursing and relactation; and the methods of sevrage are all aspects of breastfeeding style with possible health and nutritional implications. Colostrum is a good illustration of this point.

In a process oriented interpretation, colostrum may not be recognized or differentiated from "real" breastmilk at all and emphasis is on the initiation of breastfeeding: At the beginning, the process is slow and sluggish, as if passages are clogged. When the passages are cleared, milk can flow freely. This process of clearing the passages may need to be done at every feed, particularly if the mother has been out working. There may be reasons for wanting to delay the process of initiating breastfeeding for a few hours or a day until the mother is stronger, or until it is certain that the infant will live.

This contrasts with a more product oriented interpretation where colostrum is identified as a separate product with certain properties. To the health professional, it is a particularly valuable product conferring immunological protection on a newborn. To a mother, it may be the thick yellow substance similar to pus or other infected substances. In these contexts, we might expect ideas that colostrum is witch's milk or poison as Morse (ch. 14) describes in Fiji. It is this product which is being evaluated by the mother. Perhaps the variation in approach to colostrum in the South Pacific might be explained by developing this process/product contrast in breastfeeding style.

These two approaches to breastfeeding style are usefully separated to search for meaningful explanations of infant feeding practices. In actuality, process and product are two ways of looking at the same thing, and both interpretations may be present in the same community or even in the mind of one mother. In the West, the resurgence of breastfeeding may have been spearheaded by a return to a more holistic process oriented approach to health and life style. Yet in the Western world, many mothers may choose to breastfeed because they have accepted the biomedical evidence that breastmilk is indeed the best product for their infants. Similarly, in the South Pacific and elsewhere, both product and process interpretations may coexist. However, it is clearly the product interpretation which is reinforced in the clinic and hospital context.

Feeding "In Style"

Finally infant feeding is a dynamic changing process, not a set of static traditional rules. Do we have in these papers any evidence for the notion of infant feeding "in style"? Conton (ch. 6) mentions the Madang women's orientation to consumer goods; Chowning (ch. 10) notes the emulation of Europeans in West New Britain; Counts (ch. 9) observed the more educated Kaliai women's shift to wearing brassieres. We have suggestions that teachers and women exposed to Western education may be "style setters," favoring commercial foods, infant formula if available, and a shorter duration of breastfeeding, but we have little indication of why education and the change in the ways of thinking it encourages has this affect on infant feeding.

Of course, a concept such as style alone cannot account for infant feeding practices in the South Pacific. Personal and shared styles interact with organizational and institutional structures in a variety of ecological contexts. In the study of infant feeding practices, we are particularly concerned with the interaction of style with structures such as health care institutions and marketing institutions. Women also differ in physical circumstances, individual biology and temperament, and in individual experiences and personal histories. Even if they had all been exposed to the identical infant feeding style by similar role models (an unlikely circumstance), women would differ in their mothering skills. For an analyst, then, style is an heuristic device which acknowledges variation in individual experiences while recognizing the broad similarities linking individual differences.

FUTURE DIRECTIONS

These papers confirm the value of ethnographic fieldwork for understanding health and nutrition problems. They offer examples of successful inter-disciplinary cooperation and provide specific illustrations of how a variety of methods can be usefully combined. They remind us that ethnographic fieldwork and survey methods may not always be measuring the same thing, and therefore should be used to complement and not replace each other. The papers provide numerous insights which could inform nutrition education programs and infant feeding policy. The great variation in infant feeding patterns serves as a reminder that baseline data from one area cannot be widely generalized. Nor can we generalize about traditional infant feeding practices as if they were somehow homogeneous and unchanging. In fact, in some cases traditional infant feeding practices that we sometimes over-idealize may not always be uniformly supportive of breastfeeding.

Most of these papers also show how much can be gained from having a detailed knowledge of the local language. This allows for more semantic analysis than is usually found in infant feeding studies. A semantic component allows the analyst to answer certain key questions:

— Is infant feeding part of the domain of meals and food? child care? illness and medicine?

— What categories are used to describe family and infant foods?

— What categories are used to define stage of infant growth?

— How do people recognize healthy and ill infants, and how are these symptoms treated?

— How are body proportions of infants and adults described and evaluated?

These papers present convincing evidence for the importance of using *emic* categories in the collection of information on infant feeding and for developing culturally appropriate nutrition education programs (in particular, Jenkins, Orr-Ewing and Heywood, ch. 3; and Montague, ch. 5).

From the perspective of cultural anthropology, it is apparent that anthropological studies of infant feeding raise some of the broadest and most important questions in anthropology today. From issues like biocultural adaptation and the status of women, to the meaning and transformation of key symbols, infant feeding research is embedded in a series of large and complex problems worthy of attention by anthropologists of all theoretical persuasions. This collection should encourage others to choose health and nutrition related research questions not only for their applied value, but also for what they can teach us about human nature and processes of key importance to anthropologists — socialization, adaptation, and change.

REFERENCES

Daly, M. and M. Wilson (1978). *Sex, Evolution, and Behavior.* Duxbury Press, North Scituate, MA, pp. 157-158.
Douglas, M. (1979). *The World of Goods.* Basic Books, New York, pp. 119-124.
Kroeber, A.L. (1923). *Anthropology.* Harcourt Brace Jovanovich, New York.
Redfield, R. (1960). *The Little Community and Peasant Society and Culture.* University of Chicago Press, Chicago.

INDEX